With Bitter Herbs They Shall Eat It

With Bitter Herbs They Shall Eat It

CHEMICAL ECOLOGY AND THE ORIGINS OF HUMAN DIET AND MEDICINE

Timothy Johns

THE UNIVERSITY OF ARIZONA PRESS   TUCSON

The University of Arizona Press
Copyright © 1990
The Arizona Board of Regents
All Rights Reserved

This book was set in Linotype CRT Trump
∞ This book is printed on acid-free, archival-quality paper.
Manufactured in the United States of America.

94 93 92 91 90   5 4 3 2 1

LIBRARY OF CONGRESS CATALOGING-IN-PUBLICATION DATA
Johns, Timothy, 1950–
    With bitter herbs they shall eat it : chemical ecology and the
origins of human diet and medicine / Timothy Johns.
        p.   cm. — (Arizona studies in human ecology)
    Includes bibliographical references.
    ISBN 0-8165-1023-7 (alk. paper)
        1. Ethnobotany.   2. Food, Prehistoric.   3. Plant toxins.
4. Medicine, Primitive.   I. Title.   II. Title: Chemical ecology and
the origins of human diet and medicine.   III. Series.
GN476.73.J64   1990
613.2—dc20   90-31694
                                                              CIP

British Library Cataloguing in Publication data are available.

For my parents:

Together they instructed me of the virtues
of a balanced diet and a balanced mind

Contents

Illustrations and Tables

TABLES

Preface

They shall eat the flesh that night roasted, with unleavened
bread and bitter herbs they shall it eat.

EXODUS 12:8

And I have come down . . . to bring them to . . . a land flowing
with milk and honey.

EXODUS 3:8

Without recognition of the knowledge and accomplishments of humans in the past and in other cultures, Western scientific discovery is often only rediscovery. The loss of the empirical wisdom developed from human interactions with the environment over millennia is the loss of a fundamental part of our human heritage.

Scientific inquiry is seemingly self-sufficient in its quest to understand the nature of the world, and modern technological accomplishments are often used to justify the efforts of science. Moreover, scientists are little concerned with the limitations of science or with its historical roots, let alone with ties to our cultural past. However, science and the essentially materialistic vision offered by technology offer little in themselves for the sustenance of a meaningful human existence. The accelerated rate of technological change in the modern world is rapidly eroding traditional methods of interaction with the environment, traditional life-styles, and traditional cultural values. Our future is insecure if we do not recognize the social and intellectual limitations imposed by our biological origins. On the other hand, a meaningful human existence requires continuity with past cultural solutions to the questions of life.

While we often look romantically to the past for simple solutions to life's problems, science has the capacity to offer understanding of the real complexity of life. Science and technology offer many positive benefits, but the future for all peoples is dependent on successful merging of the past with the present. As the world becomes increasingly interconnected, this merger demands the recognition of our biological oneness as well as the embracing of the richness of our cultural diversity. The study of traditional biological and cultural methods of human subsistence provides a link to the past with practical and philosophical benefits for all humankind.

The quest for basic subsistence is the primary human need, and one that permeates ecological, economic, political, and other aspects of human interaction with the environment. Today, as always, the lives of both individuals and groups are dominated by a striving for a share of "the land of milk and honey." This imagery of the promised land permeates both religious and secular aspects of Western literature and language, for example, in Coleridge's *Kubla Khan:*

> For he on honey-dew hath fed
> and drunk the milk of Paradise.

Such images are a profound, time-transcending expression of our aspirations, both practical and symbolic. Taken literally, the desire for foods rich in calories, fat, and protein, and sweet in taste is consistent with what scientific study has shown to be the motivation underlying human dietary preferences. Human societies around the world have consistently considered bee honey one of life's greatest delicacies, often of divine origin or the food of the gods (Crane 1980).

A greater portion of the world's population than ever before is able to live in a world flowing with milk, honey, and other nutritionally dense foods. However, the consequences of this successful attainment of a state of abundance are not all positive, and the nutritional and medical drawbacks of a diet high in fat, protein, and refined carbohydrates are receiving increasing attention. It is clear that the land of milk and honey and the Garden of Eden are not one and the same place.

The historical events coinciding with the biblical holding forth of the promised land were marked by extreme stress and deprivation. At the start of the Exodus from Egypt God's command was, "That night the meat is to be roasted and eaten with bitter herbs and with bread made without yeast." From the point of view of understanding the human diet, the three items mentioned here have as much symbolic importance as the milk and honey of the dream of affluence. Most of this book will concentrate on the "bitter herbs." I will consider the evolution of the human use of plants, the ways in which humans obtain foods from among the myriad poisonous and unpalatable components of the plant kingdom, and the consequences of this history for understanding the basis for the human diet and the origins of human medicine.

Although bitter herbs were likely much more significant contributors to early human sustenance than either milk or honey, it is significant that neither of these two most-desired foods are plants. Vegetable products are often something to be avoided. Nonetheless, at a time when many people are able to indulge in foods their ancestors usually only wished for, it is clear that we were designed not for the reward, but for the struggle. Our physiological and dietary requirements have been affected by our evolution.

In order to understand the diet to which we are best suited, it is necessary to retrace our steps. The path backward is obscure and often confusing. Our ambiguous attitudes to plants make our relations with them especially confusing. The biological and cultural aspects of this part of the path form the focus of this book.

In retracing this path I discovered a few things I did not anticipate. Most important was a recognition that in our ambivalent attitudes toward plants can be found the origins of human medicine. The properties of plants that make them unpalatable and toxic are the same properties that make them useful pharmacologically. In exploiting plant foods it is impossible to avoid their defensive chemicals, and I believe that in adapting to them our species has made them an essential part of our internal ecology.

As this study attempts to look at the interactions of two distinct parts of the natural world—human and plant—it also attempts to fuse two academic disciplines—ethnobotany and chemical ecology. Interdisciplinary studies that attempt new mergers of categories of thought are bound to generate considerable friction. Some of the heat comes from trying to merge the unmergeable; some of it comes from the resistance of human minds to breaking out of set patterns of thought and behavior. However, sparks of imagination, not heat, are the key to a new synthesis and a new understanding.

My previous attempts to develop the ties between ethnobotany and chemical ecology have been largely de facto rather than explicitly defined. By focusing on a specific problem of plant-human interactions from an interdisciplinary perspective, I sought in a heuristic manner to tie together many of the seemingly diffuse aspects of human relations with the natural world (Johns 1985). In this book I attempt to redefine some of these insights in a more rational and systematic manner. Specifically, I have recast them into a theoretical model that offers more generalizable insights into the interactions of humans and plant chemicals, but also lends itself to defining testable hypotheses with applications in a broad forum.

The specific problem around which I developed my model, and which forms a large part of the content of this book, concerns the domestication of the potato. While it appears that selection by humans for reduced levels in glycoalkaloids must have been a crucial part of the domestication of this crop, potatoes with potentially toxic levels of these compounds continue to be used by indigenous peoples in the Andes Mountains today. I have attempted a chemical-ecological approach to the interactions of the Aymara-speaking people of Bolivia with plant secondary chemicals, in both the past and the present.

Neither the role of the Aymara in directing the domestication of the potato nor their chemical-ecological adaptations could be studied without meeting the people themselves. I initially approached the Aymara with some trepidation, and it was only through a number of unpredicted circumstances that I found myself living among them rather than with the Quechua-speaking people with whom I had had many previous positive experiences. The Aymara have a reputation for aggressiveness that has persisted in the anthropological literature in spite of voices to the contrary. As recently as sixteen years ago they were described as "the meanest and most unlikable people on the earth—the classic example of an extreme personality type dominated by excessive hostility and aggression" (Trotter 1973). More recently the reputation of the Aymara has been the subject of a particularly nasty debate in the anthropological literature (Lewellen 1981; Bolton 1984) over whether their supposed aggressiveness is due to hypoglycemia brought on by malnutrition and chronic lack of oxygen. As my data on the glucose preferences of the Aymara suggest (chapter 5), these issues are not irrelevant in relation to my own investigation.

I never had any intention of entering into the confusing debate on the Aymara personality. I found, and probably not surprisingly, that the Aymara Indians of the Province of Pacajes among whom I worked most closely do have some particular culturally related personality traits, but all in all they were just as good and bad, just as friendly, generous, suspicious, honest, and deceitful as any group of people I have known, my own English-speaking Canadians from Ontario not excepted. The Aymara personality is undoubtedly affected by their fight for survival. They are a very stoic people and incredibly tough physically and emotionally.

With strangers the Aymara are characteristically shy and unaggressive, but when the ice is broken they are as curious, friendly, and forthright as anyone. Within the regimen of daily life they are serious and controlled, but in social situations, particularly those related to various ceremonial occasions, they open up considerably. They enjoy drinking during these few annual events, and undoubtedly alcohol helps catalyze personality changes in them. The Aymara do not become aggressive under the influence of alcohol; instead they are more friendly.

They are intelligent and perceptive observers of their own environment, and among the Aymara are found individuals who are as knowledgeable and articulate about the workings of the world as persons anywhere.

The Aymara reputation for taciturnity in response to harsh environmental conditions probably comes from the experiences of anthropologists among the most acculturated groups rather than among those who might be seen to be the most traditional and the most materially deprived. In Bolivian areas around Lake Titicaca there is longstanding hostility among Indians and Europeans that dates from the oppressive control of the big hacienda owners prior to 1953. Many of these areas are reputed to be ongoing seats of political rumblings. It is true that the Aymara live a hard existence in an environment that is often inadequate to meet their basic needs. While it may also be true that some cases of aggression among the Aymara are triggered by hypoglycemia, any hostility and aggression exhibited by the Aymara likely has multifaceted biological and social causes related to their history of oppression and deprivation. One point that has been clarified in the recent aggression-hypoglycemia debate is that the Aymara as a group have a complex and well-rounded personality (Bolton 1984). The perpetuation of a negative stereotype determined by a simplistic biological theory is unfair to the Aymara and does no service to scientific inquiry. Clearly, I am speaking both personally and as a scientist on this issue. To me the Aymara are more than standard deviations on a graph. I owe them a great personal debt and among them number persons whom I call my peers and my friends.

Many other individuals have assisted me in translating my experiences in the field and the laboratory into publishable form. I acknowledged the contributions of many of these persons in my doctoral dissertation and several journal publications upon which this book is based. I have not forgotten their assistance, but I have undoubtedly overlooked the contributions to this current volume of many others. Those that I have not forgotten include Ebi Anyamba,

Stephen Baird, Gary K. Beauchamp, Sarah Booth, Stephen B. Brush, Reginald F. Chapman, J. Desmond Clark, Gerrit de Boer, Martin Duquette, Stanley Falkow, Richard I. Ford, Fredrick J. Hanke, Christine R. Johns, Daniel B. Johns, Ernest L. Johns, the late Volney H. Jones, Susan L. Keen, Ebi Kalahi Kimanani, Isao Kubo, Stan Kubow, Harriet V. Kuhnlein, Gary T. Marshall, Katharine V. Milton, Vincent Saarich, Marilyn Scott, Tim D. White, and Shelley Williams. I would also like to thank the Faculty of Graduate Studies and Research, McGill University, for a grant toward publication costs and Mindy Conner for her fine editorial assistance.

1 A Model of Human Chemical Ecology

INTRODUCTION

Plants play an important role in the lives of all humans, and our dependence on plants for survival hardly needs to be stated. Without the energy derived from photosynthesis neither the plants and animals we eat nor we ourselves would exist. However, since many plants are poisonous both to us and to other animals, a judicious strategy used by many, if not most, people is to assume that unless known otherwise, plants are not edible. With such a hands-off approach, how did humankind learn which plants were edible?

For a normal, healthy person, the problem is not usually what to eat and what not to eat. The decisions as to what constitutes food are usually made for us and form part of our cultural heritage. For most of us an acceptable range of edible foods is established by the time we are adults. However, even in a modern urban environment the issue is not a closed one for children. Childhood explorations leading to encounters with plants are particularly threatening. The mishaps of children may provide a large part of our communal knowledge about noxious plants. We learn from others or through firsthand experiences that many mushrooms and plants with names like "deadly nightshade" and "poison hemlock" are to be avoided.

For the rural ancestors of people in Western industrialized societies, and for the majority of the population living in the so-called developing[1] regions of the world today, hands-on exploitation of plants formed and still forms an important part of life and subsistence.

---

1. "Developing" is used here for lack of a better term. While it is unsatisfactory because of its implication that all countries must follow the path of the industrial nations, terms such as underdeveloped and Third World are equally problematic.

Traditional[2] societies rely to a great extent on wild and cultivated plants for food and medicines, something many less-experienced city dwellers have tried to emulate in their rural excursions. While this sometimes leads to disastrous consequences when toxic plants are mistakenly eaten, this persistent interest in exploring the natural world enforces the notion that foraging behavior is a trait of the human species. There is increasing awareness that knowledge of plant use is an important element of the human heritage which has determined our physiological, morphological, and behavioral (cf. Schleidt et al. 1988; Ulrich 1984) makeup. Plant exploitation has been at the root of our dietary and medical practices. It is this heritage that we must retrace as carefully as possible.

This book addresses the relationship of humans and plants from the perspective of the problem plant chemicals impose on this interaction. An image of humans struggling to find sustenance from a jungle of poisonous plants is a gross exaggeration. Plant consumption is not really such a life-and-death problem, but neither is it obvious how humans have learned to deal with potentially threatening plants. The key to understanding how humans have faced this problem is to recognize the wealth of knowledge that all cultural groups possess of the plant world, and, most important, to identify the biological and cultural components of the dynamic process by which that knowledge is obtained and then maintained within a human community.

The extent of the knowledge specific cultural groups have of dietary and medicinal plants is well documented in the expansive lists of plant use that make up the bulk of the ethnobotanical literature. However, the process by which humans first came to use these economic plants and to avoid others appears to be lost in history. The often unstated assumption of most ethnobotanical studies is that knowledge of the values of plants has been acquired through empirical processes, yet any understanding of how such learning was acquired is considered irrelevant, or perhaps simply inaccessible. As recorded in modern scientific literature the interaction between humans and plants is frustratingly static.

Ethnobotany as an academic discipline has gradually expanded its

---

2. "Traditional" refers to those societies that are engaged in a nonindustrial way of life, usually indigenous in a particular geographic area, and characterized by recognizable cultural practices relating to their social and biological needs.

perspectives to look at such interactions as a dynamic process. Ford (1981) regards the quest for the origins of empirical plant knowledge as an important focus for ethnobotany. I emphasize the dynamic interaction of plants and humans by attempting to give the process an evolutionary foundation that involves both biological and cultural aspects. An understanding of the acquisition and retention of knowledge by peoples in the past and the active interaction of extant peoples with the plant world requires a multidisciplinary approach to human biology.

In the following chapters I explore a specific model and various experimental approaches in this search for the synthesis of past and present interactions between humans and plant chemicals. Efforts are made to elucidate the ways humans select and choose plants on the basis of chemical constituents, a mechanism I will refer to as *chemical selection.* The basic model is familiar to investigators in the area of chemical ecology and is based on the premise that human ingestion of plant chemicals is part of an integrated adaptive response that has both biological and cultural components.

CHEMICAL ECOLOGY

Biologists' awareness of the chemical basis for many interactions between organisms has increased over the past twenty years. The theoretical basis of the field of chemical ecology has developed primarily out of studies in entomology. Chemical communication, pollination, chemical defense, and chemical aspects of herbivory are among the topics that comprise the field (Harborne 1988). A theory of biochemical coevolution arose from studies of the feeding behavior of insects, specifically butterflies (Ehrlich and Raven 1964): plant chemical constituents evolve to combat attack by insects, while insects respond by evolving new methods of overcoming these defensive compounds. Interactions between insects and plants continue to provide the focus for most of the research in chemical ecology. These interactions determine many of the paradigms of ongoing research into the interactions of mammals and other animals with plants. The feeding ecology of a number of herbivorous mammals has been considered (Freeland and Janzen 1974; Montgomery 1978; Pederson and Welch 1985), and primate feeding preferences will be discussed in greater detail in chapter 7.

Although chemical-ecological studies of humans per se are generally lacking, there is a vast body of knowledge that provides insight into the biology of our species. Much of this research has direct relevance to a chemical-ecological consideration of human evolution. For example, investigations of possible human pheromones (Sokolov et al. 1976) are a direct offshoot of entomological and mammalian studies in chemical communication.

Indirect information relevant to human-plant interactions has come from fields such as nutrition, toxicology, and pharmacology, and from studies of human taste and smell. Investigations in physiological and behavioral psychology have specifically considered human dietary preferences. Ethnobotanical studies look directly at human use of plants, and ethnopharmacological and ethnonutritional studies look for a rational basis for many of these uses. A few theoretical considerations of human dietary history have drawn on insights from chemical ecology by specifically considering the role of secondary substances in human dietary evolution (Leopold and Ardrey 1972; Stahl 1984). Coevolutionary theory has helped define scientific conceptions of the role of defensive chemicals in animal and plant interactions. Such an evolutionary model can provide a focus for tying together the diverse studies of human diet and biology. Studies of animal-plant interactions have produced several themes that are transferable to the human situation.

Plant defensive chemicals are often referred to as secondary compounds, or allelochemicals. *Primary metabolites* of plants are those chemicals associated with photosynthesis, respiration, and growth. These include chlorophyll, amino acids and proteins, simple sugars, carbohydrates, vitamins, minerals, and plant hormones. Although plants vary in their amounts of particular primary metabolites, most of these compounds are common to all plants, and they provide important nutrients for animals. Most plants produce other chemicals with no apparent metabolic function. The structures of these so-called secondary metabolites vary widely across the plant kingdom, and their roles in plants are still debated. However, it is clear that they play an important defensive role in the relationships between plants and herbivores and pathogens. These highly variable compounds create a mosaic from which humans and other animals must separate the beneficial from the harmful. The term *secondary* tends to underestimate the importance of these chemicals. On the other hand, *allelochemical* is a neutral term that recognizes the primary role of these compounds in ecological interactions, including those involving humans. Allelochemicals are "chemicals by which organisms of one species affect the growth, health, behavior or population biology of another species (excluding substances used only as food by the second species)" (Whittaker and Feeny 1971).

For insects (Harborne 1988) and also mammals (McKey et al. 1981) a balance between nutrient requirements and the metabolic cost of dealing with defensive chemicals from the plant is generally important in determining whether or not the plant is eaten. Some animals, at least, are able to handle high levels of toxins if they are compensated by a high nutrient—particularly protein—content in their food (McKey 1978; Moore 1983). Allelochemicals are important factors in determining whether or not most insects will accept a potential food. However, for some animals, nutrient quality is paramount and allelochemicals seem irrelevant. In considering animal preferences, the terms *stimulants* (attractants) and *deterrents* (repellents) are widely used to explain the chemical basis for food acceptance or rejection. The ways in which animals respond to chemical compounds have considerable interspecific differences and reflect how evolution has determined the feeding strategy of the organism.

Animals show various degrees of adaptation aimed at taking advantage of available resources, but, stated as simply as possible, most can be considered either specialists or generalists. Among plant-eating insects the more specific terms *polyphagous, oligophagous,* and *monophagous* are widely employed in relation to the variety of plant species they use. The majority of phytophagous insects are oligophagous. While they differ in their behavior from monophagous insects (which feed on only a single plant species), they are selective as far as which plant families or genera are acceptable. Other factors besides nutritional composition of the plant are important in guiding their behavior. Specialist animals exploit a particular resource: perhaps a particular plant part such as leaves or fruit, or perhaps a single plant species.

Where a coevolutionary relationship exists, specialists likely show feeding preferences for the characteristic allelochemistry of the host plant. Such is the case with moths like the silkworm (*Bombyx mori*), which requires monoterpenes and particular flavonoid stimulants before it will feed (Hanamura 1970). Specialized insects may have efficient detoxification mechanisms to deal with particular compounds that are deterrents to other species. Cabbage butterflies specialize on plants containing glucosinolates, compounds that are toxic to nonspecialized species (Erickson and Feeny 1974). Koala bears, which feed exclusively on monoterpene-containing *Eucalyptus*, and mountain viscacha, which feed primarily on pyrrolizidine alkaloid–containing *Senecio* spp., are good examples of specialized mammals.

By contrast, humans are extreme generalists. Like other omnivores such as rats, bears, pigs, and many primates, we eat a range of animal and plant foods. For omnivores, not only sensory perception but also the lessons learned from the physiological consequences of ingestion are important in determining food selection (Westoby 1974). Adaptations of humans and other omnivores for dealing with plant toxins will be discussed in detail in subsequent chapters.

ANIMAL AND PLANT FOODS IN CHEMICAL ECOLOGY

The importance of animals as food sources cannot be dismissed from chemical-ecological discussions. Animal products, or microbial contaminants in them, may present toxicological problems, some of which are discussed in chapter 2. Biological and cultural adaptations that are required for humans to exploit animals as dietary sources are, however, generally ignored in this volume. And because of the emphasis on plant-herbivore interactions in chemical ecology, animal foods take a secondary place in the model of human chemical ecology. Although chemicals that are present as nutrients or non-nutrients in animal foods are an essential part of this model, the specific testing of the model in succeeding chapters focuses only on plants.

In relation to our diet, the tissues of other animals are most important for the nutrients they provide. Certainly they can be very significant in contributing to the balance between nutrients and toxins described above. In this book, which attempts to build on the insights of chemical ecology, human use of meat is considered primarily in these terms and in juxtaposition to human interactions with plants. As I discuss in chapters 7 and 8, the changing contributions of animal products in the diet through time may have had profound importance in determining our interactions with plants.

Consider, then, two interrelated problems: (1) What are the means by which humans have interacted with plants and plant chemicals? (2) Were our hominid ancestors able to overcome the chemical defenses of plants while gaining sufficient nutritional benefit to make the exploitation of plants worthwhile? The pathway to answering these fundamental questions in human history circumscribes the model of human chemical ecology. These questions reflect two areas of philosophical and practical concern to humans: the evolution of human dietary behavior and the origin of human medicine.

EVOLUTION OF HUMAN DIETARY BEHAVIOR

Many of the major chronic diseases of modern industrialized societies, such as heart disease, hypertension, obesity, adult-onset diabetes, dental caries, and some types of cancer, have a dietary basis (Eaton and Konner 1985). Nutritional deficiencies continue to result in suffering in both poor and affluent societies. Fundamental to understanding the causes and control of diseases of both affluence and scarcity is consideration of the influence of evolution on the human diet. Human interactions with chemical constituents have determined the acceptance (or avoidance) of particular plants for food and medicine, and were probably important in determining the genetic constitution of human beings.

Human dietary physiology is assumed to have evolved little since the appearance of modern *Homo sapiens* some forty thousand years ago, and also to reflect an evolutionary continuum with our hominid and prehominid ancestors (Eaton and Konner 1985). Considerable effort has been made to reconstruct the diet of early hominids. Apart from the obvious archaeological studies, investigative emphasis has also been placed on feeding ecology of primates and gatherer/hunter groups.

Within the discussion of the diet of early hominids, the question of the importance of meat versus plant-derived foods has continued to generate a debate with considerable philosophical, economic, political, and moral overtones. Not only what our ancestors ate but how they obtained it has important implications. The idea of "man the hunter" sustains an image of man as the dominant force in the biological world that is difficult to relinquish. In actuality, an emphasis on the hunting model in investigations of human subsistence is perhaps an artifact of the role of males in hunting and the social status attached to this activity. Dietary procurement is a fundamental determinant of biology, behavior, and social structure, and indeed has relevance for helping us understand human social relations. However, the incompleteness of the data and the strong biases of various sides in the discussion combine to give a cloudy picture.

Discussions of the defensive chemistry of plants have been rightly introduced into the debate (Leopold and Ardrey 1972; Stahl 1984).

Leopold and Ardrey (1972) assumed that plant foods were unavailable to early hominid gatherer/hunters because they did not yet possess the processing techniques for removing plant toxins. Alternatively, this hypothesis suggests that early hominids must have relied heavily on meat. Studies of present-day gatherer/hunters show that these foragers[3] can subsist quite well on a mainly plant diet. Hayden (1981b) takes this argument to the extreme in essentially discounting any limitation placed on food procurement by plant toxins. However, the controlled use of fire and other processing techniques by extant gatherer/hunters is important in detoxifying plant foods and cannot be discounted as a major development in human evolution. Proteinaceous plant defenses such as proteinase inhibitors and lectins (hemagglutinins) are common in plant foodstuffs and are not detectable by taste. They do affect the well-being of humans and other mammals, but they can usually be destroyed by cooking.

As I discuss below and in chapter 2, under normal circumstances humans do have the capabilities to perceive and avoid many other kinds of toxins. However, periodic fluctuations in resource availability would not always give our ancestors the full choice of possible plant foods. The natural world is a complicated mosaic. It is hazardous to make generalizations about the nature of plants, and it must have been just as complex an issue for early hominids to obtain appreciable nutrition from plant sources. Undoubtedly they did. Most evidence points to the conclusion that humans are by nature omnivores—opportunistic consumers of a range of animal and plant products.

The cultivation of plants made certain plant foods more available to humans and has contributed to profound postagricultural changes in the human diet (Eaton and Konner 1985; Gordon 1987). We have adapted to dietary change primarily through cultural means. Changes in levels of the enzyme lactase in response to increased milk consumption associated with dairying is the best-known example of human biological adaptation to dietary change (Flatz and Rotthauwe 1973). However, in general, genetic changes in response to diet since the emergence of our species are probably minor. It is because the present-day diets of many people are out of synchrony with our biological makeup that modern diet-related diseases exist.

---

3. "Foraging" refers to the act of obtaining animal and plant foods from the wild as opposed to from domesticated or cultivated food sources. Thus it encompasses both hunting and gathering.

ORIGINS OF MEDICINE

The practitioners of present-day Western medicine have their coun-
terparts in the medicine men and herbalists of nonindustrial societies. In both settings the administration of substances called medicines is synonymous with the profession called medicine. Many
modern pharmaceuticals have in fact come from herbal remedies.
Some 25 percent of drugs used today are either plant extracts or
active principles prepared from plants (Farnsworth et al. 1985). Many
modern drugs are either synthetic equivalents of plant chemicals or
synthetic analogues of chemicals present in medicinal plants. Several of these plants were used in ancient times in European cultures.
For example, aspirin, or acetylsalicylic acid, was derived from sali-
cylic acid, the active metabolite of salicin. Salicin is widespread in
plants such as willow (*Salix alba*), which was used by the Greeks to
relieve pain and fever (Gilman et al. 1985). More recently Western
science has sought drugs from among the medicinal plants of other
cultures whose traditions parallel those of Western society. Reserpine, the first major tranquilizer and an important hypertensive
agent, was developed from *Rauwolfia serpentina*, an important treatment in Indian ayurvedic medicine for mental disorders. Quinine,
the prototype of most malarial treatments used worldwide, is the
active component of *Cinchona* spp., an important antimalarial remedy used by Amerindians on the eastern slopes of the Andes Mountains. It is perhaps a universal human trait to use plants and other
substances for medicine, and most cultures have extensive phar-
macopeias and medicinal lore.

Understanding the origin of human medicine is a perplexing problem. While we tend to view medicine as something unique to our
species, chemical compounds in plants are consumed by animals as
well, and these compounds may have similar effects on their physiology and behavior. When these compounds are deliberately administered by humans, we refer to their pharmacological properties, but
we use other terms such as *toxins* or *biologically active agents* when
the chemicals are encountered by animals or humans under other
circumstances. It is logical to assume that the plants we use for
medicines were first encountered in our quest for nourishment.
Since then, humans and other animals have used these chemicals for
their beneficial properties. A recent study of chimpanzees in Tanzania suggests that human medicine does have its analogues in animal behavior (Wrangham and Nishida 1983) (see chapter 8).

The dichotomy between food and medicine is taken for granted in

Western society. This distinction is not strong or even present in many cultural traditions. From a chemical-ecological perspective, the interactions of humans with plant chemicals appear more as a continuum. The distinctions between foods, beverages, condiments, medicines, stimulants, psychoactive agents, and toxins are not clear-cut. Often the difference between medicine and toxin is simply a matter of concentration of the ingested chemicals or the circumstances under which they are ingested. Substances like vitamins are nutrients that have therapeutic roles as well, but they can be toxic in excess quantities.

As medicine and food are linked, so do the origins of human medicine appear to be closely entwined with the evolution of human dietary behavior. In relation to the question of whether our ancestors used plants extensively for food, one might argue (although circularly) that they must have. They could not have formulated such elaborate medicinal lore if they were not actively involved in seeking sustenance from botanical sources.

INTERACTIONS BETWEEN HUMANS AND PLANT CHEMICALS

While the interaction of humans with both nutrient and nonnutrient plant chemicals is a dynamic process that extends to the beginning of time, the central focus of this book is on immediate and ongoing phenomena. In relation to the theoretical model, I develop experimental approaches to current field situations that illuminate the complexities of biological and cultural aspects of this interaction. Evolutionary consequences of this focus will be explored in chapters 7 and 8 in a more thorough examination of facts and theories on the history of human diet and medicine.

The ongoing interaction of humans and plants falls within the sphere of ethnobotany. Ethnobotany has many guises, but it has been an important component of the cultural-ecological tradition in anthropology. Likewise, ecology is a word that means different things to different people, but it is at least a starting concept that ties together chemical ecology and ethnobotany.

ETHNOBOTANY

For the purpose of this discussion I have expanded Ford's (1978) definition of ethnobotany to be *the study of the direct interrelations between humans and plants, and their evolutionary consequences.* Whether carried out by botanists, anthropologists, geographers, or others, ethnobotanical studies are often secondary to another purpose; only rarely are they carried out for their own sake (Jones 1941). Few ethnobotanical studies have satisfied the essential dual focus on plants and humans that is implicit in the above definition. Most studies have been linear in nature, lacking the interactive aspect that is inherent to the ecological perspective underlying most definitions of ethnobotany. The most essentially ethnobotanical studies have been those that have consciously attempted to study ethnobotany for its own sake. Various doctoral dissertations such as those by Yarnell (1964), Ford (1968), Bye (1976), and Alcorn (1982, 1984) reflect the developing synthesis of biological and anthropological perspectives that have characterized ethnobotany over the years since Jones (1941) provided the first comprehensive definition of this field.

By taking a diversity of approaches to the question of human-chemical interactions, I attempt to consider an ethnobotanical problem with the balanced treatment of both humans and plants that is implicit in the definitions of ethnobotany. Increasingly, practitioners of ethnobotany have recognized that both historical and contemporary ethnobotanical studies often fail to capture the diversity of plant knowledge that is possessed by a community or an individual. The social differentiation of roles that determine plant knowledge within a community, and the differentiation of recollection, synthesis, and utilization of knowledge that occurs under different seasonal, climatic, social, and other circumstances deserve increased recognition. In practice, many ethnobotanical studies present a record of plant uses by all members of a particular cultural group, for all time. My reformulation of the definition of ethnobotany attempts to underline the dynamic interaction of plants and humans implicit in previous definitions by deliberately giving this interaction an evolutionary foundation. It is a recognition of the past, present, and future importance of diversity and of change in the interrelations of humans and plants, and the emerging awareness on the part of ethnobotanists of the relevance of these considerations.

Ethnobotany by definition encompasses both biological and cultural concerns. My model of human interactions with plant chemicals reflects this dual perspective.

APPROACHES TO THE PROBLEM

The manner in which prescientific peoples select and manipulate plants for their medicinal and edible qualities can be approached from such disciplines as ethnobotany, phytochemistry, pharmacology, nutrition, physiology, and experimental psychology. The properties of plants as internal medicines or food are attributable to the chemical constituents of the ingested material. Studies that correlate therapeutic and dietary values of plants with their chemical compositions can increase our insight into human choice of plants for medicine and/or food. In many cases such investigations can support a posteriori a rational empirical and adaptive basis of such use. Both primary and secondary metabolites can contribute to human well-being; both classes of compounds are often perceived through the senses of taste and smell, which allow the user to directly associate compounds with their purported uses. Homeostatic mechanisms may be responsible for the recognition, and utilization or avoidance, of specific compounds in other cases. Consideration of chemical constituents, their perception, and the associations made by humans provide a direct means of understanding plant selection as a combination of biological and cultural adaptations.

The keystone in the development of the model of human chemical ecology presented in this volume and in tests of this model is the domestication of plants. As discussed above, the development of agriculture caused profound changes in human subsistence and diet which, while recent and relatively accessible for study, are evolutionary in scope. While the genetic effects in humans have been minimal, humans as selective agents during the domestication of plants and animals have brought about evolutionary change in these organisms. Domesticates, in fact, are the only plants that reflect detectable evolutionary change during the last ten thousand years. That this process of change has been greatly accelerated from an evolutionary perspective does not disqualify it as evolutionary change. Charles Darwin (1890) and others since his day recognized that this recent phenomenon provides excellent case studies for studying the evolutionary process. Selection for changes in secondary chemical constituents has been an essential aspect of the domestication of food plants (Johns 1986b).

The study of cultivated plants can show human selection for chemicals and other plant properties because it retains a historical as well as an ongoing component. Selection in this sense has a double meaning; as well as relating to aspects of plant evolution, it demonstrates the choices made by humans. At the community level, plants are affected by human protection and encouragement of preferred species. At the plant population level, human selection, both unconscious and deliberate (Harlan et al. 1973), can result in genetic changes that emphasize specific characters, even to the extent that speciation results (Pickersgill 1977).

The basis for selection in both the unconscious and deliberate cases involves cognitive and behavioral aspects having both a biological and a cultural basis. Domesticates are often the plants with the greatest significance in both economic and cultural terms. As Brush (1986) says, "crops are products and symbols of cultural evolution." Because they reflect the explicit concerns of their domesticators, they provide particularly interesting cases for study. The genetic changes in domesticated plants can be studied through methods of systematic botany. The possibility that humans manipulate plants for chemical constituents can be tested by studying specific examples of plant domestication. The domestication of plants encompasses both the biological and cultural components of the chemical-ecological model.

Attention to nutrients, particularly caloric content and protein quality (Harborne 1988), has been an interest of plant breeders both past and present. Hegnauer (1975) discusses several examples of plants whose cultigens reflect variation in allelochemistry brought about by human selection. Many of these reflect conscious manipulations by modern plant breeders. Examples of selection for chemical constituents during the domestication process are considered in chapter 4. Whether the resultant changes reflect unconscious or deliberate manipulations, it is a basic assumption of this model that selection for phytochemicals is an adaptive end in itself rather than secondary to other selection processes.

Although selection for changes in allelochemicals has been an essential aspect of the domestication of many food plants, how the wild progenitors of staple cultigens with appreciable levels of toxins would have been acceptable initially for human consumption, and therefore subject to genetic manipulation, is problematic. Technological and behavioral advancements in conjunction with domestication were clearly important in overcoming toxicity in our crop plants. Detoxification mechanisms are used worldwide in the preparation of many plant foods. They undoubtedly have a long history of use (see chapter 3) and may have been essential for the domestication of some plants. I present evidence to suggest that geophagy, the consumption of edible clays, has a detoxification function and may have been a crucial first step in the domestication process.

A MODEL OF HUMAN CHEMICAL ECOLOGY

Traditional uses of plants (and animals) for food and medicine have evolutionary antecedents. The selection and ingestion of plants by humans can be viewed as an integrated response to the chemical environment with both biological and cultural components. The chemically mediated interactions of humans and plants are part of a symbiotic process that is manifest in the domestication process.

COMPONENTS OF THE MODEL

The basic premises of the model are presented below. Figure 1-1 attempts to provide perspective on the way these various components interact.

1. *Humans seek to maintain physiological homeostasis through maximizing the beneficial effects of ingested components and minimizing the effects of potential toxins.* Like other members of the animal kingdom humans face an environmental array of chemical compounds. Many of these compounds are nutritional or health-promoting agents; many others are potentially toxic if ingested in even minute quantities. Survival necessitates the detection and differentiation of both classes of compounds and appropriate behavior in response.

2. *Natural selection has produced the interrelated physiological and behavioral mechanisms that allow humans to deal with environmental chemicals.* Three general mechanisms allow humans to deal with chemicals. First, organisms have evolved taste senses specific to the role of the particular compounds relevant to their survival. Sweetness usually evokes a sense of pleasure and is associated with chemicals essential for basic energy metabolism and growth such as sugars and some amino acids (Schiffman et al. 1982; Shallenberger and Acree 1971). Bitter compounds such as alkaloids are generally perceived as unpleasant, and these are often highly toxic (Garcia and Hankins 1974). The experience of stimuli as pleasurable or displeasurable provides motivation for responding to a stimulus (Cabanac 1979). Second, the vast majority of chemicals—whether bene-

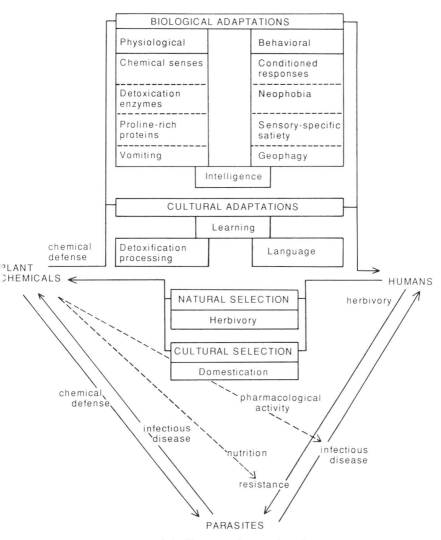

FIG. 1-1. Diagrammatic model of human chemical ecology.

ficial, detrimental, or neutral—have either no taste or no taste relevant to their effect on survival. Alternatively, then, animals respond to chemicals by learned aversions or preferences based primarily on associations of dominant tastes with general effects on well-being. These effects are mediated through various physiological and neurological processes. The association may be specific for a taste that may be attributable to some chemical constituent other than the one affecting well-being. Third, potentially toxic chemicals cannot be totally avoided, and all animals have an aresenal of detoxication and detoxification mechanisms for dealing with toxins that are obtained from the environment and for expediting the excretion of these toxins back into the environment. These mechanisms are primarily biochemical and physiological.

Phylogenetic Considerations. Studies of animals provide useful analogies to understanding human evolution. Comparative studies offer insights into the differences and similarities of animals' responses to chemicals which reflect variation in ecological adaptation and ancestral history. Omnivores are similar to each other in their food-selection behavior (Rozin 1976), and omnivores such as rats (Rozin 1976) and primates (Glaser 1972) are valuable subjects for studies of feeding behavior and for physiological experiments that provide comparison with humans. Numerous studies have examined animal feeding behavior in the field and in captivity. Animals have been extensively employed in experiments that examine their ability to digest and absorb nutrients and avoid toxins. Animals serve as models for understanding human dietary requirements and diseases and are often studied explicitly with this intent. Rats, mice, guinea pigs, and primates are the most common subjects of study.

Studies on primate food procurement are important in themselves for adding to our understanding of the biology of these animals, many of which are threatened by the activities of humans. Our closest relatives among the apes provide the best analogies to understanding human diet as well as other aspects of human nature, and considerable experimental effort continues in this area. Patterns of food procurement can be related to body size, energetics, metabolism, locomotion, and functional morphology of teeth and the gastrointestinal tract. Diets of our closest relatives, the chimpanzees, and of other primates contain large amounts of leaves, many of which are defended by secondary compounds. The behavioral mechanisms exhibited by these animals for dealing with these plant toxins are analogous to our own.

Mechanisms for Dealing with Environmental Chemicals. Many of the items included here are discussed in chapter 2, which provides details in relation to interactions with allelochemicals. This introduction considers general principles and provides details only in particular aspects related to nutrient attainment that will not be considered elsewhere.

Sensory perception. Although four basic tastes—sweet, sour, salt, and bitter—are generally recognized by sensory physiologists (Bartoshuk 1978), the perception of a compound depends on the nature of the compound, its concentration (Shallenberger and Acree 1971), and its admixture with other compounds (Erickson 1982). Thresholds of perception have been worked out for many compounds (Fazzalari 1978). Olfactory senses are less well understood: although they mediate (and confound) taste perception and quality, they are probably less fundamental in determining behavior responses.

Physiological mechanisms. Two interconnected responses regulate the interaction of animals with environmental chemicals. The first is mediated by internal information, while the second involves taste and smell perception and changes in hedonic response (the subjective level of the pleasantness with which a stimulus is perceived). Perception depends on neurophysiological responses, but also on interpretation of the stimulus. Hedonic responses are complicated by physiological and psychological factors.

Appetite plays a critical role in motivating appropriate feeding behavior in animals. It is a complex behavior that depends on physiological, perceptual, neurological, and cognitive factors.

Efficient homeostatic and behavioral mechanisms respond to the nutritional needs of the internal environment and stresses from ingested toxins. Animals fed intragastrically, that is, without input from senses of taste and smell, are capable of accurately regulating caloric intake (Ramirez 1986), and satiety, in relation to energy at least, does relate to the internal state of the organism. Moreover, humans develop specific appetites related to nutritional deficiencies such as for salt (Richter 1936), iron, and vitamins (Rozin 1976). The basis for this apparent adaptive response is unclear. The internal state, or "milieu interieur," of an organism is regulated in response to internal receptors (Garcia et al. 1974), and both the threshold of perception and the preference for compounds are affected by the milieu interieur. A given stimulus may arouse pleasure or displeasure under different circumstances (Cabanac 1979), and the pleasure of a taste depends on its usefulness as determined by internal signals. Sweet and salt tastes are more pleasurable in states of nutritional deficiency; satiated subjects find sweet stimuli less pleasant.

Variety plays an important role in determining the food intake of humans and animals. In controlled experiments humans fed a varied diet consumed more than control subjects (Rolls et al. 1981). Rats fed a diet of controlled composition that contained varied nonnutritive flavors consumed more than rats provided with food all of a standard taste (Treit et al. 1981). Rolls et al. (1986) introduced the concept of "sensory-specific satiety." Satiety may develop for some foods while not for others. Taste, appearance, smell, and texture are all factors that determine satiety to particular foods. Changing palatability for different food items may be an adaptive response to ensuring a varied diet containing a spectrum of required nutrients (Naim et al. 1986). Alternatively, this behavior may ensure that large amounts of particular toxins are avoided. Sensory-specific satiety will be reintroduced in this regard in chapter 2.

The classic work of Davis on the self-selection of diet by young children demonstrated that human children are capable of selecting a nutritionally balanced diet, but only if they are provided with only nutritious and unsweet foods (Story and Brown 1987). Children show a particularly strong preference for sugar-containing substances, and when presented with highly palatable but nutritionally poor foods they cannot be expected to maintain adequate nutrition. Our dietary preferences are adapted to a "primitive diet" rather than to the processed foods of the modern world.

Bitter and sour substances, which may be toxic compounds or pharmaceuticals, do not appear to show the same short-term hedonic shifts as salt and sweet, which are nutritionally relevant tastes. However, the relevance of biologically active compounds to survival changes with an organism's ability to respond to them. Particular plant secondary compounds may have positive physiological effects that might be expected to lead to enhanced preferences, and induction of detoxification enzymes will decrease the potential detrimental effects of compounds with time. However, in the case of nonnutrients, any changes in hedonic response appear to be mediated more strongly by long-term experience and the conditioned mechanisms discussed in chapter 2 than by short-term physiological responses.

It is inevitable that nonnutrient and potentially harmful compounds will be ingested in the quest for adequate nutrition. Physiological mechanisms exist to prevent the absorption of these compounds and to hasten their excretion. Most foreign compounds are excreted in the bile and the urine. Enzyme-controlled detoxication systems that act to reduce toxicity of dangerous compounds and make them easier to excrete from the body will be discussed in greater detail in the next chapter.

Behavioral responses. The popular fallacy that humans learn to recognize toxic plants from watching the deaths of a few of their fellows overstates the case by condensing to a few generations a process that dates back to much earlier stages of the evolutionary process. Humans, like other animals, have different behavioral as well as biochemical ways for avoiding toxins. The distinction of chemicals as feeding stimulants or feeding deterrents clearly relates to human sensory perception and preferences for chemicals. Behavioral characteristics, specifically neophobia and conditioned responses, also play a role in the avoidance of toxins. These behaviors will also be discussed in chapter 2.

3. *Humans have unique cultural traits that play a role in their interactions with plant (and animal) chemical constituents. Language and technological innovations, including plant domestication, are particularly powerful forces.*

Memory, Communication, and Culture. Taste perception in humans is easy to study because the subject can communicate directly, but it is complicated by ambiguous responses (Bate-Smith 1972; Cabanac 1979). Bitterness evokes particularly ambivalent responses in some subjects. The interrelationships between innate biological restraints, the learned aversions already discussed, and cultural values all interact to produce variation in taste acceptabilities. For example, exposure to distasteful items such as chili pepper (Rozin and Schiller 1980) can lead to strong preferences for them in some cultures. Studies on food evaluation and food preference in humans can provide relevant data for understanding human chemical selection (Rozin 1980).

Communication. Human sensory experiences and learned responses to chemical agents—good or bad—are, of course, mediated cognitively through language and culture. These unique characteristics provide a means of communication and retaining knowledge, and in turn determine the way humans will approach and interpret new stimuli. Ethnobotanical practices reflect a memory of historical observation and intelligent reflection passed on by generations of users. The association of plants with particular effects is immediate and ongoing, and it combines sensory perception, physiological responses, and cognitive memory. The cultural record of plant use can be communicated verbally and is contained in the classification schemes of the language. Knowledge may also be communicated nonverbally by example and observation in the course of day-to-day subsistence.

However, culture can act as a conservative force in the acquisition of new knowledge, and the process of changes in plant use is probably quite gradual. Human interactions with novel phytochemicals may be limited to only a few individuals or may not even occur in every generation. Whether toxins are encountered inadvertently or deliberately, the memory of the event would be essentially the same. The majority of cases of poisoning from plants involve children (Klaassen 1985). They are the most active experimenters with plant toxicity, and their individual lessons can be integrated into the codified wisdom of a culture.

Humans, while cautious experimenters, can be keen observers of the natural world. Various sources have suggested that humans learn much about the edibility of plants by watching animals (see chapter 8). Knowledge of medicinal plant use can be elaborated into integrated systems with philosophical and empirical foundations. Traditional Chinese medicine, for example, provides an integrated method of diagnosing and treating disease that relies heavily on the use of herbs. Within this self-contained system, empirical knowledge of the properties of plants has been compiled through careful observation. Information on the appropriate plant medicine to use in a given situation is intricately detailed in a literature that dates back more than two thousand years to the Han dynasty (Reid 1987). Written language can be the greatest conservative force in limiting human exploration of the plant world. Chinese anthelmintic remedies, for example, have changed little over the last two thousand years (Chan and Croll 1979).

While language provides a means for the direct transmission of knowledge, many of its messages about plant properties are communicated indirectly and subtly. Myth and oral traditions can communicate information about the significant features of plants. Plant names are more than simply labels for plants. Plant nomenclatures that contain words like the English names "deadly nightshade," "death angel," "poison hemlock," or "skunk cabbage" offer powerful controls to limit human encounters with harmful or unpleasant plants.

Consider *Zygadenus venenosus,* an extremely poisonous plant containing veratrum alkaloids that is commonly called "death camus" or "poison onion" in English. In the language of the Okanagan-Colville Indians of western North America it is called *yiẃístn* (literally, causing twitching). Eating the bulbs is said to cause twitching, foaming at the mouth, and eventually death. The Okanagan-Colville plant nomenclature contains many more names that communicate pharmacochemical information (Turner et al. 1980) either in references to the effects of the plant or to its taste or smell. The name "corpse's plant" (*stemtemnihílph*) in reference to *Symphoricarpos albus* suggests the poisonous properties of this plant. It was also consumed as medicine, as was the highly valued x̱ásx̱es (literally, always good) (*Ligusticum canbyi*). Information about chemical properties of plants perceived by taste and smell is communicated in plant names such as *tx̱ax̱agáẏkst* (menthol-tasting leaves), *nek'nek'tílhp* (rotten bush), *tk̲wik̲wágẏkst* (skunk's scent-bag), *yuyugwálhk̲w* (strong odor), *ktextxikst* (bitter-tasting leaves), *kts'eⱦts'eⱦíkst* (sour-tasting leaves), and *yititemnílhp* (bitter-tasting plant).

Folk taxonomies of taste and smell are likely to reflect human experience with the chemical world. Examples of the manner in which chemical sensations are classified and communicated are discussed in chapter 5.

Technological innovation. Technological manipulations of the environment are potent forces that determine food procurement as well as innumerable aspects of environmental and social interactions.

Simple processing of plant foods can be seen in the feeding behavior of many animals. Gulls and crows drop shellfish and nuts onto rocks or pavement in order to crack open their shells. Squirrels, primates, and other animals peel fruits and other plant parts to obtain the edible portion. The archaeological record provides evidence suggesting that hominids used stone tools for woodworking and cutting soft vegetation as early as 1.5 million years ago (Keeley and Toth 1981). Evidence for the controlled use of fire for cooking food appears as early as 500,000 years ago (Clark and Harris, 1985; Isaac 1984) in the archaeological record. Processing techniques make food more digestible and more palatable in various ways. Detoxification is a specific function of many processing techniques, including cooking, peeling, leaching, fermentation, and grinding (see chapter 3). Geophagy, or the consumption of edible clays, is a detoxification mechanism for both animals and humans (Johns 1986a). While this practice has a strong biological and behavioral basis, its cultural elaboration may have made it an important technique in the human exploitation of plant resources.

Domestication of plants is a human cultural practice that represents a zenith in environmental manipulation. By controlling the evolutionary process itself, considerable advancements were made in detoxifying plant secondary compounds.

The technological manipulations of the diet that developed through several hundred thousand years of hominid history have taken on new dimensions since the industrial revolution. Processing and artificial breeding profoundly affect human use of plant resources today and will in the future.

4. *The nature of human interactions with plant (and animal) chemicals occurs within, and is influenced by, a broader ecological framework.* Fluctuations in environmental conditions play a role in the likelihood that humans will encounter and be affected by toxic chemicals. The ability of humans to avoid toxins in their food is determined to a considerable extent by availability. Periodic scarcity may bring humans into greater contact with noxious secondary compounds in their search for nutrients. The chemical composition of plants varies in response to factors such as climate, plant maturity, or attack by herbivores (including humans), and humans utilizing a plant as a food or medicine must adjust to the changes.

Stress may alter humans' physiological state and affect their interactions with chemicals. In states of nutritional deficiencies human biochemical and behavioral responses may be impaired. Extreme environmental conditions such as cold may alter human physiology in particular ways.

Humans are subject to attack by pathogens. The health of the human organism may determine the incidence and severity of infection and other illness. Exogenous chemicals that have no role in primary metabolism may limit or increase the severity of attack. The three-way interaction between body, pathogen, and plant chemical is the probable basis for herbal medicine and modern pharmacology (see chapter 8).

Undoubtedly, no plant chemical defenses evolved specifically to deter humans. They are much older than that. Their presence in crops and other plants is a response to threats from microorganisms, insects, plants, and other organisms. Human selection for reduced plant toxicity during domestication must be considered in a dynamic relationship with other ecological factors. In selecting for more palatable genotypes, humans make plants more susceptible to pests and pathogens, and these biological forces act as constraints.

The parasitic agents that attack both humans and plants are shown at the same pole of the basic triangle in Figure 1-1. While the intent of this model is to discuss the interaction between the other two poles—plant chemicals and humans—this cannot be done without an appreciation that plant chemicals have a fundamental role in the interaction of both humans and plants with pests and disease.

TESTING THE MODEL

It is difficult to test any model of human adaptations to deal with plant chemicals and the acquisition of knowledge concerning plant chemicals that is based on the physiological and behavioral considerations described above. Domestication potentially offers a unique opportunity for considering the dynamic evolutionary nature of the direct interaction of humans with specific compounds. Ethnobotanical field studies of an interactive and observational nature (cf. Alcorn 1984) can show the dynamic process of plant use in a broad sense, but they are likely to be frustrated by the subtleness of the evolutionary process at the physiological level. Evaluation of the empirical record of the medicinal and dietary use of plants, and the correlation of phytochemical constituents with biomedical and toxicity data, can help us understand why humans came to utilize particular plants. However, neither of these methods approaches the direct interaction of humans and chemical constituents. They cannot answer the question of how the specific knowledge was gained.

Domestication of plants and animals is the most recent and rapid evolutionary process available for study. Because the process involves humans, it provides a focus for considering humans within an evolutionary context. Humans direct plant domestication, and in turn are affected by and required to adjust to the changes that occur. Studies of toxic plant domestication can focus on the historical record of human-directed chemical change as well as on the ongoing dynamics of humans, plants, and environment in determining plant chemistry.

I have drawn nine testable hypotheses from the model outlined above. While they focus on the domestication process as a component of the quest of humans for food, the insights generated are of broad significance for understanding human relationships with plants through time.

1. The phylogenetic relations of domesticated plants and their wild relatives will reflect chemical changes occurring during the domestication process.

2. Chemical change will be for either reduced levels of allelochemicals (in order to improve flavor or to lessen toxicity) or for greater levels of allelochemicals (in order to increase resistance of the plant to pests or to enhance pharmacological properties).

3. Genetic change is determined by factors in plant reproductive biology in combination with human treatment of the plant populations.

    a. Farming practices will determine the extent to which genetic recombination occurs and the proliferation of particular chemotypes.

    b. Plant reproductive methods will determine patterns of chemical variability. Reproduction involving sexual reproduction and seeds offers greater potential for bringing about plant change than vegetative propagation. Different patterns of chemical variability will occur according to mode of reproduction.

4. Humans can perceive differences in chemical concentration and will show preferences for chemicals consistent with changes in plant chemical makeup seen during domestication.

5. Human language concerning taste, smell, and other chemical properties will reflect human concerns relevant to the chemicals under selection as well as to the chemical environment.

6. Geophagy and other processing techniques as practiced by humans have an evolutionary role in detoxification of plant secondary compounds.

7. Toxic plants used in human diets provide nutrients that would otherwise be deficient.

8. Processing techniques used by humans are effective in removing toxicity while retaining nutrients.

9. High levels of secondary compounds are retained in the diet for ecological reasons.

    a. Chemicals contribute to plants' natural resistance to herbivores and pathogens, and toxic plants may be higher yielding.

    b. Humans may favor high levels of biologically active compounds in some plants if these are recognized as having positive phar-macological activity.

Subsequent chapters will address many of these hypotheses, both by outlining data provided by experiments that test them and by drawing insights from other sources. The domestication of the po-tato, and specifically the interactions of Aymara Indians in Bolivia with the potato, provide a common focus directly related to the model.

POTATO DOMESTICATION AS A MODEL OF CHEMICAL SELECTION

Flavor is a prime factor used by Andean peoples in determining potato nature and quality. In the broadest sense potatoes are classi-fied as bitter or sweet. The level of bitterness appears to be at least partly a direct result of glycoalkaloid content, and bitter potatoes are potentially toxic without elaborate processing. Glycoalkaloids have been shown to be distasteful to humans (North Americans) in con-centrations above 14 mg/100 g fresh weight (Sinden et al. 1976). Although it has been suggested that human selection has reduced levels of these compounds in most cultigens (Dodds 1965; Simmonds 1976; Harborne 1988), including the primitive domesticate *Solanum stenotomum*, high glycoalkaloid content is found in some frost-resistant species that grow above 3,600 meters above sea level on the central Andean altiplano. These species are hybrids between *S. steno-tomum* and wild potatoes such as *S. megistacrolobum* and *S. acaule*, which are adapted to the frigid and arid conditions of this environ-ment. Various approaches and methods described in the following chapters consider the interactions of Aymara Indian cultivators with glycoalkaloids in these potatoes.

Each chapter focuses on a different aspect of the manner in which humans have adapted to deterrent and toxic plant chemicals, using the specific interactions of the Aymara with potato glycoalkaloids as an example. The general themes of chapters 2–5, respectively, are the biological bases for the interaction of humans and other organisms with plant chemicals; solutions offered by processing technologies to detoxify plant toxins; domestication as a solution; and the interac-tion of perception, cognition, and language in determining human behavior toward chemicals.

In specific reference to the domestication of potatoes, chapter 2 considers the toxicity and perception of glycoalkaloids as only one of many examples of plant secondary compounds encountered in the human quest for food. Chapter 3 examines technological solutions to the initial critical impasse to domestication posed by toxic levels of glycoalkaloids characteristic of wild potatoes. The detoxification role of geophagy in the evolution of cultivated potatoes is a major focus of the discussion.

Chapter 4 examines the biosystematic and chemotaxonomic relationships of *S. stenotomum, S. acaule,* and *S. megistacrolobum* with the hybrid species *S.* × *ajanhuiri* and *S.* × *juzepczukii.* Human selection of glycoalkaloid constituents is correlated with the phylogeny of the group. In addition, the dynamic role of humans in effecting genetic change and evolution is examined in relation to current agricultural practices.

Chapter 5 considers the manner in which perceptions of taste are categorized and transmitted by traditional Aymara farmers. The cognitive and perceptual data are considered in relation to the capability of humans to select for changes in the glycoalkaloid composition of potatoes in relation to domestication trends.

This volume consciously attempts to use the case study outlined here as a heuristic device providing insight into a problem of broader interest. Many of the general connections between the case study and the larger picture are not explicitly belabored through most of the book. In chapters 6, 7, and 8 hypotheses drawn from the model are reconsidered, and insights relevant to a chemical-ecological consideration of the human diet and human medicine are discussed.

AYMARA INDIANS AND THE ALTIPLANO ENVIRONMENT

The Aymara-speaking people are discussed rather two-dimensionally and tangentially throughout most of this volume. They are an intricate part of the equation comprising plants and humans that is considered in this study, and as such they are an important component of the chemical-ecological model. Viewing them so narrowly, however, does a disservice to them as people and limits our understanding of the breadth of their adaptation to an exceptional environment.

The central Andean region of Peru and Bolivia actually comprises several environments that have together determined human history in this part of the world. The western coastal desert, the altiplano, the moist eastern slopes and valleys, and the upper Amazon jungle, while separated by distance and by vast physical and biological differences, have been interconnected by human movement and economic activities. The Inca Empire was successful in exploiting all of these zones and represented a zenith in Andean economic and cultural achievement.

Recent studies of Andean cultural ecology have stressed verticality as the primary feature in Andean subsistence and culture (based on the model of Murra 1975). Verticality is an Andean means to achieve self-sufficiency by exploiting multiple agricultural zones along an altitudinal gradient. The altiplano and eastern Andean slopes are of central importance to this study. These are the regions where the potato was domesticated and the regions occupied most extensively by Aymara and Quechua farmers today. The moist inter-Andean valleys are a rich biological area; their ample rainfall and moderate temperatures make them highly productive of a number of crops. The rugged terrain provides the greatest challenge to human existence in this area, but the problems posed by steep slopes are overcome by the endurance of Andeans and by the elaborate terraces constructed during Inca and pre-Inca times. Altitudes ranging from 3,800 meters to below 1,500 meters facilitate the production of crops adapted to different climatic regions. Maize is the dominant staple in the lower reaches of the valleys, while potatoes and other tubers predominate in the cooler temperate regions between 3,000 and 3,800 meters. *Solanum stenotomum* and *S. tuberosum* were probably domesticated in this region.

The majority of the Aymara live on the altiplano of what is today western Bolivia and southern Peru. Lake Titicaca is a geographical and economic focal point in the north of the region occupied by the Aymara, while to the south and west extends a vast treeless plain delineated on the west and east by two mountain ranges, the Cordillera Occidental and the Cordillera Real, respectively. Lake Titicaca, with an elevation of 3,810 meters, has no outlet to the ocean, and its waters drain southward via the Rio Desaguadero to Lake Poopo and eventually evaporate or disappear underground. Saline flats such as the Solares of Uyuni and Coipasa and many smaller salt deposits are a common feature of western Bolivia. The altiplano gets progressively more frigid and arid as one moves south and west away from the moderating effects of Lake Titicaca and the moist air rising from the Amazon basin to the east.

The high plateau is by no means a uniform plain. It is broken by numerous hills that provide a variety of habitats facilitating the exploitation of various cultivated and wild resources. The area of Caquiaviri is typical of the Departments of La Paz and Oruro where I worked. The extensive pampas (flat areas of grasses and woody shrubs) are predominantly devoted to grazing sheep and the American camels, alpaca and llama. Wild vicuñas are fairly common in the area today. Cultivation of frost-resistant potatoes along with the grain crops quinoa, *cañahua,* and barley is carried out in this region. Less frost-resistant potatoes such as *S. stenotomum* and *S. tuberosum* and other tubers are planted in protected valleys and on hillside slopes which are characterized by cold-air drainage and are less subject to frost on cold nights.

Cold and water availability are the predominant factors that determine plant life in this area. Temperatures are relatively stable throughout the year, but dramatic daily ranges from highs of 16°C to below freezing are common. Frost can occur at any time of the year. Agricultural productivity is highly dependent on rainfall, and although the rainy season extends from October to May with average seasonal highs in January and February, the amounts and patterns of rainfall vary considerably from year to year. Periodic droughts are a condition of life here, and are tied in with El Niño, a phenomenon that is increasingly recognized as a major component of variable weather conditions in the Southern Hemisphere and around the world (Brock 1984). Most of the research described in succeeding chapters was carried out in 1982–83 in the midst of an unusual and very severe El Niño–precipitated drought (Brock 1984).

Human survival in this extreme environment is determined by cold, hypoxia, and the availability of food. Atmospheric oxygen is approximately 60 percent of that at sea level, and adjustments in respiration are required of the Aymara and their visitors. As well, the reduced atmosphere exposes people in the area to greater amounts of solar radiation than are experienced elsewhere. Because of their adaptations to cold, hypoxia, and high solar radiation, the Aymara have been subjects of extensive biomedical and ethnographic studies. While they show numerous physiological adaptations to their extreme environment, evidence for genetic changes that enable their survival is equivocal. The direct discussion of physiological, behavioral, and technological adaptations of the Aymara and other Andeans is beyond the scope of this book, but details can be found in various sources (Baker and Little 1976; Bastien and Donahue 1981; Buck et al. 1968; Gade 1975).

Human beings first came to the altiplano as hunters of camelids and other animals following the last ice age. Domestication of camelids probably preceded plant cultivation in the area, with cultivation gradually expanding from its base at lower altitudes and the favorable areas around Lake Titicaca. The Tihuanaco culture (A.D. 100–1200), which was centered on the lake, exerted a strong influence on cultural and political developments in the central Andean highlands. The modern Aymara are believed to be descendants of these people. The Aymara remained an intact, although subjugated, society within the Inca Empire, which ruled the Andean region extending from southern Columbia to northern Chile and Argentina from A.D. 1430 until the Spanish conquest in A.D. 1532. The language of the Aymara is distinct from Quechua, the tongue of the Incas of Cuzco, and the two groups continue to coexist in Bolivia and in the Department of Puno, Peru. The Aymara are the largest ethnic group in Bolivia, making up about one-third of a population of five million inhabitants, and another half a million Aymara live in the Department of Puno. Although the Bolivian Aymara continue to be dominated by the Spanish-speaking and largely urban minority, their numbers make them an important force in the national identity. Access to universal bilingual education since the agricultural reform of 1952 has brought them increasingly into the economic and political mainstream of Bolivian life, and they have been gradually exposed to technological and cultural influences from outside. Urban migration, particularly into La Paz, has increased dramatically in the last twenty-five years, and many Aymara have joined the mercantile class.

Because of their numbers, their predominance in the countryside of the altiplano, and their relative exclusion from economic and political power, the Aymara have been conservative in incorporating change, and thus the majority have retained a traditional life-style. Ecological conditions of the altiplano also contribute in a major way to the lack of change in the Aymara life-style. Modern technology can offer the Aymara very little that supersedes their own technological and behavioral adaptations developed over millennia. For instance, high-yielding varieties of potatoes that have been introduced from breeding programs into other areas of the Andes do not thrive in this region and have had virtually no impact.

The Aymara have been seduced at times by the material benefits offered by modern technologies and economic systems. Change has been positive where its economic base is self-sustaining. However, taking advantage of the benefits of modern civilization requires orienting oneself toward city life. Lacking access to solid capital, the uncertainty of the Bolivian economy continues to make the Aymara dependent on a subsistence life-style. The constraining forces of cold and drought continue to determine their well-being, and it is in the face of these forces that traditional Aymara ways of life remain superior. The world order being what it is, they themselves are often inclined to acquiesce to the too-often patronizing admonitions of outside experts and technocrats. However, it is they who come closest to being masters in this harsh environment, and scientists must first learn from them. The merging of traditional technologies with modern scientifically based advances is inevitable and not undesirable. However, it must be based on careful understanding of the Aymara and their way of life, in their terms. This study has attempted to contribute to that understanding.

The betterment of the lives of people like the Aymara is the goal of much of the effort in international development. Such groups are often innocent victims of political and technological changes on the road to progress. Their well-being suffers both in relation to failure to meet their material needs and in the loss of control over their own lives. Direct efforts to assist traditional groups often fail, no matter how well meaning. The reasons they do so may be unclear. What is clear is that transfer of technology and cultural values must be done within the context of the localized cultural traditions and biological adaptations. Meaningful efforts in development must be accompanied by basic sociological and biological research.

EVOLUTION IN HUMAN AFFAIRS

In this volume I attempt to examine both human biological and human cultural solutions to the same problem. Although human responses to plant secondary compounds have involved concurrent biological and cultural developments, cultural evolution and biological evolution are not one and the same. In mingling the two, one must proceed with considerable caution. Culture innovation undoubtedly does feed back to evoke genetically determined changes in human biology (Tobias 1981), and the interconnections of biology and culture have become more complex with time.

Plant domestication, which is central to the model developed here, has been viewed in recent work by Rindos (1984) as a case of coevolution involving plants and humans. While that model is useful conceptually and provides new insights into human-directed evolution of plants, it is not helpful in linking the coevolution of plants along biological lines to humans along cultural lines. Human intent must be considered as part of the domestication process; otherwise, the role of plants in domesticating us is as great as our role in domesticating them (K. V. Flannery, personal communication).

Intent is a crucial element that must be considered in relation to human history. Consciousness leaves no archaeological record, and discussions about intent and consciousness versus unconsciousness in evolution are primarily philosophical in nature. Nonetheless, if consciousness and intent are important aspects of human interactions with the environment today, then they were just as important in the past. Our intellectual capacities are a product of biological evolution and have a genetic basis. While the cultural manifestations of these intellectual capabilities have been compounded over time, little suggests that these capabilities themselves have evolved significantly in many thousands of years. Who will say that the contemplation shared by reader and author through a book, or the modern elaboration of science and technology, are not representative of the reality and force of human conscious intent? To deny intellect, conscious awareness, or intent—whether these be qualities of people of other cultures, other races, or of *Homo sapiens* in the past—is to deny the humanness of others and ourselves. Certainly culture is a product of and intercurrent with biological evolution, but it is my assumption that through conscious self-awareness, or the reflective function, it is also independent from it.

Understanding the interactions of our culturally determined lifestyle with our biology are essential for the well-being of all members and groups of the human race. Central to this interaction is food procurement, which has been recognized as a factor in many aspects of our biology and social structure. The biological basis for our sometimes aggressive nature must be understood if we are to cope successfully with the problems of human social relations in an increasingly complex and interdependent world. Similarly, understanding the biological determinants of nutrition are essential for understanding many of the diseases of modern life.

Our ability to supersede culturally the constraints that biological evolution places on our behavior and physiological adaptability depends on our conscious awareness of the biological, philosophical, and ethical implications of our nature and past. Science can offer humankind this conscious awareness of who we are. Ironically, whether or not the human community reconciles itself to its place in nature is the true test of our capacity to rise above and be independent from our biological origins.

2 Biological Adaptations for Dealing with Plant Toxins

Animals seek the constituents necessary for their biochemical processes from the complex mixtures of chemicals that make up their food. For herbivores, every meal may contain potentially hazardous plant allelochemicals, and like other organisms they are challenged to absorb the compounds essential for metabolism and growth while eliminating others that are nonuseful and potentially harmful. Even nutrients can occur in excess, and a balance must be maintained. Many plant allelochemicals and other natural and synthetic chemicals are highly reactive compounds that can interact with body constituents in dynamic, sometimes destructive, ways. Toxins and pharmacological agents have common mechanisms of action, and in using drugs therapeutically it is essential to employ a concentration that is effective but nontoxic.

Biologically active chemicals act in a multitude of ways, and organisms have evolved a variety of physiological, biochemical, and behavioral mechanisms to minimize the potential damage from such compounds. This chapter discusses basic biological mechanisms that are employed by animals for avoiding the toxic effects of naturally occurring constituents of food or for overcoming toxicity. In reality these mechanisms, outlined in Table 2-1, act not independently but as a coordinated concert of reactions to problems of toxicity.

TASTE AND SMELL PERCEPTION OF PLANT ALLELOCHEMICALS

The task of obtaining adequate nutrition while avoiding toxic effects of food is made much simpler if wholesome foods are consumed and toxic substances are avoided. Herbivores are capable of rejecting plants containing potentially harmful chemicals, and the chemical senses—e.g., olfaction and gustation—play a fundamental role in the

Table 2-1. Human Biological Adaptations for Dealing with Environmental Toxins

---

Physiological
    Chemical senses
        Gustation
        Olfaction
        Urtication
    Detoxication enzymes
    Bacterial transformations
    Vomiting
    Proline-rich proteins

Behavioral
    Conditioned responses
    Neophobia
    Sensory-specific satiety
    Geophagy

---

interactions of all animals with their chemical environment. The sensory capabilities of animals, including humans, have evolved in line with their particular survival needs. When an animal regularly encounters a chemical that has specific significance to its survival, there is selective pressure for it to evolve specific sensory capabilities to recognize that chemical. The perceptual apparatuses of specialist animals that consume a limited range of foods need only be capable of discriminating salient characteristics of the food from other materials lacking these characteristics.

Receptor cells specific to particular classes of chemicals are known in insects. For example, the cabbage butterfly (*Pieris brassicae*), which feeds only on plants of the family Brassicaceae, has receptors specifically sensitive to the glucosinolates that characterize this family (Schoonhoven 1967). Comparable specialization is uncommon in mammals, with the exception of a few such as koala bears and mountain viscachas (see chapter 1). It has not been shown, however, that these animals have specific chemoreceptors. Omnivores, such as humans, must make choices among a wide range of possible food items. Their chemical senses respond to a wide range of chemicals of various classes as well as function in coordination with various learning mechanisms.

The sensation of taste in vertebrate animals, including humans, is

mediated by chemoreceptors in the oral cavity, primarily on the surface of the tongue. Smell receptors are located in the nasal cavity,
where they are stimulated by odorous molecules during respiration. Taste buds comprise gustatory receptor cells and accompanying
cells. The axons of gustatory cells connect via synapses with the
fibers of interneurons, whose axons synapse in turn with cells of the
central nervous system. In contrast, the olfactory receptors occur
singly in the nasal epithelium, and their axons make direct connections with the cells of the central nervous system.

Odorous molecules are trapped in the mucus covering the olfactory epithelium and then stimulate the olfactory receptors. Gustatory receptor cells have microvillar projections into the apical pore of
the taste bud, where they may encounter stimulant molecules that
likewise are trapped in a mucous covering. Stimulatory molecules
interact with molecules in the membranes of receptor cells, causing
changes in the ion permeability of the membrane that result in a
depolarization of the membrane. A depolarization of sufficient magnitude produces an action potential that is transmitted along the
membrane surface of the entire receptor cell. The interactions of
organic chemicals with receptor macromolecules on the surface of
taste and olfactory receptor cells depend on steric, ionic, and/or
polarity properties of the chemical. A receptor may be stimulated or
inhibited by different molecules, and a molecule may stimulate or
inhibit different receptors.

Chemical perception is complicated further by the trigeminal
nerve's sensitivity to chemical stimuli. The trigeminal innervates
the epithelia of the head, including the oral and nasal cavities, and is
responsible for pain, touch, and temperature as well as chemical perception (Silver 1987). Sensations such as astringent, harsh, pungent,
and metallic may be produced by stimulation of the trigeminal nerve
(Lawless 1987). Irritation of the trigeminal nerve is involved in the
perception of naturally occurring compounds such as capsaicin (from
chili pepper) and piperine (from black pepper).

The action potentials from gustatory, olfactory, and trigeminal
receptors are processed in the central nervous system in such a way
that the animal recognizes the qualities and quantity of the chemical
stimulus. In vertebrates many taste receptor cells may contact a single neuron, and a fundamental question in understanding perception
is how the chemosensory information is coded by the receptors and
subsequently interpreted by the central nervous system. The hy-

pothesis of across-fiber pattern coding maintains that the precise recognition of a chemical (and a mixture of chemicals in a food item) depends on specific interactions of the chemical(s) with different receptors and the pattern of action potentials thus generated (Maes and Erickson 1984). The activity of any single neuron does not signal a separate message. Generalist receptor cells respond to a range of chemicals but differ in the sensitivity and nature of their response to particular chemicals. Different cells respond to a spectrum of stimuli and may overlap with other cells in the portion of the sensory spectrum they recognize (Boeckh 1980). The overall pattern of sensory output generated by a particular chemical in all the receptor fibers leading to the brain is unique, and recognition of a chemical then depends on the interpretation of this total pattern within the central nervous system (Chapman and Blaney 1979).

Human verbal communication facilitates understanding of human perception of chemicals, and while different languages recognize varying sensory sensations (see chapter 5), sweet, salt, sour, and bitter are often recognized as distinct sensations by scientists and laypersons alike in Western societies. The Japanese concept of umami, associated with consumption of monosodium glutamate (MSG) (Kawamura and Kare 1987), may also correspond to a specific physiologically based perception. However, it is questionable whether distinct types of taste receptors exist (O'Mahony and Ishii 1987).

In view of the general response of most receptor cells, it is difficult to project these human taste categories onto animals. In any animal a particular range of receptors with a continuum of receptor sensitivities may exist. Sweet, salty, sour, and bitter may have specific meaning for humans in that receptors with the capabilities to recognize and discriminate certain types of stimuli relevant to our nutritional needs are found in higher frequency and may be grouped together. An animal's perception will probably reflect its nutritional needs, and a different nutritionally relevant pattern of perceptual stimuli will probably be found (Maes and Erickson 1984) in each animal species.

Olfactory receptors appear to reflect an even greater continuum of sensitivities to stimuli than taste receptors, and the identity of discrete categories of odor is highly problematic. There are probably no fundamental odor categories; and to a large degree efforts to create categories of odor reflect factors in human cognition (Cain 1980). Food flavor is a subjective perception of the combined effects of all chemicals on both taste and olfactory senses.

Perception of plant allelochemicals does not involve only taste and smell. As Swain (1979) points out, no textbook dealing with sensory perception mentions tannins, but they can be sensed. Astringent compounds like tannins exert their effects on proteins of oral membranes and probably on the trigeminal nerve. Urtication is caused by the effects of the chemicals found in stinging hairs of a number of plants. For example, *Urtica dioica* (Urticaceae), the stinging nettle, contains histamine, serotonin, and acetylcholine (Hegnauer 1973), which cause pain and itching in the affected area.

The perception of certain compounds should evoke appropriate behavioral responses in an organism. Human preferences for various taste stimuli are discussed in chapter 5 and can be associated with motivation to consume or not to consume a particular substance. In general, sweetness signals energy-rich compounds, while bitterness correlates best with toxic compounds that are generally aversive. While the nature of many responses is genetically determined, many other responses are learned.

Chemical perception is a fundamental characteristic of all organisms. Chemical threats shared by animals have remained essentially unchanged over time, and perhaps because of this constancy much of the gustatory and olfactory processes occur in anatomically primitive areas of the brainstem (Scott 1980). Our responses at this level are genetically determined, while learned experiences such as that associated with conditioned taste learning (see below) are mediated at the level of the forebrain (Scott 1980).

Temporal aspects of chemical recognition are important (Maes and Erickson 1984) in the way an animal responds to an ecologically relevant stimulus. Rats can make a fast approximation of taste quality of some stimuli in less than 0.10 second (Beidler 1987), while fine discrimination of the intensity of some substances may take considerably longer. Humans can make similar fast and simple qualitative discriminations in 0.4 to 2.2 seconds. Rapid recognition of approximate taste quality appears to have a function in facilitating the evaluation and rejection of unpalatable and toxic substances. Slower integration of taste stimuli may involve information learned through experience.

Chapman and Blaney (1979) discuss the specific perception of plant allelochemicals by animals. Insects have been studied most extensively in this regard, while the majority of studies with vertebrates have focused primarily on quinine as a representative bitter substance. In general, the characteristics of the electrophysiological responses of receptor cells to secondary plant compounds are no different from responses to any other chemical. While receptors for specific compounds are unlikely in vertebrates, some receptors may show greater sensitivity for certain plant chemicals or perhaps for certain types of functional groups. Even though quinine has no ecological significance to rats and macaques, these animals show some sensitivity at the neurophysiological level, with certain nerve fibers in the tongue of each animal responding more strongly to stimulation by quinine (Chapman and Blaney 1979). Although the output of these neurons is derived from a number of receptor cells, there is some correspondence between the receptor cells on the tongue and neurons responding to quinine. This suggests a function of the receptors in mediating behavioral responses of animals to this compound or to similar compounds that they might encounter in their normal foraging activities. Caffeine, gymnemic acid, brucine, nicotine (Sato 1980), and flavonoids such as naringin are among other compounds of natural origin that have been employed in taste experiments with vertebrates.

Exposure to some secondary chemicals may modify the response of receptor cells to stimulation by other classes of compounds, and these effects could be an important mechanism by which plants containing these compounds influence the behavior of animals (Chapman and Blaney 1979). The alkaloids sparteine, nicotine, and quinine are feeding deterrents for the red turnip beetle (*Entomoscelis americana*). These compounds stimulate a cell that is also responsive to glucosinolates, which act as feeding stimulants in this animal. However, it is more significant in relation to the behavior of the insect and the defense of the plant that they also inhibit the response of the sugar-sensitive cell (Mitchell and Sutcliffe 1984). This example suggests that plants may have evolved chemicals that inhibit an animal's recognition of sugar as a defense against herbivores.

In vertebrates quinine and other bitter-tasting substances, various local anesthetics (Sato 1980), and the related gymnemic acids and ziziphins (Kennedy and Halpern 1980) have been shown to reduce the sensory responses of some animals and not others. Gymnemic acids, which are found in the leaves of *Gymnema sylvestre* (Asclepiadaceae), and ziziphins from leaves of *Ziziphus jujuba* (Rhamnaceae) are triterpenoidal saponins that depress sweetness perception in humans. Gymnemic acids have a similar effect on dogs, hamsters, and some fly species, but they have no effect on the perception of any of the nonhuman primates that have been tested (Glaser et al. 1984). Gymnemic acids are highly surface active and therefore produce effects on the cell membrane (DeSimone et al. 1980). Studies of chemicals like these are excellent tools for attempting to unravel the complexities of taste perception.

Three naturally occurring proteins from different tropical fruits have been reported to have sweetness-enhancing effects (Cagan 1973). The glycoprotein miraculin from the miracle fruit (*Synsepalum dulcificum* [Sapotaceae]) acts as a taste modifier by making sour substances taste sweet. While it has no effect in rats, guinea pigs, pigs, dogs, or rabbits, nonhuman primates (with the exception of prosimians) do detect sweetness (Hellekant et al. 1974, 1981). The proteins thaumatin, from *Thaumatococcus danielli* (Marantaceae), and monellin [*Dioscoreophyllum cumminsii* (serendipity berry, Menispermaceae)] produce intensely sweet tastes in humans (Cagan 1973). Thaumatin, for example, is two thousand times as sweet as sucrose on a per weight basis.

Both of these proteins elicit a sweet response in catarrhine primates, that is, the infraorder that includes Old World monkeys and higher primates such as chimpanzees, gorillas, and orangutans. A few prosimians and South American primates respond to monellin but not thaumatin. Thaumatin is not perceived by nonprimates. Glaser et al. (1978) concluded that the capability to taste thaumatin developed as long as thirty-eight million years ago.

Gymnemic acid found in leaves may have a natural function as a feeding deterrent, while the taste-modifying proteins found in fruit may serve as feeding stimulants to encourage consumption and seed dispersal. Where responsiveness to a compound such as thaumatin is restricted to a certain phylogenetic group of animals, it is likely that there is an evolutionary relationship between the animals and the plants producing the compound. Sweet-tasting proteins may be produced by the plants at less metabolic cost than an amount of sucrose or other sugars that produce equivalent sweetness. Certainly the perception of secondary compounds is important in the food procurement of animals and may have an important evolutionary role in determining the nature of chemical perception in general. The perception of these compounds needs to be investigated more directly.

The ability to taste phenylthiocarbamate (PTC) or phenylthiourea is not shared by all humans, and its incidence varies in frequency among human populations. It is well known as a marker in studies of population genetics. We share this trait with chimpanzees and rodents (Greene 1974), but general distribution in the animal kingdom is unknown.

PTC is a synthetic compound that is goitrogenic through its interference with thyroid function. Naturally occurring thioureas, thiocyanates, and other compounds released from glucosinolates (see below) have been shown to be goitrogenic as well. It is suggested that our ability to taste PTC is a reflection of the adaptive importance of detecting (and avoiding) this class of compounds (Greene 1974, 1980).

NEOPHOBIA

It is almost always safe to consume familiar items, and neophobia plays an important role in the food selection of most animals. Omnivores exhibit intense neophobia when dealing with biologically relevant tastes (Rozin 1976). They explore gingerly and allow time for a physiological reaction before ingesting more than an exploratory mouthful. Depending on the effects of a sample of a novel food, they may reject it or eat larger amounts of it. I have observed similar behavior by humans sampling new foods (wild fruits) in the upper Amazon in Peru.

Humans are generally conservative in dietary habits and after childhood show little or no tendency toward experimenting with novel foods. Interestingly, the ontogeny of our feeding behavior closely parallels that of mountain gorillas (Watts 1985). During the postweaning period of early childhood children must accept many new foods that are essential for growth and development. Neophobia is reduced in children of this age, and they are more willing to try new foods and accept them after relatively little exposure (Birch and Marlin 1982). The high incidence of poisoning from plants and other sources among children of this age (Klaassen 1985) is consistent with minimal neophobia. The role of adults in introducing children to new foods and protecting them from toxins during this period is clearly important. Conversely, learned associations involving social-cultural dimensions made during this period can be very strong and can last throughout a person's life.

SENSORY-SPECIFIC SATIETY

The concept that humans and other animals develop selective satiety to specific foods as an adaptive means for ensuring the consumption of a nutritionally balanced diet was discussed in chapter 1. This so-called sensory-specific satiety (Rolls 1986) has parallels in the feeding behavior of some animals (Freeland and Janzen 1974). By moving from one plant species to another, animals may avoid consuming large amounts of any particular toxin. This behavior may have a similar basis in relation to human behavior.

CONDITIONED RESPONSES

Animals learn to avoid toxic foods by associating toxicity with a particular taste, and taste aversions in rats and other animals can be evoked by pairing tastes with subsequent illness. In conditioned taste aversion experiments, ingestion of a novel-tasting food or fluid is followed by drug treatment, radiation, or other stress. Taste aversions are unique compared to other associative learning because the conditioned stimulus (novel taste or food) need be administered only once, and it can be separated from the unconditioned stimulus (toxicity) by a considerable length of time. Garcia et al. (1974) list three principles developed from conditioned aversion experiments: (1) all other conditions being equal, the stronger the taste of the food or drink, the greater the aversion induced by illness; (2) given a constant taste, the more severe the illness, the stronger the aversion for the taste; and (3) given a constant taste and equivalent illness severity, the strength of the aversion is inversely related to the span of time between consumption and illness.

Although taste is the most powerful novel stimulus in conditioning aversions to food, other secondary stimuli, including olfactory, tactile, and visual properties of food, are significant in aversion learning. While odor is a poor conditioned stimulus for flavor aversions, when odor and taste are part of a compound stimulus the odor becomes even more aversive than taste (Bermúdez-Rattoni et al. 1986). Conditioned reactions extend to include similar tastes, although along a continuum reflecting the degree of similarity and the nature of the conditioning (Nachman 1963; Domjan 1975). In summary, an animal will avoid a food source after a single encounter that invokes illness. This avoidance need not involve the toxic chemical itself, but other qualities, particularly taste qualities, that the animal associates with it.

Although ethical considerations prohibit carrying out similar experiments on humans, many instances of "natural" or accidental experiments where aversions are associated with gastrointestinal toxicity have been documented (Garb and Stunkard 1974; Logue et al. 1981; Midkiff and Bernstein 1985; Pelchat and Rozin 1982). In addition, in more controlled situations aversions produced by chemotherapeutic agents given to cancer patients have been demonstrated (Bernstein 1978; Bernstein and Sigmundi 1980).

Nausea plays a special role in the acquisition of taste aversions and food dislikes by humans (Pelchat and Rozin 1982). In animal studies taste avoidances are greater when tastes are followed by internal malaise as opposed to peripheral pain. Stimuli coming from the upper gastrointestinal tract (i.e., the stomach) are the most potent. Chemotherapeutic drugs which produce gastrointestinal toxicity are more effective in producing conditioned rejections of tastes than those that do not (Bernstein et al. 1980). This learned reaction to adverse chemicals is clearly adaptive in that it is related to the primary defense against toxicity—nausea and vomiting.

Other physiological effects associated with toxicity may play a lesser role in acquired aversions to foods and drugs. Physiological responses associated with illness, such as altered blood pressure (Kresel and Barofsky 1979) and alterations in sweating and body temperature, may play a role in some cases. These effects, while perhaps not so important to the survival of an organism, may play an important part in the way an organism learns about the physiological effects of nonlethal chemicals that may have either pharmacological or toxic properties. Psychoactive drugs such as those affecting brain serotonin levels have been shown to induce conditioned taste aversions, although these aversions are not as strong or as lasting as those mediated by gastrointestinal effects (Fletcher 1986). These weak conditioning effects may not alter the palatability of tastes, but they may still be important in teaching the negative consequences of a taste.

Pelchat and Rozin (1982) distinguish between food rejection based on taste and that based on anticipated danger. The former, in which sensory properties of a stimulus are altered, may be based on conditioned aversions resulting from nausea, while the latter may result from experiences of other kinds. Human learning and cultural transmission of information from either type of experience can determine our interactions with plant chemicals.

Particularly instructive for understanding the acquisition of dietary nutrients and the human pharmacological use of plants are positive associations made between beneficial effects and conditioned stimuli. Specific appetites develop during nutritional deficiencies for salt, iron, and vitamins. These appetites likely are in response only to the internal state of the organism and do not involve learning. However, previous dietary experiences (Rozin 1976; cf. Woods et al. 1977) may mediate resultant behavioral changes with many nutrient and nonnutrient compounds. When rats were sickened by administering apomorphine, a distinct flavor such as grape or milk offered during the recovery period was subsequently preferred (Green and Garcia 1971). Other conditioned taste preferences in rats have involved pairing novel-tasting substances with recovery from thiamine deficiency (Garcia et al. 1967; Zahorik et al. 1974) and morphine withdrawal (Parker et al. 1973), and with self-stimulation with intracranial electrodes (Ettenberg and White 1978). Rapid satiation produces strong preferences in animals and humans (Booth 1982).

Dietary constituents affect brain neurochemistry and behavior, and in turn neurochemistry can affect feeding and satiety (Wurtman and Wurtman 1988). Of particular interest is the role of precursors in affecting neurotransmitter synthesis. Consumption of tryptophan or carbohydrates that enhance the uptake of tryptophan by the brain can increase the synthesis of serotonin. On the other hand, meals rich in protein tend to raise tyrosine concentration and promote catecholamine synthesis while decreasing the availability of tryptophan serotonin and therefore reducing the synthesis of serotonin (Anderson and Johnston 1983; Spring 1986). The interactions of such effects with learned preferences for food have not been studied.

Conditioned taste preferences have not been as extensively studied as conditioned aversions and have been variously interpreted (Zahorik et al. 1974; Zahorik 1977). As might be expected in relation to the acute effect that specific toxins can have on survival, preferences are not as strong, or as easy to condition, as aversions (Rozin 1976). It is not clear that the same mechanisms apply in the positive and negative conditioning situations. Pelchat and Rozin (1982) ask, "Is there an 'opposite' of nausea (perhaps satiation, or the termination of nausea) which is particularly effective in establishing likes for associated foods?" It seems that acquired likes for good tastes do not result in improved taste but instead parallel Pelchat and Rozin's (1982) "danger" situation in being learned recognitions of probable positive consequences of particular tastes or food. These authors suggest that "in the light of the fact that strong acquired likes are much more characteristic of humans than other organisms, one would be more inclined to look in the direction of cultural or social factors" in producing food preferences.

DYNAMICS OF BIOLOGICALLY ACTIVE COMPOUNDS INGESTED BY ANIMALS

Ingestion of potentially toxic chemicals is inevitable, and such compounds must be transformed and/or eliminated from the body. To understand the strategies of organisms in countering foreign compounds it is necessary also to consider normal biochemical and physiological processes. While much of the focus of toxicology and pharmacology is on the problem of man-made organic chemicals, the same principles apply to understanding the dangers posed by naturally occurring compounds. Indeed, our mechanisms for dealing with environmental toxins must have evolved in relation to naturally occurring plant chemicals. Foreign compounds that play no role in the primary metabolism of an organism, whether they be natural or man-made, are collectively termed *xenobiotics*.

A vast number of biologically active chemicals occur naturally in plants, and each plant produces its own complement of secondary compounds. The phytochemical complexity of plants is such that it is often difficult to determine what chemical is responsible for a particular property. Plant toxins are usually categorized by chemical class. Certain plant taxonomic groupings are characterized by particular chemical types (see chapter 4), and a close parallel may exist between the taxonomy of toxic plants and phytochemicals common to a group. Among the major plant allelochemicals are alkaloids, saponins, cardiac and cyanogenic glycosides, various terpenoids, nonprotein amino acids, and proteinaceous phytohemagglutinins and proteinase inhibitors. However, specific members of these classes may or may not be toxic to humans or other organisms.

Studies of structure/activity relationships correlate the actions of plant allelochemicals with their chemical structure. The toxicity of individual compounds, while not totally unpredictable, can often only be established by looking empirically at the symptoms of intoxication within a particular test organism. The complex interactions of exogenous chemicals with the tissues and biochemical constituents of an organism provide a challenge to the understanding of xenobiotics' mechanisms of action. A comprehensive discussion of phytotoxicity is beyond the scope of this book. It is equally difficult to draw simple patterns from the complex of harmful interactions that occur. Instead, I will give a few examples that look at the toxicology of classes of compounds that have some specific relevance to the problems considered in this book. More detailed botanical discussions of plant poisons can be found in Kingsbury (1964) and Lewis and Elvin-Lewis (1977). Chemically oriented discussions of naturally occurring toxins in plant foods can be found in Jelliffe and Jelliffe (1982), Liener (1980), National Academy of Sciences (1973), and Rosenthal and Janzen (1979), while toxicology and pharmacology texts such as Klaassen et al. (1986) and Gilman et al. (1985) provide relevant background on the dynamics of the action of particular molecules.

Often the interactions of biologically active compounds with constituents of the body are highly specific to tissue and cell types. Chemicals may target particular cellular compounds such as nucleic acids, proteins, lipids, and carbohydrates. They may interact with specific receptors and upset the function of the receptor in normal biochemical processes. These interactions may bring about general damage and failure of various tissues. In general, xenobiotics affect physiological processes in the following ways: (1) alteration of DNA replication, RNA transcription, and protein synthesis; (2) alteration of membrane transport processes; (3) enzyme inhibition and activation; (4) blocking of receptor sites for endogenous chemical transmitters; and (5) affecting the conformation of various macromolecules (other than those already mentioned) (Robinson 1979).

Toxicological studies both document characteristic illness patterns as a consequence of toxin exposure and look for biochemical mechanisms that explain the symptoms observed. Individual organisms and persons exhibit variability in their responses to biologically active compounds, which may reflect genetic or environmental differences, including nutritional status (see chapter 7).

BIOCHEMICAL BASIS OF DETOXICATION

Mechanisms by which organisms protect themselves against toxic compounds are referred to as *detoxication*, which is different from *detoxification*, referring to the process of correcting a state of toxicity. Although the principal means of avoiding toxicity have been studied mostly in animals, the same mechanisms generally apply equally to humans. Most enzymatic systems of detoxication are common to all animals, although the relative rate and capacity of enzymatic pathways, and the time required to degrade particular compounds, may vary considerably from species to species and may reflect evolutionary history (Caldwell 1980). Primates have evolved a few detoxication enzymes that distinguish them from other animals (Caldwell 1980).

Detoxication involves several interrelated components, generally classified as absorption, distribution, biotransformation, and excretion (Fig. 2-1) (Klaassen et al. 1986). The absorption of chemicals from the gastrointestinal tract and the distribution and transport of xenobiotics within the body mediate the degree of toxicity of specific compounds. In addition, the constant shedding of the surface layer of cells of the digestive tract acts as a defense against the direct toxicity of chemicals to digestive and absorptive tissues. Foreign compounds that are absorbed from the gastrointestinal tract, or via other routes, are efficiently eliminated in the bile and through the urinary tract. Pharmacological agents are subject to the same processes—processes referred to as pharmacokinetics in medical contexts (Gilman et al. 1985). In order to understand the action of a particular drug, it is necessary to establish how it is handled and altered by the body.

Both the urinary tract and the bile system are effective excretors of water-soluble, or hydrophilic, compounds, and potentially toxic hydrophilic xenobiotics pose a relatively minimal threat to the organism because they can be rapidly eliminated. A more serious threat comes from fat-soluble, or lipophilic, compounds, and organisms have evolved important biochemical mechanisms for dealing with lipophilic xenobiotics. Metabolism of foreign compounds involves two basic components. First, enzymatically catalyzed reactions involving processes of oxidation, reduction, hydrolysis, and rearrangement change xenobiotics into compounds that can be readily excreted or linked with an endogenous molecule which makes them excretable. The latter process, called conjugation, is the second important component of detoxication. Conjugation may be either enzymatically controlled or spontaneous. The originally ingested compounds as well as their metabolites may be prepared for excretion through conjugation.

The enzyme systems involved in the biotransformations of many compounds are concentrated in the endoplasmic reticulum of liver and kidney cells, with less activity found in the intestinal mucosa, lungs, skin, testes, and thyroid (Freeland and Janzen 1974). The best-known and perhaps most important of these systems are the oxida-

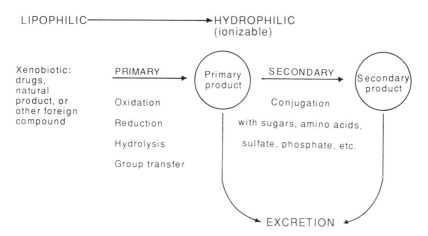

FIG. 2-1. Physiological mechanisms of detoxication.

tive enzymes called the cytochrome P-450 mono-oxygenases, or mixed-function oxygenases. Quite significantly, many of the products of this and other enzyme systems which control the metabolism of xenobiotics are more potent than the parent compound. Highly reactive nucleophilic (electron-rich) and electrophilic (electron-deficient) agents and free radicals are common products of biotransformation. Specific enzymes such as superoxide dismutase play an important role in the detoxication of free radicals produced in the metabolism of primary and xenobiotic compounds (Reed 1985). Even though detoxication enzymes often produce highly reactive intermediates, this is only the first step toward excretion after subsequent conjugation. A balance must exist in each tissue between enzymes that form toxic intermediates and processes that protect the cell and neutralize these highly reactive metabolites.

The most important metabolic conjugation reaction involves the enzyme UDP-glucuronosyltransferase. Glucuronic acid is conjugated to nucleophilic atoms of lipophilic molecules, thus increasing their polarity. Under most physiological conditions this bound acid forms a salt, which greatly increases the ease of its excretion in the bile or urine. The process of glucuronidation is regarded as functioning primarily in the detoxication of harmful compounds. Noncatalytic conjugations also occur between nucleophilic and electrophilic compounds in which one is exogenous and the other is of intracellular origin. Glutathione, a major intracellular source of nucleophilic activity, is an important conjugation agent for defending cells against reactive radicals or electron-deficient intermediates. Dietary antioxidants such as vitamins C and E, β-carotene, and flavonoids play similar protective roles (Reed 1985; Robak and Gryglewski 1988).

Besides cytochrome P-450 mono-oxygenases, other enzymes that metabolize xenobiotics are reductases, hydrolases, and group transfer enzymes (Klaassen et al. 1986). Epoxide hydrases play an important role in the secondary metabolism of epoxide intermediates formed by mono-oxygenases and act in close concert in the inactivation of foreign compounds. Conjugation reactions besides glucuronidation and glutathione transfer involve the addition of amino acids, sulfate, and phosphate groups to make xenobiotics more hydrophilic (Klaassen et al. 1986). Methylation and acetylation of compounds can also change their biological activity. Individual compounds may be metabolized in many ways (even in the same cell) to lead to their ultimate excretion. A few important routes of detoxication involving specific natural compounds are discussed below.

BACTERIAL TRANSFORMATIONS

Chemical compounds may be metabolized in the gastrointestinal tract prior to absorption, or they may be altered in such a way as to prevent absorption from occurring. Although stomach acid conditions may assist the hydrolysis of some compounds, the bacterial flora has been increasingly recognized as playing an important role in biotransformation (Rowland et al. 1985). In humans the highest bacterial populations and the largest amount of biotransforming activity are in the large intestine. Bacterial biotransformations have been implicated in producing toxic and carcinogenic metabolites as well as in detoxifying xenobiotics. Again, most investigations in this area have been concerned with synthetic chemicals. Nonetheless, in humans and other mammals bacteria have been shown to decarboxylate amino acids and transform plant steroids and lignans. Gastrointestinal bacteria are involved in the ring fission and metabolism of flavonoids (Hackett 1986) and have β-glucosidase activity, which hydrolyzes various naturally occurring glycosides to produce their aglycones. Cyanogenic glycosides are hydrolyzed to release hydrocyanic acid, and methylazo-oxymethanol glycosides are hydrolyzed to release their highly toxic aglycones. Mutagenic activity of many plants has been ascribed to flavonoids (e.g., quercetin) that have been released by glucosidases from their corresponding glycosides (Hackett 1986).

Bacterial metabolic activity can interact with liver detoxification systems. Bacteria contain the enzyme β-glucuronidase, which hydrolyzes the compounds conjugated with glucuronic acid in the liver and excreted into the bile. Sulfate and glucuronic acid conjugates of steroidal hormones (Rowland et al. 1985), flavonoids (Hackett 1986), and other compounds may be deconjugated, transformed in various ways, and reabsorbed with altered biological activity.

Dietary modifications, particularly in the proportion of poorly digested dietary components such as fiber and plant cell walls, alter the activity of the gastrointestinal microflora of humans and mammals (Rowland et al. 1985) interacting with xenobiotics. Many studies have been done in this area in relation to the role of diet in colon cancer. Allelochemicals may also have inhibitory effects on gut microflora with implications for nutritional ecology.

Symbiotic relationships between animals and their bacterial flora are important for digestion, and in nonhuman primates with ruminant digestion they have been proposed as an important part of the mechanism by which plant allelochemicals are detoxified (McKey 1978). Humans absorb some of the volatile fatty acids that intestinal bacteria produce through the digestion of cellulose and other plant wall polysaccharides (Cummings 1984), although bacteria are less important in human digestion than in the digestive processes of many other animals. Nonetheless, the interactions of plant nutrients, plant structural polysaccharides, and plant allelochemicals with the gastrointestinal microflora have no doubt been important in the evolution of human dietary patterns.

NATURAL SUBSTRATES OF DETOXICATION ENZYMES

In order to understand the evolution of detoxification systems, we should first consider the natural substrates of detoxication enzymes. Reactive intermediates can pose a more serious threat to an organism than the original compound, and many cancer-causing agents from natural sources are activated in this way. The formation of such negative products might seem to indicate that plant allelochemicals are not natural substrates for detoxication enzymes. In fact, most of the enzymes involved in metabolizing noxious chemicals have multiple functions in basic cellular processes. This might further suggest that detoxication is not their primary function. These enzymes generally have low specificity for the compounds—xenobiotic and endogenous—that they accept as substrates.

This broad specificity of detoxication enzymes does sometimes lead to inefficiency and failure in their defensive effectiveness. This weakness should, however, be viewed in relation to the unpredictability in the myriad possible noxious chemicals that an organism might encounter. The production of specific proteins to metabolize every environmental component is not an efficient solution for an omnivorous animal. It is much more economical to produce enzymes with multiple functions that can be effective in removing a broad range of xenobiotics. It is notable that polyphagous insects appear to be more resistant to novel chemicals such as insecticides than monophagous insects (Brattsten 1979). The nonspecificity of human detoxication enzymes is consistent with the evolutionary view of humans as omnivores.

It should be noted again that most studies on detoxication deal with known toxins and carcinogens, many of them man-made. For example, the *N*-oxygenase enzymes produce many carcinogens from synthetic compounds. However, the majority of the naturally occurring alkaloids they interact with are converted to less toxic derivatives (see below). The cytochrome P-450 mono-oxygenases, glucuronidation, and the detoxication of cyanide by the enzyme rhodanese are examples of systems that have been viewed as having detoxication as a primary function (Jakoby 1980). Mono-oxygenase enzymes are known to be involved in the metabolism of such plant allelochemicals as pyrethrins, rotenoids, nicotine, lignans, and opium alkaloids (Brattsten 1979). Thus allelochemicals can be viewed as the likely natural substrates for many of the important detoxication systems.

INTERACTION OF BEHAVIOR AND DETOXICATION SYSTEMS

Enzymatically controlled pathways in living systems are typically regulated through feedback mechanisms leading to inhibition, activation, and induction of particular enzymes. Induction is important in the efficient response of organisms to unpredictable environmental toxins (Bresnick 1980). After an initial experience, biosynthesis of appropriate detoxication enzymes will enable an animal to ingest greater quantities of a potentially valuable food.

The neophobic behavior of omnivores can act in conjunction with biochemical function at the cellular level. If an animal takes only a small bite of a potentially toxic food, it may avoid negative effects associated with exceeding its detoxification capacity. Enzymes will be produced in response to toxins in food, and greater amounts can be eaten with impunity on subsequent occasions (Brattsten 1979; Freeland and Janzen 1974).

Animals vary in their induction responses (Brattsten 1979). While induction times in insects may take less than an hour, the inducing effects of plant allelochemicals may take a number of days in animals such as rats and rabbits. In short-lived insect larvae, rapid responses to environmental conditions are necessary for survival. On the other hand, mammals usually have variable feeding options and behavioral means of avoiding toxins. Their relatively slow induction responses may avoid the metabolic cost of unnecessarily responding to feeding situations that offer minimal long-term gain. Monooxygenases in mammals are induced by plant allelochemicals such as pyrethrum, α-terpineol, isobornyl acetate, isosafrole, α-pinene, spironolactone, and caffeine (Brattsten 1979).

Taste perception, neophobia, detoxication enzymes, and conditioned learning act in concert to protect an animal against the effects of biologically active compounds. Humans, like many mammals, have little difficulty surviving toxins in their food as long as a variety of foods are available to them. Our behavioral and biochemical defenses can be best understood in the evolutionary context that is discussed in greater detail in chapter 7. Toxins in plants would constrain, but by no means prohibit, the use of plants as food by early hominids. The detoxification technologies discussed in chapter 3 complement rather than replace our normal biological mechanisms for tolerating plant foods.

NATURALLY OCCURRING TOXINS

In spite of the evolutionary importance of naturally occurring compounds in determining the nature of detoxication systems, relatively few toxicological studies have involved these compounds. The following summary focuses on mechanisms of biological activity and the detoxication of several naturally occurring compounds that are discussed in various parts of this volume. Just as the diversity of secondary compounds makes it difficult for organisms to provide specific defenses, so is it difficult to include all natural plant toxins in any discussion. More than one of these toxins is likely to be present in a particular plant, and the effects on, and the responses of, organisms are thus more complicated (Kubo and Hanke 1985).

Synergistic interactions between different allelochemicals are receiving increased attention from chemical ecologists and may be coordinated components of plant defense. For example, myristicin and other compounds with a methylenedioxyphenyl substituent are mono-oxygenase inhibitors and thus may increase the toxicity of compounds that would otherwise be detoxicated (Berenbaum 1985). In reality, the organism as a whole responds in a complex manner and employs a complement of defenses to the chemical onslaught contained in a particular meal. Izaddoost and Robinson (1987) discuss several examples of medicinal plants in which pharmacologically active alkaloids interact synergistically or antagonistically with other active phytochemicals.

CYANOGENIC GLYCOSIDES

The bitter-tasting cyanogenic glycosides are widespread in the plant kingdom (Conn 1979). Amygdalin, for example, is well known from seeds of apples, apricots, bitter almonds, and other members of the Rosaceae. Harborne (1988) summarizes evidence supporting the view that these compounds provide protection to the plant from predation from a broad range of animal species.

These compounds are glycosides of α-hydroxynitrile (cyanohydrin). Enzymes present in the plant or in intestinal bacteria act to release hydrocyanic acid (HCN) from the cyanogenic glycoside (Fig. 2-2). Cassava (*Manihot esculenta*) contains linamarin as its major cyanogenic glycoside. The enzyme linamarase (a β-glucosidase) hydrolyzes linamarin to form glucose and 2-hydroxyisobutyronitrile. In a second step the hydroxynitrile dissociates to form acetone and HCN. *Phaseolus lunatus* and *Sorghum bicolor* also contain cyanogenic glycosides.

Toxicity of plants containing cyanogenic glycosides is directly attributable to HCN, which may have both acute and chronic effects, although other products of the metabolism of cyanogenic glycosides may also have deleterious effects (Conn 1979). Acute toxicity results from the affinity of the cyanide ion CN− for metal ions in enzymes that are fundamental to the respiratory process. By inactivating cytochrome oxidase, the respiratory enzyme which links atmospheric oxygen with metabolic respiration, cyanide brings about asphyxiation at the cellular level. Chronic toxicity from the consumption of low concentrations of cyanogenic glycosides poses a more serious problem worldwide to people dependent on cassava and other cyanogenic foodstuffs for their subsistence.

Although HCN is rapidly absorbed into the body, acute cases of

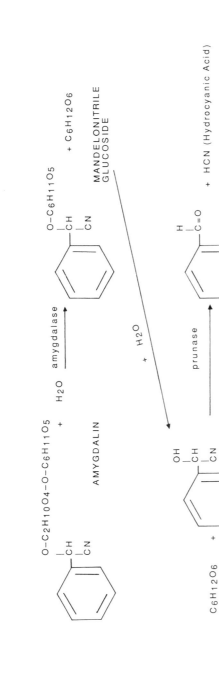

FIG. 2-2. Breakdown of cyanogenic glycosides.

toxicity in humans are rare under normal circumstances because of the efficiency of cyanide detoxication mechanisms. However, severe poisoning from cassava and other foods does occur during periods of famine.

Physiological detoxication is catalyzed by the enzyme rhodanese (Westley 1980), which transfers sulfur to the cyanide ion to form thiocyanate. An available store of sulfur in the form of sulfane is thus essential for detoxication. Thiocyanate competes with iodide in thyroid tissue and leads to thyroid insufficiency. Thus, while the rhodanese system is effective in preventing acute toxicity, its actions add to the chronic effects of cyanide. This detoxication mechanism risks chronic poisoning in return for rapid detoxication of potentially lethal levels of cyanide.

Cyanide detoxication has been viewed as the primary function of rhodanese, although there is no direct evidence to support this assumption. The enzyme may have a much broader role in physiological regulation of sulfane sulfur (Westley 1980). The high concentration of rhodanese in the liver (and kidney) does make it well suited for detoxifying cyanide released from food. The prevalence of cyanide and cyanogenic glycosides in dietary plant materials underlies the need for an effective detoxication system for these compounds, and it is tempting to argue that cyanide is the natural substrate of rhodanese, and further that this enzyme has evolved for the primary purpose of countering the threat of cyanogenic glycosides.

GLUCOSINOLATES

Glucosinolates, which characterize a small number of plant families, including the Brassicaceae, Capparaceae, and Tropaeolaceae, are responsible for the pungent flavors of horseradish, mustard, capers, and a number of common vegetables. In this volume glucosinolates are discussed in relation to *Tropaeolum tuberosum*, *Lepidium meyenii*, and the Capparaceae. Although they contribute positively to food flavor, high amounts of these compounds make plants unpalatable. Chronic effects of glucosinolate-containing plants, particularly endemic goiter, have been associated with metabolites of glucosinolates. Like cyanogenic glycosides, glucosinolates undergo enzymatic hydrolysis when the plant tissue is damaged. The thioglucosidase enzyme (myrosinase) releases glucose, sulfate ion, and an aglycone portion. The latter entity converts to one or more possible products, depending on the nature of the particular glucosinolate and the reac-

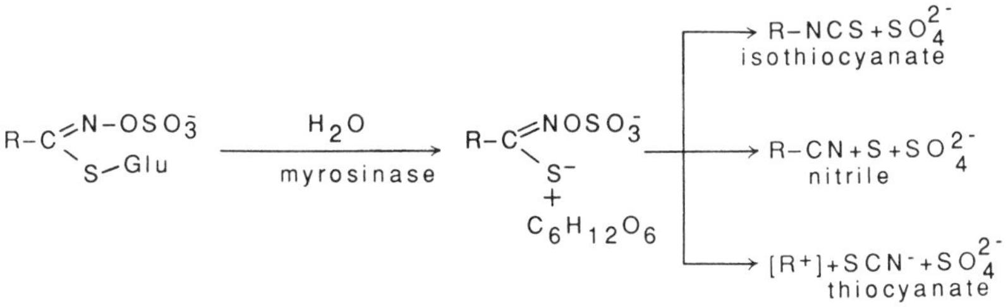

$$R-C\underset{S-Glu}{\overset{N-OSO_3^-}{\lessgtr}} \xrightarrow[\text{myrosinase}]{H_2O} R-C\underset{\substack{S^- \\ + \\ C_6H_{12}O_6}}{\overset{NOSO_3^-}{\lessgtr}}$$

$\longrightarrow R-NCS + SO_4^{2-}$
isothiocyanate

$\longrightarrow R-CN + S + SO_4^{2-}$
nitrile

$\longrightarrow [R^+] + SCN^- + SO_4^{2-}$
thiocyanate

FIG. 2-3. Breakdown of glucosinolates.

tion conditions (Fig. 2-3). Isothiocyanates, or mustard oils, produce the pungent flavors, while thiocyanates, nitriles, and oxazolidine-thiones are responsible for the goitrogenic and other toxic effects of glucosinolate-containing plants. Thioureas resulting from the conjugation of isothiocyanates (Johns and Towers 1981) or the reaction of basic amino acids and small peptides with isothiocyanates may also be responsible for the goitrogenic properties of glucosinolate-containing plants (Benn 1977).

Naturally occurring isothiocyanates differ among themselves in the way they are metabolized, although mercapturic acids formed from conjugates of these compounds with the amino acid cysteine are commonly excreted in the urine of humans and animals (Mennicke et al. 1983).

Isothiocyanates, nitriles, and thiocyanates play important roles in plants' defense against insect herbivores and have been the focus of classic work in the area of plant-insect relations (Rodman and Chew 1980). Isothiocyanates are also antibiotic (Johns et al. 1982).

ALKALOIDS

Alkaloids are a very diverse and widely distributed group of plant secondary compounds. They are broadly classed as compounds derived biosynthetically from amino acids that contain nitrogen as part of a heterocyclic ring. The nitrogen atom makes most alkaloids basic. Hence their actions in physiological processes, digestive processes in particular, can be affected by changes in pH.

The diversity of alkaloid types makes it extremely difficult to make generalizations about the biological importance of these compounds. Robinson (1979) wrestles with the ecological role of alkaloids from the perspective of plant-herbivore interactions. The modes of action of alkaloids and their metabolism within biological systems encompass almost a complete overview of xenodynamics in

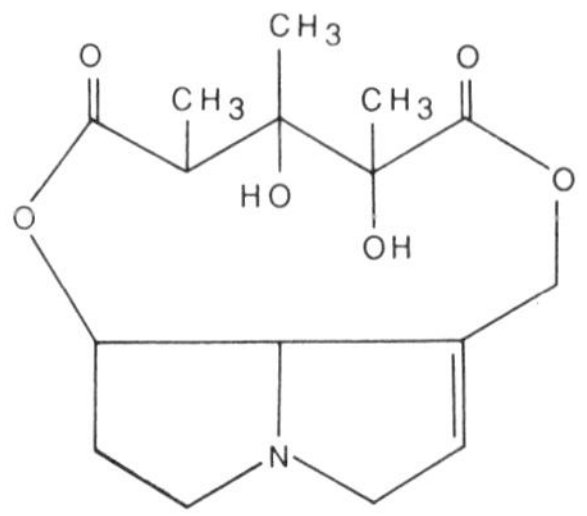

Monocrotaline

FIG. 2-4. Pyrrolizidine alkaloids.

general. Many alkaloids are not toxic to humans. Others, such as quinine and reserpine, are important drugs. They vary in their taste properties; many, such as quinine, are extremely bitter, while others, such as capsaicin, irritate the trigeminal nerve. Pyrrolizidine alkaloids (Fig. 2-4), which are found in *Crotalaria* spp., *Senecio* spp., and other plants have received particular attention because of the hepatotoxicity and carcinogenicity reported for many of these compounds (Mattocks 1968).

Detoxication of alkaloids proceeds by a complex variety of routes and again is best studied in relation to individual classes of alkaloids. A reaction that is characteristic of these molecules is *N*-oxygenation of the heterocyclic nitrogen. Many amine compounds are converted to highly reactive intermediates by *N*-oxygenation enzymes. Conjugations with sulfate, glucuronic acid, and acetyl groups are important in alkaloid metabolism (Robinson 1979). Interestingly, most naturally occurring alkaloids yield less toxic derivatives (Bickel 1969). *N*-oxygenation of alkaloids appears to be an effective route for which alkaloids may be natural substrates.

GLYCOSIDES OF TRITERPENES

Terpenoid compounds are the largest and most structurally diverse class of allelochemicals (Geissman and Crout 1969). They have a common biosynthetic origin as derivatives of mevalonic acid. Common classes of terpenoids occur as multiples of five carbons. Monoterpenes (C_{10}) are often volatile compounds that are part of many fragrant plant materials and essential oils. They are biologically active as well. Iridoids and their glycosides are biologically important molecules derived from monoterpenes. Among other important biologically active terpenoids are sesquiterpenes and sesquiterpene lactones (C_{15}) and derivatives of diterpenes (C_{20}).

The compounds discussed in detail below are glycosides of compounds with a triterpenoid (C_{30}) origin, except for cucurbitacins,

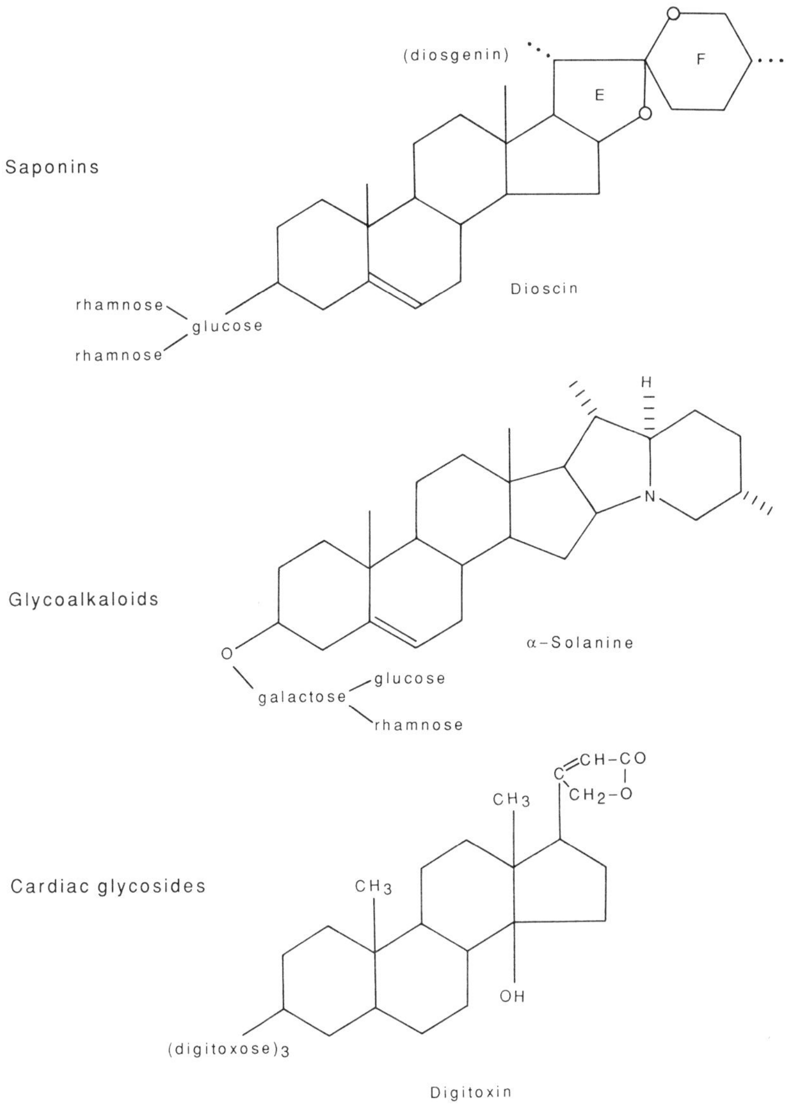

FIG. 2-5. Structures of the triterpene glycosides diosin, α-solanine, and digitoxin.

which are nonglycosidal triterpenoids. Degradation and structural modifications of a triterpene intermediate give rise to sterols, of which cholesterol is the most important, as well as other biologically active compounds. Included in the classes of glycosides with a steroidal or triterpenoid aglycone are saponins, glycoalkaloids, cardiac glycosides, and withanolides (Fig. 2-5). Because of their structural relationships to cholesterol and other membrane constituents of mammals, the aglycone portions of these compounds generally exert their actions on cell membranes. Their sugar components increase their solubility in aqueous environments.

Saponins. Saponins are glycosides of both steroidal and triterpenoid compounds and are widespread in the plant kingdom (Applebaum and Birk 1979). Their presence in *Chenopodium quinoa, Dioscorea* spp., and potatoes is discussed in this volume. Saponins find widespread folk use as fish poisons, arrow poisons, soaps, and medicines. They generally have a bitter taste and are characterized by their properties of foaming in aqueous solutions and their ability to hemolyze red blood cells. Hemolytic activity against red blood cells, intestinal wall cells, and other cells probably accounts for the major part of the toxicity of saponins (Birk and Peri 1980). The wide range of biological activities of saponins includes effects on basic metabolism and on the cardiovascular system (Chandel and Rastogi 1980; Mahato et al. 1982). Much of their biological activity probably results from interactions with cholesterol and other membrane components. Saponins also have antinutritional effects in mammals. By complexing with cholesterol, they make this steroid unavailable for absorption and can noncompetitively inhibit dietary enzymes.

Some saponins are active antifungal and antibacterial agents. The variety of activities exhibited by individual saponins makes it difficult to generalize about their role in the plant beyond saying that saponins as a class (but not necessarily individual compounds) act as defenses against a broad range of pathogens and herbivores.

Saponins are generally unpalatable and when ingested are poorly absorbed, if at all, from the gastrointestinal tract. Their primary activity is against intestinal tissue, and the irritation they induce leads to vomiting and diarrhea. While they can cause damage to humans and other animals, they are rarely life threatening when eaten. They are deadly arrow poisons because they enter the bloodstream directly.

Complexing with cholesterol may act as a form of detoxication of saponins in the gut (Applebaum and Birk 1979). Saponins are metabolized not in the small intestine but in the large intestine, where they are hydrolyzed to nontoxic aglycones, probably through the action of microbial enzymes.

Cardiac Glycosides. Cardiac glycosides have steroidal aglycones and are generally limited to the Apocynaceae, Asclepiadaceae, Liliaceae, Moraceae, Ranunculaceae, and Scrophulariaceae. They are not characteristic of any major human food plants but are noteworthy for their pharmacological activity as well as their toxicity.

As their name implies, these compounds are cardiotonic, that is, they increase the contractility and improve the tone of heart muscle (Gilman et al. 1985). Digitoxin and digoxin derived from *Digitalis purpurea* and *D. lanata* are among the cardiac glycosides that are widely used therapeutically in treating congestive heart failure. Their therapeutic dose is often close to the toxic dose (Gilman et al. 1985), and the use of cardiac glycosides as medicine thus poses some risk. Both the positive and toxic effects of cardiac glycosides arise from the inhibition of the enzyme Na,K-ATPase, which is involved in the transport of sodium and potassium ions across cardiac cell membranes. These glycosides are readily absorbed from the gastrointestinal tract. Digoxin is excreted unchanged in the urine, although digitoxin is hydrolyzed in the liver to its aglycone and subsequently metabolized before excretion.

Glycoalkaloids. Glycoalkaloids are glycosides of nitrogen-containing steroidal compounds. Because they are not derived from amino acids they are often considered pseudoalkaloids. They are often chemical analogues of saponins and may co-occur with related saponins in the same plant.

Taste perception of glycoalkaloids has not been studied physiologically, but it appears to result at least partly from an astringent effect on oral membranes, perhaps in combination with direct effects on taste receptors. The bitter taste of glycoalkaloids develops slowly (within fifteen to thirty seconds) and is described as a hot/burning, persistent irritation at the side of the tongue and the back of the mouth (Woolfe 1987).

Solanine, the glycoalkaloid common in most potatoes, is the most widely studied toxin in this class. It is considered to be toxic to humans in quantities above 20 mg/100 g fresh weight (Gregory 1984; Jadhav et al. 1981), although cases of solanine poisoning are rare. Glycoalkaloid poisoning involves gastrointestinal and neurological disturbances similar to those caused by the related classes of steroid glycosides, saponins, and cardiac glycosides (Baker et al. 1987; Jadhav et al. 1981). Glycoalkaloids produce hemolytic effects similar to saponins. Their neurological effects are at least partly due to their activity as inhibitors of the enzyme acetylcholinesterase (Jadhav et al. 1981).

Like saponins, α-solanine and other glycoalkaloids are poorly absorbed from the gastrointestinal tract and pose minimal threat in low concentrations such as those found in most cultivated potatoes (Jadhav et al. 1981). They are hydrolyzed in the gastrointestinal tract to their aglycones, which are less soluble, less likely to be absorbed, and less toxic. The hemolysis and irritation of intestinal cells by glycoalkaloids brings about vomiting and diarrhea, which further limit the chance that absorption will occur. In higher concentrations glycoalkaloids concentrate in the spleen, kidneys, heart, brain, and blood. α-Solanine is readily excreted in the urine and feces, although other glycoalkaloids—for example, α-chaconine—appear to be less readily eliminated from the body (Woolfe 1987).

CUCURBITACINS

Cucurbitacins are a group of closely related triterpene compounds found in the Cucurbitaceae. Humans find these compounds to be the most bitter chemicals in the plant kingdom. They are usually too intensely bitter to allow a toxic dose to be ingested. Cucurbitacins do, however, play a role in the interactions of plants and insect herbivores (Harborne 1988).

Cucurbitacins will be discussed in chapter 4 in relation to chemical selection during the domestication of a number of important food plants.

METHYLAZOXYMETHANOL GLYCOSIDES

Cycads, a primitive, pantropical group of plants, all contain a closely related series of glycosides with the nitrogen-containing methylazoxymethanol (MAM) aglycone but differing in the sugar moiety. Enzymatic hydrolysis by β-glucosidase liberates the toxic MAM aglycone, a highly reactive molecule that has been primarily studied as a carcinogen (Feinberg and Zedeck 1980). Glucosidase produced by intestinal bacteria plays an important role in the toxicity of MAM compounds (Freeland and Janzen 1974). The acute toxicity of MAM aglycones likewise results from the reactive nature of the compounds (Freeland and Janzen 1974). Typically, unprocessed cycad plant material results in severe gastrointestinal disturbance, fever, numerous other symptoms of distress, and often death. Whiting

(1963) presents an extensive summary of historic and scientific data related to the use and dangers of cycads as food.

PHYTOHEMAGGLUTININS (LECTINS)

Phytohemagglutinins, or lectins, are proteins characterized by their ability to agglutinate red blood cells (Jaffé 1980; Liener 1979). They possess a specific affinity for certain sugar molecules. Carbohydrate moieties that occur in most animal cell membranes may act as specific receptor sites for lectins. If the phytohemagglutinin has at least two active groups, it can act to agglutinate red blood cells and other cell types. Acute toxicity may result from the impairment of the function of various tissues. Absorption of nutrients from the intestine is reduced in rats fed phytohemagglutinins, and chronic growth retardation and death may result. Damage to the intestine may be lethal in severe cases.

Phytohemagglutinins are found in a number of higher plants, particularly in legume seeds. Beans (*Phaseolus* spp.), faba (broad) beans, soybeans, lentils, and peas all contain lectins, which are also responsible for the toxic effects of *Abrus precatorius,* the rosary pea. Although this is one of the most poisonous plants known, it can be eaten after sufficient processing. Phytohemagglutinins are also present in potatoes and various cereals. They may play a role in the resistance of plants to insect herbivores (Janzen et al. 1976).

Phytohemagglutinins are not absorbed from the intestine. They produce gastroenteritis after a period of time which may be as long as three days. Toxicity is limited because the body can replace damaged cells at a rate faster than they are destroyed.

INHIBITORS OF DIGESTIVE ENZYMES

The toxins discussed above act directly on the tissues. Tannins (see below) and inhibitors of digestive enzymes are sometimes distinguished from toxins because they act as digestibility-reducing substances (Rhoades and Cates 1976). Digestive inhibitors, which are themselves proteins, bind to and inhibit the activity of proteolytic enzymes (Liener and Kakade 1980) as well as amylase, the pancreatic enzyme that functions in starch digestion (Gallaher and Schneeman 1986).

Proteinase inhibitors are widespread in plant foodstuffs. Trypsin and chymotrypsin inhibitors have been extensively studied from soybeans (*Glycine max*), other legumes, cereals, potatoes, sweet potatoes, eggplants, and other important crop plants. Their role in affecting the nutritional status of animals ingesting plant foods is a complex matter. There are considerable interspecies differences in animals' susceptibility to particular inhibitors and physiological responses to them. Primates generally appear to be minimally affected by low levels of proteinase inhibitors (Harwood et al. 1986; Schneeman and Gallaher 1986).

Growth depression and enlargement of the pancreas are the two metabolic responses most commonly studied in relation to trypsin inhibitors (Schneeman and Gallaher 1986). Short-term detoxication mechanisms are not totally satisfactory. An increase in enzyme production may restore proteolytic function but is limited by the cost that can be tolerated in terms of the loss of protein and energy. Pancreatic enlargement occurs when the organism excretes more pancreatic enzymes in response to regulatory feedback signals from the digestive tract. While human enzymes are certainly affected by proteinase inhibitors (Weder 1986), the capacity of the human pancreas to adapt to dietary inhibition is unknown (Schneeman and Gallaher 1986). In general, the importance of proteinase inhibitors in vivo under normal circumstances is not known (Rothman 1986), and although proteinase inhibitors have been shown to have some role in resistance to insects (Ryan 1979), their specific protective role has not been directly tested (Nelson et al. 1983).

TANNINS AND OTHER PHENOLICS

Flavonoids are the most ubiquitous plant allelochemicals and are a constituent of most plant foodstuffs. Over three thousand flavonoids of a wide range of structural types are known. They appear to serve a variety of functions in plants (Swain 1986), including defense against animal predators. However, perhaps because they are so widespread and of such antiquity in the plant kingdom (Swain 1986), humans and other animals have evolved efficient enzymatic systems for their detoxication.

Hackett (1986) recently reviewed flavonoid metabolism in mammals. Bacterial enzymes in the gut are important in the ring fission of specific flavonoids and in the demethylation and dehydroxylation of phenolic acid metabolites, as well as in the hydrolysis of flavonoid glycosides to form aglycones. Mono-oxygenases are involved in the oxidation of flavonoids, and reduction, methylation, and conjugation reactions with glucuronic acid and sulfate have also been reported.

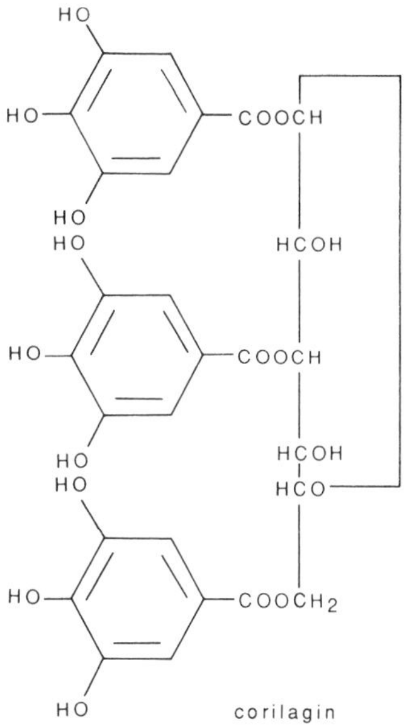

FIG. 2-6. Structure of a hydrolyzable tannin.

Although flavonoids are considered nonpoisonous to humans, many of them are biologically active and have important potential as therapeutic agents (cf. Cody et al. 1986). Indeed, many medicinal plants owe their activity to flavonoids. Flavonoids are highly variable in taste: many are tasteless, while bitter compounds such as naringin and highly sweet synthetic derivatives such as various dihydrochalcones have been extensively studied (Horowitz 1986).

Tannins are large-molecular-weight polyphenolic compounds (Fig. 2-6). Hydrolyzable tannins can be enzymatically or spontaneously degraded to yield ellagic or gallic acids. Condensed tannins, which are polymeric flavonoids, do not break down readily under physiological conditions.

Tannins form complexes with proteins and other molecules, which is the characteristic associated with their traditional use in leather making. They are highly astringent when ingested and are unpalatable in levels that are acutely toxic to humans. Tannins act as strong feeding deterrents and cause growth depression in many animals (Butler et al. 1986). Decreased digestion of food is probably caused by tannins interacting with digestive enzymes, including trysin, α-amylase, and lipase (Griffiths 1986), as well as through binding of dietary protein into an indigestible form. However, the ecological importance of tannins in plant-herbivore relations is unclear (Mole and Waterman 1985). Swain (1979) suggested that the major evolutionary importance of tannins is in protecting plants against fungal and bacterial attack.

Tannic acid, a hydrolyzable tannin, is absorbed from the gastrointestinal tract and can cause liver damage in humans (Gilman et al. 1985). An 8 percent tannic acid diet is fatal to rats (Freeland and Janzen 1974); in general, concentrations higher than 2 percent tannic acid deter feeding in herbivores. Acorns (*Quercus* spp.) and some varieties of sorghum are high in tannins. Acorn species vary considerably in their content of tannic acid. Ofcarcik and Burns (1971) reported a range of 0.7–8.8 percent tannic acid among twelve species of acorns growing in Texas.

Recent reports that sorghum high in tannins stimulates the production of proline-rich glycoproteins by rat and mice salivary glands (Butler et al. 1986) are intriguing. Hamsters, which do not synthesize these proteins, are much more sensitive than rats or mice to the deleterious effects of tannins, and it is suggested that by complexing with tannins these compounds play a role in the detoxication of naturally occurring tannins. Proline-rich proteins may play a role in human interactions with plant allelochemicals (see Table 2-1).

Proline-rich proteins comprise about 70 percent of the protein content of human parotid gland secretions and may be important in enabling humans to utilize tannin-containing plant foods. Humans have been postulated to have a taste for tannins, and Butler et al. (1986) suggest that perhaps as a consequence humans may select diets containing sufficient tannins to maintain the parotid gland in the induced state.

At low concentrations tannins actually stimulate the hydrolysis of proteins by the mammalian enzyme trypsin (Mole and Waterman 1985), which suggests that herbivores may be adapted to a certain quantity of these compounds in the diet. Equally intriguing is evidence showing that some hydrolyzable tannins inhibit the growth of *Streptococcus mutans*, a bacteria that causes tooth decay (Kakiuchi et al. 1986). Perhaps these compounds play an anticariogenic role as dietary constituents.

Flavonoids have antioxidant properties. The ability of many of these compounds to scavenge superoxide anions may be responsible for their hepatoprotective and vasoprotective properties (Robak and

Gryglewski 1988). Additionally flavonoids may spare vitamin C and thus in times of dietary stress help prevent the onset of scurvy (Carpenter 1986).

The possible role of phenolics as normal positive constituents of human physiological ecology has important implications for the model of the evolution of medicine that is presented in chapter 8.

OXALATES AND PHYTATES

Oxalic acid and phytates are common nonnutritional substances that form chelates with various minerals. Excessive intake of phytates, which are components of many plant fibers, has been linked with nutritional deficiencies of zinc, copper, and iron (Hambraeus 1982). Oxalates, which are common in various plant products, also make various minerals unavailable and may have adverse effects on renal function. Calcium oxalate crystals, which are widespread in plants of the Araceae and other families, are involved in the oral irritation caused by ingestion of these plants.

In spite of their detrimental aspects, phytates, like tannins, may interact in a beneficial way with digestive processes (Thompson et al. 1987) and might be considered a normal part of our nutritional ecology.

BACTERIAL AND FUNGAL CONTAMINANTS OF FOOD

Bacterial and fungal contamination of foods can have unpredicted and dramatic consequences and has played a historical and perhaps evolutionary role in human food procurement. Mycotoxins such as ergot alkaloids and aflatoxins likely became health problems after the development of agriculture, and as food contaminants are associated with particular cultivation practices, methods of food storage, and preparation of plant products. While ergot can produce dramatic short-term health problems, aflatoxin (derived from the mold *Aspergillus flavus*) and other mycotoxins such as those from *Penicillium* and *Fusarium* fungi produce more subtle, longer-term effects for which organisms have few adaptive remedies (Hambraeus 1982). Although aflatoxin may produce acute health problems (Srikantia 1982), it is best known as a very potent carcinogen. Ergotism results from derivatives of the alkaloid lysergic acid that are produced by *Claviceps purpurea*, a parasitic fungi on cereals, particularly rye and wheat. Outbreaks of ergotism were a periodic affliction in Europe during the Middle Ages. Ergot alkaloids cause vasoconstriction, which leads to convulsions and gangrene of the extremities. The convulsions, which lead to painful spasms of the limb muscles and to epileptic-type seizures, were known as Saint Anthony's fire during medieval times.

Microbial exotoxins—proteins that are released by microorganisms—are the chemicals associated with food poisoning. Some exotoxins, such as the botulism neurotoxin, are produced prior to the ingestion of contaminated foods. Botulism is a potentially fatal neurological disease caused by *Clostridium botulinum,* an anaerobic bacterium found in the soil and often associated with animal carcasses and improperly canned food (Smith 1982). Severe diarrhea, nausea, vomiting, fever, and dehydration result from toxins released into food by organisms such as *Staphlococcus aureus* and *Clostridium perfringens.*

Other food-borne infections proliferate in the gut and the body's tissues and cause food-poisoning-like symptoms in humans. These include the *Salmonella* species that are responsible for salmonellosis, *Salmonella typhi* (typhoid fever), *Shigella* spp., *Vibrio parachaemolyticus, Vibrio cholerae* (cholera), *Yersinina enterocolitica, Campylobacter jejuni,* and *Escherichia coli* (Corry et al. 1982). *Vibrio cholerae, Shigella, E. coli,* and the exotoxic organisms discussed above all release enterotoxins—protein exotoxins which act on the surface of the epithelial mucosa of the intestine. The other organisms mentioned above, along with *Shigella* and *E. coli,* invade host tissue and produce lipopolysaccharide endotoxins within host cells which induce fever and related disease symptoms. Humans are the prime reservoirs of *Staphlococcus aureus, Salmonella typhi, Shigella, Vibrio cholerae,* and *E. coli,* and poor hygiene and improper food preparation and storage likely are responsible for most problems with these organisms. *Clostridium perfringens, Salmonella* organisms causing salmonellosis, *Yersinina enterocolitica,* and *Campylobacter jejuni* are widespread among mammalian and avian hosts, and foods of animal origin are the most frequent source of human infection. *Vibrio parachaemolyticus* is found in seafood.

Humans can develop only limited and transient immunity to enterotoxins. Although effective immunity can be developed to endotoxin-producing organisms, over two thousand types of *Salmonella,* as well as numerous types of *Shigella* and *Campylobacter jejuni,* are known. Thus the chances of encountering infectious microorganisms are high.

Herbivores likely encounter deleterious quantities of microorgan-

isms only rarely. Our ancestors, who evolved from plant-eating primates, may have had minimal exposure to bacterial contaminates
and the toxins they produce until relatively late in their evolutionary
history. Since these organisms and their toxins have no specifically
associated taste, they are difficult to avoid. Thus organisms such as
Salmonella, Campylobacter jejuni, and *Clostridium botulinum*
may have presented severe problems to early hominids eating raw
meat obtained through scavenging and hunting until cooking was applied to control them (cf. Washburn and Lancaster 1968). Fortunately,
cooking is an effective way of controlling most food-poisoning toxins
and pathogens.

MECHANISMS OF DETOXIFICATION

When the detoxication capacity of an organism has been exceeded,
methods exist for preventing further harm and for hastening the
process of recovery. The biochemical mechanisms described above
will continue to reduce the concentration of toxins in the blood and
other tissues. At the same time toxicity is recognized and responded
to at the neurophysiological level through symptoms such as hypo-
thermia, increased blood pressure, and general malaise, which are
typical of illness states of various kinds. Two very effective behaviors
for dealing with toxins are vomiting and geophagy. The former expels
toxins directly, the latter binds them so that they are not absorbed as
they pass through the gastrointestinal tract. Both mechanisms are
triggered by the direct gastrointestinal disturbance caused by many
compounds, as well as by more complex mechanisms involving
other physiological and psychological processes.

VOMITING

Vomiting provides relief for many animals from gastrointestinal distress caused by poison-induced illness. Dogs and cats induce vomiting by eating grass and other unusual substances, perhaps after eating
contaminated food. Human vomiting is often a response to a food
that is unacceptable for some reason. It is an important biological
survival mechanism, but its practice can be inhibited by cultural
influence. Someone who seldom vomits is said to have a "strong
stomach," implying that he or she is resistant to the negative effects
of some undesirable substance. There is clearly a psychological component to vomiting that ties in closely with the conditioned aversion
response and with cultural norms.

The detoxification function of vomiting is recognized by most people. In fact, most traditional folk remedies for poisoning act to induce vomiting. Emetics, substances that induce vomiting, are widespread in folk medicine and have an important place in modern Western forms of therapy. The emetic of choice in clinical treatment of poisoning is ipecac, the alkaloid-containing roots and rhizomes of *Cephaelis ipecacuanha* from Brazil (Windholz 1976). Lewis and Elvin-Lewis (1977) list a few examples of remedies explicitly stated to be antidotes to internal poisoning. As these authors point out, many of the substances used to cause vomiting are themselves very toxic.

The concept of cleansing attributed to vomiting extends from the purely functional cleansing of the gastrointestinal tract to more symbolic and spiritual forms of ridding a person of an undesirable condition. Ritual vomiting has important significance in many cultures. For example, *Ilex vomitoria*, the black drink of Indians of the southeastern United States, was used ritually to invoke vomiting in order to achieve a state of purity (Hudson 1979). The religious recognition of something that relates in such a basic way to biological survival shows how seriously humans have understood the threats of dietary poisons.

GEOPHAGY

Interestingly, the second detoxification behavior that I consider is also often practiced by humans in a ritualistic manner. For example, in Central America geophagy, or the consumption of mineral substances, is a religion-endorsed practice combining Amerindian, Spanish, and African cultural traditions (Hunter and de Kleine 1984). Geophagy is a complex human activity that is common worldwide and crosses all geographical, racial, and ethnic boundaries.

Soil consumption has been observed in many animals, including invertebrates (Shachak et al. 1976), reptiles, birds, and mammals (Kreulen and Jager 1984). Although clays may be important nutritionally, geophagy also may have a detoxification function in several mammal species (Kreulen 1985). The best experimental evidence supporting this hypothesis comes from studies with rats, which consume clay when subjected to chemical toxins in laboratory settings. Mitchell et al. (1977) demonstrated that if a rat develops a conditioned taste aversion to a novel (and nontoxic) flavor, it will respond to this conditioned stimulus by consuming clay, even if clay was not administered during the original conditioning episode.

Geophagy in rats is thought to have evolved in relation to the gastrointestinal illness that occurs after eating toxic substances. Rats are notoriously difficult to control by poisoning, and Mitchell has concluded (personal communication) that geophagy plays a big part in their capacity to survive in an often unpredictable dietary environment. Interestingly, rats lack the capacity to vomit; thus geophagy and aversion conditioning may be more important responses to poisons in this species than in other animals.

Geophagy has been observed in many nonhuman primates that include toxic plant material in their diets. For example, specific clays, often from termite mounds, are eaten daily, sometimes seasonally, by chimpanzees (Hladik and Gueguen 1974; Uehara 1982; Wrangham 1977). The nature of geophagy has led primate ecologists to suggest a detoxification function in relation to plant secondary compounds (Davies and Baillie 1988; Oates 1978). However, this hypothesis has been difficult to test experimentally and is not universally accepted. In chapter 3 I will demonstrate the detoxification function of clay consumption by humans, as well as the importance of the cultural elaboration of the practice in human dietary history.

POTATO TOXICITY

Potato tubers contain a number of chemicals that act as defenses against attack by herbivores and pathological microorganisms. Glycoalkaloids, saponins (Kaneko et al. 1977; Osman et al. 1982), a phytohemagglutinin (Woolfe 1987), proteinase inhibitors, sesquiterpene phytoalexins (Tjamos and Kuc 1982), and phenols (Woolfe 1987) have all been reported from potato tubers. Tuber-bearing species of *Solanum* vary in the identity and quantity of the glycoalkaloids contained (Gregory et al. 1981; Johns and Osman 1986; Osman et al. 1978). Most cultivated potatoes have less than 10 mg/100 g fresh weight of glycoalkaloids. While wild species generally have a higher glycoalkaloid content, quantities vary greatly, with some clones having levels of several hundred mg/100 g. While the major cultigen, *S. tuberosum*, contains primarily α-solanine and α-chaconine, wild species and cultigens formed from hybrids with wild species contain a variety of glycoalkaloids with the demissidine, solanidine, tomatidine, tomatidenol, and leptinidine aglycones. Proteinase inhibitors can make up more than 15 percent of the soluble protein of mature potato tubers. At least thirteen different inhibitors have been identified from the potato, including an unusual inhibitor of the carboxypeptidase enzyme (Woolfe 1987).

The cohort of chemical defenses in potatoes act in concert to protect the plants from herbivores and pathogens; conversely, they present complex detoxication problems for animals exploiting potatoes. On the other hand, potatoes offer a high-quality food for humans and other animals. They provide abundant energy in the form of carbohydrate, 2 percent protein, appreciable quantities of ascorbic acid and other water-soluble vitamins, and minerals such as iron, magnesium, and phosphorus. Wild potato tubers contain as much as 4 percent protein. In spite of the nutritional value of this plant, any organism exploiting potatoes must maintain a positive balance between the nutritional benefit and the toxic cost. The quality of the potato as a human food is greatly enhanced by cooking. Cooking results in a high-quality, digestible protein which is moderately limited in sulfur-containing amino acids but contains substantially more lysine than cereal staples. Woolfe (1987) has recently summarized information on the nutritional quality of the potato from a human perspective.

The levels of glycoalkaloids, proteinase inhibitors, and possibly the phytohemagglutinin in unprocessed wild potatoes generally make them unacceptable as a human food. The phytohemagglutinin and the proteinase inhibitors, with the notable exception of the carboxypeptidase inhibitor, are generally heat labile. The acute toxicity and unpalatability of wild potatoes are a result of their glycoalkaloid content. Severe gastrointestinal disturbances and vomiting are the preliminary effects of potato poisoning. Glycoalkaloids, as large-molecular-weight cations (under stomach conditions), are very effectively bound by clays. Significantly, geophagy has been employed extensively in the detoxification of bitter potatoes.

3 Technological Methods of Detoxification

Technological advances have made humans less susceptible to the dangers of dietary toxins and less dependent on the biological defenses discussed in chapter 2. Consequently, cultural innovations have made plant foods more available to humans than they are to much of the animal kingdom, or than they would have been to our prehominid ancestors. Cooking, the controlled use of fire to alter the character of food, is the most obvious method by which humans expanded their food resources. Evidence for fire utilization dates back at least 500,000 years (Stahl 1984), and its effects on human resources must be considered in any discussion of the evolution of the human diet. The use of heat, however, is only one of the methods employed to improve food quality. Leaching, fermentation, grating, the use of lye, and drying are all techniques that may have considerable antiquity.

This chapter considers the nature of these techniques as they are applied by people around the world today to detoxify plant foods. Most important, it seeks to provide insight into the importance and significance of these processes through time. The final portion of the chapter is devoted to an extensive discussion of geophagy, both in general terms and in relation to the utilization of potatoes by humans. Geophagy is the one of the detoxification techniques that is the most difficult to understand. As a behavior with a biological basis, its cultural elaboration may provide a link to the way our earliest ancestors dealt with plant toxins that extends through time.

Although the elimination of toxic and bitter constituents in plant foods is the focus of this discussion, we should also recognize that some processing techniques can improve foods by making them more digestible or more palatable in several ways. Cooking, soaking, grating, and the addition of lye are used widely to soften foods.

Comminution increases surface areas of foods on which digestive enzymes can act. Increasing the digestibility of foods makes nutrients more available.

However, processing techniques may remove or destroy nutrients along with toxins and thus fail to improve the metabolic balance between these two qualities of plants. Balance can also be considered in terms of return for effort. Processing requires an expenditure of energy that must be returned in added caloric value (Stahl 1989).

Whatever the purpose of processing techniques, several key problems arise. First is the question of the origins of these techniques. Their behavioral basis can be seen in the simple manipulation of plants through peeling and the simple use of clay as a detoxicant as practiced by some animals. Unfortunately, technologies other than cooking leave a poor archaeological record. However, the fact that the same basic processing methods are practiced worldwide today by traditional peoples suggests that they have considerable antiquity. Important techniques such as boiling and grinding may predate the origin of agriculture by only ten or fifteen thousand years (Washburn and Lancaster 1968). Second is the question of how humans learn these various techniques. Since detoxification processes developed by humans are directly related to plant chemistry, they reflect the interaction of humans with these compounds. Third, were the same methods of food processing discovered repeatedly worldwide? How has the basic knowledge been disseminated? And fourth, are the methods utilized worldwide specific to particular plants or to the same classes of chemical constituents?

A survey of processing techniques (Johns and Kubo 1988) shows that humans worldwide have considerable sophistication in detoxifying plant foods, and the methods employed are often highly specific to particular plants. However, the basic mechanisms by which detoxification is achieved are universal and can be categorized. The processing methods described in the following section have specific physical-chemical functions in detoxifying plant foods. Often these basic methods are used in concert rather than individually. These various combinations in particular reflect the sophistication of individual traditional groups in processing plants. Appendix 1 presents a classification scheme of detoxification techniques developed around the world. While the following discussion assumes an understanding of elementary chemical concepts, everyday experience and common sense should make most of these techniques as sensible to the reader as they were to the preliterate peoples who developed them.

HEATING

Heating provides energy to drive chemical reactions and can be applied to plant foods through a number of cooking techniques. Each plant contains a complex mixture of constituents, and many different chemical reactions are likely to occur within it. Toxic organic compounds may participate in these reactions, and in converting or degrading to other compounds these chemicals may lose their toxic properties. On the surface, where food is exposed to oxygen in the air, compounds can be heavily oxidized, and often follow common pathways such as browning. In the interior of a cooking food less oxygen is present, the temperature is usually less due to the temperature-limiting effect of the water present, and more complex reactions among various constituents may take place. Because most foods are composed primarily of water, the interior reactive environment will be less than 100°C. Many plant toxins are highly stable at this temperature, and thus heating is not always an effective means of detoxification.

Rather than acting on the poisonous principle itself, heat may also act by denaturing enzymes that are necessary to liberate certain active compounds such as glucosinolates and cyanogenic glycosides. Some proteins such as proteinaceous lectins and proteinase inhibitors (see chapter 2) are themselves toxic. In addition, the exotoxins excreted by bacteria, as well as bacteria themselves, are usually destroyed by heat. The activity of proteins is usually dependent on the secondary structure of the molecule, and heating deactivates these bacteria toxins and plant enzymes by destroying their natural conformation.

Boiling and some form of roasting or baking are the most common cooking techniques used worldwide (Stahl 1989). Although many plant foods are eaten raw, most are cooked in some way. However, detoxification is usually not the explicit function of the cooking process.

Roasting was the first cooking technique and perhaps the only one used for most of human history. Fragile plant parts such as leaves are likely to be destroyed by roasting. Not surprisingly, they are rarely detoxified with this technique (cf. Johns and Kubo 1988), and the use of fire probably provided little innovation for early hominid use of leaves.

Boiling requires the availability of watertight containers. Although woven baskets and clay pots can be used for boiling foods, metal pots have greatly increased the distribution of this technique. Many peoples solved the problem of applying heat to water by placing heated rocks directly within the container with its contents.

While boiling is a means of applying heat, it can also function as a means of increasing the rate and solubility of some compounds and leaching them out of a material. In the classification scheme presented in Appendix 1, boiling is included under detoxification by solution as well as detoxification by heat. It is difficult to separate the two functions, and classifying a particular case is a matter of judgment as to the most important part of the process. Although water boils at 100°C, the addition of solute to the solution will elevate the boiling point several degrees. The greater the resultant temperature, the faster the chemical reactions involved will take place. Perhaps for this reason people often boil foods in salt water.

SOLUTION

Water is a solvent that is readily available to humans in most parts of the world, and it is widely used as a way of removing toxins and bitter compounds. Although this technique takes many forms, it basically involves dissolving the toxic compounds in the water and leaching them from the food. Compounds vary in their solubility in water, and the solubilizing techniques will vary in efficiency depending on the chemicals involved. Some organic compounds are already in a soluble state within a tissue. Less soluble constituents require greater processing time.

The process is enhanced in specific and often sophisticated ways. The contribution of heat has been discussed above. When solubility of a toxin is low, a turnover of water, either by repeated pouring off and replacing, by placing the object in running water (Fig. 3-1), or by passing water through a food, will help. Salt increases the polarity of the aqueous environment and can help make certain compounds more soluble. Any process that causes more tissue to be exposed to the water or liberates plant constituents by destroying the integrity of the plant cells will speed up leaching. Many techniques combine the breaking down of tissues through grinding, grating, pounding, freezing, and so on with the use of water. These techniques, which are collectively termed *comminution*, allow cellular contents to be leached out more rapidly. A high or low pH may increase the degradation of tissue constituents that make up cell walls. A change in pH can sometimes affect the solubility of certain compounds, specifi-

FIG. 3-1. Leaching *Oxalis tuberosa,* "oca."

cally alkaloids. Changing the pH through the addition of ashes or acids is discussed below.

FERMENTATION

The alteration of the chemical composition of plant and animal materials by yeast, bacteria, and fungi has achieved considerable refinement in the processing of dairy products and soybeans, in the baking of bread, and in the production of alcoholic beverages. Microorganisms are ubiquitous, and fermentation proceeds spontaneously under appropriate conditions.

Useful fermentation conditions foster the growth of nonpathogenic microorganisms while excluding those responsible for food poisoning. Temperature and an anaerobic environment are important in many controlled fermentation processes. Simpler fermentation techniques are part of the human repertoire of detoxification worldwide. The basic techniques employed involve burying a plant food or enclosing it in some kind of container. Comminution will enhance the fermentation. Microorganisms involved in fermentation may employ detoxification and digestive enzymes similar to those of the intestinal microorganisms described in chapter 2.

ADSORPTION

Chemical constituents in foods may be bound by physical and chemical processes to other substances. Components of dietary fiber can interact with organic compounds and eliminate their positive or negative interaction with the absorptive tissues of the gastrointestinal tract. The decreased exposure of intestinal tissue to dietary toxins has often been discussed in relation to the function of dietary fiber in decreasing colon cancer (Reddy and Spiller 1986). Charcoal is the standard substance used in cases of acute toxicity in modern clinical settings (Gilman et al. 1985). Charcoal and clay are both made up of small particles and thus have large surface areas. They undergo weak interactions with organic compounds, primarily through van der Waals and electrostatic forces. Clay mineral lattices may be charged (usually negatively), and adsorption of chemicals may also occur by ion exchange.

The pharmacokinetics of clay-organic interactions under gastrointestinal conditions have been studied (White and Hem 1983). Deliberate drug release or inadvertent drug deactivation when clays are ingested with various medications, either purposefully or accidentally, have motivated most such investigations. Pharmacological data are directly applicable for understanding adsorption of naturally occurring compounds and the role of geophagy as a detoxification technique (cf. Browne et al. 1980). Alkaloids and other naturally occurring compounds have been shown to be effectively bound by clays (Johns 1986a). Binding depends on reaction conditions and on the nature of the clays. Montmorillonite clays have higher surface areas and cation exchange capacities than kaolinites and other clay types and are superior adsorbers of organic compounds (see below).

Humans deliberately use the adsorption properties of clays and perhaps other substances such as charcoal to bind toxins in food in ways that appear to be elaborations of the geophagous behavior of animals. Although charcoal is reported to be eaten by humans as a form of pica (Lackey 1978), charcoal ingestion has rarely been reported among traditional peoples. Interestingly, coprolites believed to be those of Neanderthals were found to contain charcoal (Kliks 1978). Detoxification using clay basically involves adding the substance directly to food during processing or at the time of ingestion, or soaking the plant product in wet mud. The role of geophagy in detoxifying bitter potatoes and other plant materials is discussed in more detail below.

DRYING

Drying is likely to be an effective technique for removing volatile toxins from food. It is usually used in combination with one of the other methods of detoxification.

PHYSICAL PROCESSING

The importance of comminution methods such as grating, grinding, or pounding was mentioned above in relation to other modes of processing. In some situations such processing is a primary means of detoxification. The enzymatic release of volatile, water-soluble, or heat-labile compounds utilizes the metabolic machinery of the plant cell. Compounds such as cyanogenic glycosides and glucosinolates, while part of the defensive arsenal of plants, are themselves nontoxic until cleaved. These glycosides are stored in plant cells separately from glucosidase enzymes. When the plant is damaged and its tissues are disrupted, glucosidases contact and interact with the compounds; the sugar molecule is removed and the toxic chemicals, which act to deter attacking predators, are released (see chapter 2). Even the ingestion of the nontoxic principles is undesirable, however, because the human gut can also cleave glycosides and activate the toxins. Grating cassava (*Manihot esculenta*) is a widespread mechanism for detoxifying bitter varieties of this important staple in the world's tropics. Hydrogen cyanide is released into the atmosphere during processing rather than while the plant is being chewed or digested. Because heating destroys the glucosidase enzyme, and because acidic gut conditions and bacterial glycosidases can release HCN from cyanogenic glycosides, it is important that comminution precede cooking rather than follow it.

pH CHANGE

The importance of pH changes in affecting the solubility of chemicals has been discussed above. In addition, acidic and alkaline conditions can lead to hydrolysis of many compounds such as glycosides and amides. Cyanogenic glycosides, for example, are unstable in both dilute acid and base.

Concentrated acidic substances are not readily available in the environment and appear to be rarely employed deliberately in traditional food processing. However, acidic fermented products such as vinegar and organic acids from fruits such as tamarind (*Tamarindus indica* L.) may serve some role in the hydrolysis of secondary compounds. Pickling is carried out with some toxic plants in combination with other techniques and may play a role in producing the final nontoxic product. Acids formed during fermentation may contribute to the breakdown of toxins. Tamarind pulp is widely used in tropical regions as a flavor additive, although because of its acidity it may play other roles in altering food quality. Tartaric acid, which makes up 10 percent of the weight of tamarind pulp (Windholz 1983), is a good organic buffer. A concentrated solution of tamarind that I tested had a pH of 2.54. I am familiar with three cases where tamarind is used to detoxify food. Two of these cases involve plants in the family Araceae (App. 2), which often have high levels of calcium oxalate. Significantly, tartaric acid is effective in increasing the solubility of the highly irritating crystals (raphides) of this compound (Oke 1969). The third case of tamarind use with toxic plants involves roots of the legume *Neorautanenia mitis*. The genus is characterized by rotenoids and other toxic flavonoid derivatives (Brink et al. 1977).

Alkaline materials in the form of lye from plant ash and mineral lime are often more available than acids for processing foods. Different cultural groups often use very specific ashes, and they burn specific plants to get the desired product. Water washes of such plant ashes can be in the pH range of 10–12. Hydrolysis reactions of common chemical linkages such as esters, acetals, and hemiketals can readily occur in solutions of this pH range. Ashes are usually used in solution and are often combined with heat. Heat will greatly speed up the hydrolysis process as well as assist in cellular breakdown under alkaline conditions. Alkaline substances also improve the availability of nutrients such as niacin and amino acids (Carpenter 1981).

PATTERNS OF DETOXIFICATION

My survey of detoxification techniques (Johns and Kubo 1988) lists 137 genera and 216 species from 65 families of plants that are used after some detoxification. While this survey is comprehensive, it is not exhaustive. Although similar processing techniques are used in various other circumstances, this survey lists only those cases where it was explicitly stated that the plants were being detoxified or bitterness was being eliminated. The cycads, the families Araceae, Dioscoreaceae, and Fabaceae, plus *Quercus* spp. and *Manihot esculenta*, are especially important for the degree that detoxification has contributed to their exploitation around the world.

These plants represent a unique subset of the world's flora. With the exception of the Fabaceae and *M. esculenta,* few of the processed plants are members of the families of flowering plants with the largest number of species. Neither, with the exception of the legumes, are the heavily processed plants included among the families of major economic plants.

The classification scheme presented in Appendix 1 demonstrates the complexity of individual detoxification techniques applied by people around the world, although many more permutations could be added. The complexity and sophistication of processing is demonstrated in numerous cases. For example, in West Africa rhizomes of *Anchomanes difformis* (Araceae) are subjected to prolonged washing and cooking, with hearth ashes added to the cooking water. The plant material is then left in the water to ferment for several days before it is sun dried and stored (Burkill 1985). On the coast of East Africa *Dioscorea* yams (*D. dumetorum* or *D. sansibarensis*), which contain high levels of saponins and alkaloids, are sliced, boiled in several changes of salt water, and finally washed in fresh water (Weiss 1979).

There is no one correct way to detoxify a plant. The variety of techniques applied successfully to make certain plants edible reflect cultural variations as much as physical, chemical, and biological constraints. The best example of this is again cassava. Lancaster et al. (1982) and Miracle (1967) provide excellent discussions of the myriad mechanisms for detoxifying tubers of this plant.

With a few exceptions, the same basic methods of detoxifying plant foods are found worldwide. Of the seven major techniques listed above, heat, solution, and physical processing are part of detoxification procedures on all the inhabited continents and in Oceania. Our survey did not encounter the use of ashes in Europe, South America, or Australia. However, in each of these areas the sample size of cases is less than for Africa, Asia, and North America. The chemical properties of ashes are, in fact, used in South America with the cocaine-containing stimulant coca. In Australia, while food may be roasted in ashes, boiling was uncommon, and thus the opportunity for adding alkali to solution was virtually nonexistent. No cases were encountered where fermentation was reported for detoxification in Europe or Oceania. Nonetheless, fermentation is widely employed as a food-processing technique in both of these areas (Cox 1980; Pederson and Albury 1969). Geophagy was not encountered as a detoxification technique in Oceania, although geophagy is practiced to a limited extent in this area (Laufer 1930; Glaumont 1889). Drying as a detoxification technique was not encountered in South America, although it is important in food processing in the Andes at least.

Why do humans go to the trouble to process certain plant foods and not others? People who utilize toxic plants always exploit other plants that require little or no processing. Perhaps the answer lies in the fact that the processed plants are all of widespread distribution and produce a reliable, recognizable, and abundant food resource. The aroids, cycads, yams, acorns, and cassava are all major carbohydrate-supplying staples for various cultural groups. The legumes represent another source of abundant food. However, none of the major exploited legumes (e.g., beans, peas, and lentils), except edible lupines, require detoxification (other than cooking).

Why has detoxification technology rather than domestication provided the solution to overcoming plant chemical defenses of certain plants? In actuality, many of the plants in Appendix 2, including cassava, *Colocasia esculenta,* and *Dioscorea* spp., are important domesticated plants, and in each case some selection for chemical change has been noted. The same ecological reasons discussed in chapter 1 for chemical defenses to be maintained in crops apply here as well. Processing provides a means of exploiting toxic genotypes that may have the defensive capacity to be vigorous and high yielding in the face of predator attack.

In long-lived trees such as oaks, domestication is less feasible than in plants where selection can be exerted on each annual generation. Genetic factors controlling the inheritance of secondary compounds, and human cultural factors that affect the dynamics of wild and cultivated plants (see chapter 4), may also limit the possibility of selection leading to genetic change. Thus processing may often be a more practical method of detoxification than domestication.

The particular chemicals that are eliminated or destroyed during processing might reveal something about detoxification techniques in general, and on a broader perspective the nature of the chemicals involved might reveal something about the direct interaction of humans with allelochemicals. Are there certain chemicals that humans recognize as unfavorable and work to eliminate from food? The extensive lists in Johns and Kubo (1988) and Appendix 2 both show a lack of any clear pattern. Human techniques eliminate a large range of secondary compounds, seemingly representing a cross-section of the classes of chemicals found in plants. Appendix 2 includes plants

containing calcium oxalate, alkaloids, glucosinolates, MAM glycosides, cyanogenic glycosides, saponins, tannins, phytohemagglutinins, and nonprotein amino acids.

It is difficult to say whether humans apply certain techniques to
detoxify certain plants through recognition of the plant principles
they contain. This is conceivably the case in local situations or for
some compounds. Tannins, for example, are a widespread class of
chemicals in plant foods that are characterized by their oral as-
tringency. They are detoxified in a variety of ways, although similar
techniques are employed in different situations. Acorns were detoxified by boiling or by pulverizing and leaching in western North
America, Persia, and Japan, while clay was used with them in both
western North America and Sardinia (see below). In eastern North
America acorns are similarly detoxified by boiling in water containing ashes (App. 2). The fact that a variety of techniques are used to
detoxify the same chemicals and the same plants suggests that no
broad pattern linking detoxification techniques to chemical recognition exists. Although there may be some universal recognition of
certain chemicals in relation to processing, and although certain
compounds may only be detoxified by specialized techniques, parallel use of particular techniques could simply reflect parallels in the
empirical process leading to the most efficient solutions in each case.

Studies that explicitly evaluate the effectiveness of detoxification
processes are generally lacking. The global importance of cassava and
potatoes has stimulated assessment of methods to remove cyanogenic glycosides and glycoalkaloids, respectively, from these crops
(Lancaster et al. 1982; Christiansen and Thompson 1977). In general,
the existence of detoxifying methods indicates that they are effective
enough to warrant their ongoing application. But all of the processing
techniques discussed above have costs as well as benefits. Heating
and leaching in particular can destroy or eliminate nutrients such as
vitamins and minerals (Stahl 1989). Nonetheless, a positive balance
between overall nutritional enhancement and toxin elimination appears to result from most detoxification processes.

The intriguing question of how humans learned to detoxify plants
in particular ways is difficult to answer. The ubiquity of the various
techniques supports their antiquity. The use of clays for their adsorption properties has antecedents in animal behavior. Heating, leaching, fermentation, and drying all have simple cause-and-effect relationships with changes in food palatability that could be observed in

common events. Harris (1977) suggested that plants that are detoxified by leaching were originally placed in water as fish poisons. On the other hand, hyenas are known to store food in standing water (Kruuk 1972), and early hominids could have learned to place foodstuffs in water for similar reasons, conceivably by observing hyenas. Whatever the origin of the practice, plants that had been left in streams could be subsequently discovered to be acceptable foods.

The comminution of plants and the use of lye and salt to facilitate detoxification require greater sophistication. However, the use of tools is a long-standing human trait that would have been involved in detoxification since early in human history. Once tools were used to open nuts or other foods, refining the techniques to diminish the foods further is not a great leap. The use of salt in boiling would take place once boiling itself was established. Salt water would be used initially in coastal areas simply because of its availability. The use of lye also would follow cooking because ashes would be readily available. Perhaps hot coals or ash-covered rocks were initially added only to heat water but consequently were discovered to improve the food. The use of ashes in processing perhaps represents the most sophisticated of the basic techniques, and perhaps for this reason it is less widely distributed than the other methods.

Once a technology for detoxifying foods was established, deliberate elaboration using available resources would certainly occur. As the basic detoxification mechanisms spread, human adaptability and intelligence were involved in refining them in sophisticated ways to deal with particular plants. Humans' adaptation to new environments would require the application of detoxification techniques in order to use the available plant resources. Where human groups were in intimate association with a food resource over many generations, it is not surprising that considerable refinements took place.

TECHNIQUES OF POTATO DETOXIFICATION

Detoxification of bitter potatoes by modern Andeans is most commonly carried out through elaborate and labor-intensive leaching and freeze-drying processes (Fig. 3-2) (Werge 1979; Mamani 1981). The most common product, *tunta*, or *chuño blanco*, reportedly contains only 3 percent of the alkaloids originally present (Christiansen 1977). To produce tunta, potatoes are spread on the ground overnight at high altitudes. By morning they are frozen solid, and in this state they are trampled underfoot. In the still-frozen state, the tubers are placed in a

FIG. 3-2. Chuño production.

rock-lined well or similar depression in moving (or sometimes stagnant) water and left for a period of several weeks to more than a month. Freezing bursts the cell walls and trampling destroys the epidermis of the tuber. These steps greatly facilitate the leaching of cellular constituents. After leaching, the tubers are snowy white in color. They are dried for long-term storage by alternatively freezing them at night and squeezing them and drying them in the day under the brilliant Andean sun. The resultant product is high in carbohydrates and low in vitamins, protein, minerals, and glycoalkaloids.

Mamani (1981) discusses the production of *lojota*, or *chuño fresco*, from *Solanum × juzepczukii* at harvest time. I observed the use of this product in the Department of La Paz in June 1983. Frozen tubers are thawed and squeezed before cooking to produce a dense product,

not unlike other forms of chuño. Elimination of the juice may significantly reduce the alkaloid content of these bitter potatoes.

Peeling could potentially serve as a simple technique for eliminating the portion of the tuber with the greatest concentration of glycoalkaloids. Investigations with cultivated potatoes show that this is
generally effective, although as tubers age, glycoalkaloids diffuse
throughout the tuber (Woolfe 1987). In addition, peeling would not
necessarily remove a sufficient portion of the glycoalkaloids to detoxify many wild species. Although Andeans often peel potatoes,
they do not view this as a means to remove bitterness. The Zuni of
New Mexico, however, peel tubers of the wild *S. fendleri* for use in
boiled dishes and as an uncooked snack food (Cushing 1920; Richard
Ford, personal communication) with the intent of reducing bitterness. Such techniques warrant more detailed examination in order to
evaluate their detoxification potential.

Detoxification processes employed with potatoes have undoubtedly been important in expanding the exploitation of this cultivated
resource. However, such technological developments may not predate the initial stages of the agricultural revolution. Today's techniques requiring frost were probably not applied until agriculture
moved to higher altitudes. Clay use for the apparent detoxification of
glycoalkaloids in bitter cultigens and wild potatoes, both in the
Andes and among Indians of the American Southwest, will be discussed in detail below.

These geophagous practices are suggestive of a phenomenon that is
directly linked with the process of human-directed evolution leading
to plant domestication, and which appears to have its origins at an
earlier stage of history.

DETOXIFICATION OF PLANTS BY CLAY

As we saw in chapter 2, a detoxification function for geophagy was
not widely accepted until recently, being dismissed in favor of accepted theories on the nutritional or maladaptive importance of
the practice. Anthropologists and other observers have occasionally
pointed to the likely detoxification function of specific examples
of human clay consumption. For instance, the pre-Roman medical
practice of prescribing clay in cases of heavy metal and other poisoning persisted until the nineteenth century (Black 1956; Halstead
1968). Toxicity in plant foodstuffs has been reported to be neutralized
through the deliberate ingestion of clay. The two most suggestive

FIG. 3-3. Native Californians preserve their traditional use of acorns as food. Many Pomo like this woman from northern California still remember the preparation of acorn bread from a red clay called *ma-po* or *ma-sil*.

cases in the literature deal with alkaloids and tannins—two classes of naturally occurring compounds that are known to be effectively bound by clays (Theng 1974; Van Olphen 1977). The first case concerns glycoalkaloids in potatoes.

The second case describes clay being used by two widely separated groups of people for a similar detoxification purpose associated with acorn use. Natives of California and Sardinian villagers both used clay in the preparation of bitter acorns, probably to bind tannins (Gifford 1936; Laufer 1930; Usai 1969). The Pomo of California baked bread from ground (unleached) acorns (Fig. 3-3) and a specific type of red clay. The so-called black bread that was produced was regarded as sweet and was preferred to bread made from leached acorn meal (Barrett 1952). In Sardinia, acorns were boiled in the presence of a ferruginous clay until soft (Fig. 3-4), then made into cakes and baked in an oven (Usai 1969). Several Indian groups in California buried whole acorns in mud for an extended time (Gifford 1936). This practice may have served the same purpose as baking acorns with clay.

Analogous consumption of clay with unpalatable plants has been recorded for natives of Western Australia, Indians in Florida, and the Ainu of the North Asian Pacific. Aborigines used clay to remove the bitter taste from roots of *Haemodorum coccineum* Hook. (Laufer 1930; Grey 1841), a plant characterized by phototoxic phenalenone (Hegnauer 1986). Cabeza de Vaca (Hodge 1907; Castetter 1935) reported in 1527 that Indians of the American Southwest ate clay with the beans of mesquite, *Prosopis juliflora* DC. These seeds likely contain nonprotein amino acids. The Ainu boiled bulbs of *Corydalis ambigua* Cham. & Schlecht. (Papaveraceae) with clay to remove bitterness (Laufer 1930) probably caused by isoquinoline alkaloids (Hussain et al. 1988). In addition to its use with wild potatoes by Indians of the American Southwest, Castetter (1935) mentions the consumption of clay with fruits of *Datura inoxia* Mill. (*D. meteloides* DC) and *Lycium pallidum* Miers. The former plant contains highly toxic tropane alkaloids (Hegnauer 1973), while *L. pallidum* probably contains alkaloids or steroidal glycosides related to saponins and glycoalkaloids. In addition, Indians in the Southwest added clay to latex of *Apocynum angustifolium* Wooton and chewed it. The genus *Apocynum* is characterized by cardiac glycosides.

These cases documenting the detoxification of naturally occurring plant chemicals are similar in that all involve the exploitation of a resource that is not absolutely inedible and may have been consumed on a small scale without processing. The adsorption of weak, dose-dependent toxins such as tannins and alkaloids does make these foods more palatable and safer for large-scale consumption. Detoxification was the stated rationale for clay consumption in the cases recorded here. A careful reading of the geophagy literature reveals many other instances where detoxification, while not expressed as the basis for clay use, is probably a central reason for this practice.

Laufer (1930) and others have compiled a large number of reports of geophagy during famine. Most observers have explained this widespread practice as an attempt to achieve a sense of satiety (Eidlitz 1969; Laufer 1930; Solien 1954). While this may be one function of

FIG. 3-4. Like the Pomo of California, natives of Sardinia preserve knowledge of the preparation of acorn bread with clays rich in iron. Here two men from the area of Baunei gather clay from a source of the traditional acorn clay called *torco* or *trocco.*

the clay ingested, a few reports explicitly document the consumption of clay with famine foods such as wild roots and bark. Others talk about geophagy and famine foods in relation to the same famine episodes (Laufer 1930; Sorokin 1975), and it is likely that an essential connection between the two has been overlooked. Famine foods are usually marginally edible, often because of their high content of unpalatable allelochemicals. The specific use of clay in such stressful situations is likely an atavistic behavior held over from foraging in the resource-limited, chemically threatening environment that our hominid ancestors encountered.

In an extensive article on emergency plant foods in West Africa, Irvine (1952) mentions in passing that famine foods of mineral origin are known. He gives only one example where soil is used as a detoxicant under these conditions. Perhaps significantly, the plant involved is one of the most important famine foods in West Africa, the wild yam, *Dioscorea dumetorum.* Toxicity due to the presence of saponins and alkaloids is reportedly alleviated by burying the tubers in "black cotton soil" for three days or steeping tubers for three days in a mixture of water and cotton soil. Black cotton soils are dominated by montmorillonite (Bunting 1967), the type of clay recognized as the best binder of organic compounds (see below).

Gaud (1911) reported from the French Congo that "at the present time it is only during famines that the Manja gather the earth of termites' nests and consume it mixed with water and powdered tree-bark. This compound is said to assuage the tortures of hunger in a similar manner. We think that this effect must be attributed not only to the physical action resulting from filling the stomach, but also to the absorption of organic products existing in the clay" (trans. Laufer 1930). These African practices show remarkable similarity to reports from Europe where clay, euphemistically referred to as "mountain meal," was consumed in times of war or famine (Laufer 1930). Laufer (1930) cites a particular report from 1837 that "during the famine of 1832, the foodstuffs used in the parish Degerna (Finland) on the frontier of Lapponia contained a meal-like silicious earth mixed with real flour and tree-bark." Sorokin (1975) reported that during famines in Russia and Western Europe "the people have been brought to eating . . . all sorts of roots, grass, hay, bark, earth with a small addition of flour, clay and even human flesh. . . . In 1734–35 in Nizhegorodskaya province people ate rotten oak logs, acorns, pig-weeds and chaff, and in 1922 they ate bread made from clay, bul-rushes and pigweeds."

Clay is eaten with food in many other contexts (Anell and Lager-crantz 1958; Laufer 1930). Although widespread, the uses of clays as condiments, relishes, or spices—e.g., in Indochina, the Philippines, New Guinea, Chukchee and Gilyak of North Asian Pacific, Guatuso of Costa Rica, Guatemala, Senegambia, and the Amazon and Ori-noco basins—are more difficult to interpret. While small amounts of clay in foods may make a mineral contribution to the diet (Danford 1982) or may improve food taste, the use of clays as condiments or spices may be attributable to the detoxification function. For exam-ple, many observers (Mazess and Baker 1964; Otero 1951) refer to clays used with potatoes in South America as spices or condiments, when in fact informants in the same area today relate that clays are used specifically with bitter potato species to render them more palatable. What is at issue may be the perspective of the observer or his/her choice of words, not the detoxification function.

The widespread use of edible clays in tropical Africa has been explained on both nutritional (Hunter 1973; Vermeer 1966) and pharmacological (Vermeer and Ferrell 1985) grounds. Although clays are widely eaten by pregnant women who may require supplements of calcium and other minerals, they are also commonly eaten by men and children. Some support for a detoxification explanation of African geophagy comes from informants among the Luo and the Kisi of western Kenya, who reported to me that charcoal is eaten by the same persons who eat clay. Again, it is most commonly eaten by pregnant women. While charcoal is unlikely to provide anything nutritional, it could be very effective as a detoxification agent. The consumption of clay or charcoal by pregnant women may be related to the gastrointestinal upset, or morning sickness, associated with pregnancy, which has been attributed to hormonal changes (Weigel and Weigel 1988), dietary influences, and psychological problems (Luke 1979). Nutritional deficiencies bring about reductions in the detoxification capacity of hepatic enzymes (Klaassen et al. 1986), and as well diet affects the metabolic function of gastrointestinal microflora, including their biotransformation of steroidal hormones such as estrogen (Rowland et al. 1985). The widespread use of clays by poor women in the American South and other places may be diet related, not directly as replacement therapy for nutritional deficiency but as a result of the greater vulnerability of malnourished women to the normal metabolic stresses of pregnancy. Behavioral responses to toxicity such as conditioned aversions (see chapter 2) or geophagy are evoked through gastrointestinal distress. Gastrointestinal upset in general may trigger craving for clay.

Mineralogical and Chemical Characteristics of Clays. Details of the structures of clay minerals can be found in works more specifically devoted to this topic (Brindley and Brown 1980; Dixon et al. 1977; Theng 1974). Clays are crystalline in nature and are essentially made up of stacked layers of silica tetrahedral and aluminum (or magnesium) octahedral sheets. Organic molecules may be adsorbed onto the surfaces of clays through a number of means, including cation, anion, and ligand exchange, protonation water bridging, hydrogen bonding, and nonspecific van der Waals interactions (Sposito 1984).

It is possible for cations of similar size but different valency (usually lower) to substitute within clay structures. Usually Al^{3+} substitutes for Si^{4+} in the tetrahedral sheets, while Fe^{3+}, Fe^{2+}, and Mg^{2+} substitute for Al^{3+} in the octahedral sheets. This so-called isomorphous substitution results in a charge deficiency in the clay which is

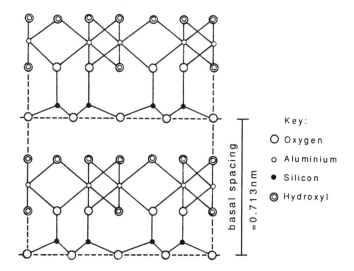

FIG. 3-5. Structure of kaolinite.

generally compensated by an exchangeable cation adsorbed onto the surface of the clay.

When clays are ingested with food, the cations associated with the clays may be exchanged with inorganic cations in the food or in digestive fluids. In this interchange there may be a net gain or loss in terms of mineral nutrition, depending on the nature of the food, the diet, and the necessity of the particular mineral to the ingestor. For example, exchange of sodium in the diet for calcium from clays might be positive, while a loss of dietary zinc or iron could be detrimental.

The exchange of organic and potentially toxic cations from food with cations from clays can be important in the detoxification function of clays.

Different clays can be identified according to the arrangement of the sheets, the layer charge, and the interlayer cation. Kaolinite, which is representative of 1:1 type minerals, is formed of two sheets (one tetrahedral and one octahedral sheet with no isomorphous substitution), while 2:1-type minerals have one octahedral sheet sandwiched between two tetrahedral sheets. Organic molecules can be adsorbed onto the surface of the clays and can penetrate into the interlayer space. In kaolinite (Fig. 3-5) the oxygen plane of one sheet is superimposed on the hydroxyl plane of the successive layer to produce interlayer hydrogen bonding. Because the forces holding ad-

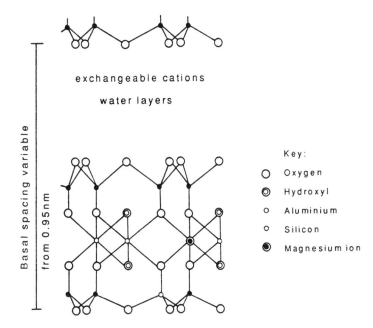

Key:

○   Oxygen

◎   Hydroxyl

o   Aluminium

o   Silicon

◉   Magnesium ion

exchangeable cations

water layers

Basal spacing variable from 0.95nm

FIG. 3-6. Structure of montmorillonite.

jacent layers together are strong, penetration by organic compounds is generally difficult and adsorption is restricted to the external surfaces. In montmorillonites (Fig. 3-6), which are 2:1 clays, the bonds holding the layers together are weak, and water, cations, and organic molecules may intercalate between them. This effectively increases the surfaces for adsorption and the potential of these clays for detoxifying organic toxins.

Cation-exchange capacity (surface charge) and surface area determine how particular organic molecules are adsorbed onto various clays. Kaolinite has a surface area of 3–15 m^2/g and a cation-exchange capacity of 1–5 mequiv/100 g, while montmorillonite has a surface area of 600–800 m^2/g and a cation-exchange capacity of 80–150 mequiv/100 g. More detailed discussions of the adsorptive properties of clays can be found in various sources (Bolt 1979; Brindley and Brown 1980; Theng 1974).

GEOPHAGY AND POTATO DOMESTICATION

Geophagy in the Andes can be viewed as part of the integrated sensory and cultural factors that relate to human exploitation of potatoes and adaptation to the high-altitude environment. I have presented clay-use data (Johns 1986a, forthcoming) associated with the gathering of potatoes from the wild, with their incipient domestication, and with their primitive cultivation in North and South America. Interpretation of the adaptive significance of geophagy associated with potato consumption and domestication is decidedly less problematic than in other cases where geophagy and organic toxins have been linked, in that detoxification is the explicit intent here. The literature and field data outline a biological activity associated with the consumption of potatoes that are high in distasteful and potentially toxic glycoalkaloids.

Indians of the American Southwest and adjacent Mexico consume clays with the wild potato species *S. jamesii* Torr. and *S. fendleri* (Johns forthcoming). The Navajo, Zuni, and other tribes gather potatoes from the wild; the Hopi protect potato plants; while the Tarahumara of Chihuahua transplant wild tubers and maintain them in fields. Tubers are exploited as a famine food and during seasonal periods of scarcity. The expressed reason for the consumption of specific clays is their effectiveness in eliminating bitterness and preventing stomach pains or vomiting that result when large quantities of these foods are eaten without clay. Clays used by the Hopi, Navajo, Zuni, and Keres are referred to in the literature, and by a handful of extant traditional practitioners, as potato clay (e.g., *tumin tcuka* [Hopi], Fewkes 1896).

Similar reports of Andean native consumption of specific clays with potatoes date from the sixteenth-century chroniclers (Johns 1985). Specific clays are still consumed with frost-resistant potato cultigens (*Solanum* × *juzepczukii* and *S.* × *curtilobum*) by the inhabitants of the central Andean altiplano, with the stated rationale of eliminating the bitter taste characteristic of these potatoes.

Clays are obtained from locally well-known sites in both North and South America. In Peru and Bolivia I have seen excavations for edible clays extending two or three meters below the surface (Fig. 3-7). Edible clays may be used medicinally, but they are distinguished from clays used for other purposes such as pottery making and whitewash. Remnants of the historical trade in edible clays persist among the Quechua and Aymara of southern Peru and Bolivia.

In the following analysis, adsorption of the glycoalkaloid tomatine by four specific clays consumed with potatoes in North and South America demonstrates the high adsorptive capacity of various clays and the astuteness of prescientific peoples in enhancing the detoxi-

FIG. 3-7. Site of the edible clay *ch'aqo*, near Acora, Peru, in the Lake Titi-caca basin.

fication effect of specific clays. The association of clay use with domestication phenomena provides a specific case for demonstrating the detoxification function of geophagy.

SOURCES OF CLAY SAMPLES

I purchased samples of southern Peruvian clays in the markets of Juliaca and Puno, Department of Puno, in July 1981. Although various samples of each type originate from different local towns, two types of clay recognized by color and organoleptic quality are differentiated under the names *ch'aqo* and *p'asalla*. The samples of ch'aqo and p'asalla used in this study came, respectively, from Acora (Fig. 3-7) and Tincopalca, Department of Puno. Clay that is similar to the Peruvian ch'aqo in appearance is referred to as *p'asa* by the Bolivian Aymara of the Department of La Paz. It is obtained from Achocalla, Province of Murillo, and is widely available in markets (cf. La Barre 1948).

Navajo clay, called *dleesh*, was obtained directly from a Navajo informant. Its source in the Chinle, Arizona, area of the Navajo reserve is consistent with an earlier reported source for the material used by the Hopi (Laufer 1930).

CHARACTERIZATION AND ADSORPTIVE CAPACITIES OF
POTATO CLAYS

Quantitative characterizations of the four field-collected clays—ch'aqo, p'asalla, p'asa, and dleesh—carried out by powder-pattern X-ray diffraction are recorded in Table 3-1. The sample of p'asa shows high interlayer swelling in water, as is expected from a clay high in smectites (montmorillonite). Many of its properties are similar to those of bentonite, a known montmorillonite clay.

Determination of the adsorptive capacities of each clay used the Langmuir adsorption isotherm as described by Johns (1986a). Adsorptive capacities were determined under physiological conditions of pH and ionic strength (Johns 1986a). Values of the adsorptive capacities for tomatine for each of the clay samples at pH 5.5 and 0.1 M ammonium acetate are shown in Table 3-1. The adsorption of tomatine by the six clay samples and standards follows a pattern consistent with the known cation-exchange capacities and intercalation properties of different clay groups (Theng 1974). The weak adsorptive capacity of kaolin clay compared to montmorillonite clays such as p'asa and bentonite is consistent with the known weak cation-exchange capacity and lower surface area of kaolin (Barr and Arnista 1957; Meshali 1982). With an adsorptive capacity of 0.50 g/g (pH 5.5, 0.1 M ammonium acetate), the Bolivian p'asa is superior to bentonite (adsorptive capacity: 0.37 g/g), one of the best commercially available adsorbents of organic compounds. At pH 4.5 and 0.01 M buffer concentration, p'asa and bentonite have capacities of 0.68 and 0.65 g/g, respectively. Adsorptive capacities of 0.61 g/g and 0.64 g/g, respectively, were recorded for unbuffered samples of the two clays. Under these conditions of low ionic strength, tomatine adsorption by bentonite is comparable to that of previously studied pharmaceutical agents (Wai and Banker 1966).

Of the four clay samples, the weakest adsorption was shown by the dleesh obtained from the Navajo reservation. Its adsorptive capacity of 0.14 g/g is nonetheless approximately three times greater than kaolin under similar conditions.

DETOXIFICATION UNDER PHYSIOLOGICAL CONDITIONS

The adsorptive capacities of tomatine by p'asa and bentonite show no significant differences over a range of pH and ionic strengths simulating gastrointestinal conditions (Tables 3-2 and 3-3) (Johns 1986a). The binding of tomatine by both p'asa and bentonite as

Table 3-1. Characterization of Clays and Comparison of Detoxification Potentials as Represented by Adsorptive Capacities for Tomatine (from Johns 1986a)

Clay	Characterization	Adsorptive capacity [g/g (r^2)]
P'asa	Predominantly smectites, mixed layered with illite; quartz. High interlayer swelling.	0.50 (0.99)
P'asalla	Mixed layer of smectites and illite; quartz.	0.28 (0.94)
Ch'aqo	Predominantly illite, mixed layered with smectites, chlorite, kaolinite; quartz.	0.19 (0.98)
Dleesh	Predominantly illite, mixed layered with smectites, kaolinite, chlorite; quartz.	0.14 (0.97)
Bentonite		0.37 (0.98)
Kaolin		0.05 (0.77)

ammonium acetate concentration is decreased below physiological ionic strength is significantly increased for both clays (Johns 1985).

Adsorption studies of a range of conditions encompassing gastrointestinal pH and ionic strength (Lentner 1981) demonstrate the ability of these edible clays to bind tomatine and hence neutralize the bitterness and eliminate the gastrointestinal irritation associated with glycoalkaloid ingestion (Jadhav et al. 1981). The quantities of clay necessary to reduce glycoalkaloids from a level of 100 mg/100 g fresh weight, typical of wild potatoes, to a recognized nontoxic level

Table 3-2. Effect of pH on Adsorptive Capacity of P'asa and Bentonite (from Johns 1986a)

pH	Ionic strength (M)	Adsorptive capacity [g/g (r^2)]		
		P'asa	Bentonite	Dleesh
5.5	0.10	0.50 (0.99)	0.37 (0.98)	
4.5	0.10	0.45 (0.98)	0.41 (0.96)	
3.5	0.10	0.50 (0.97)	0.42 (0.97)	
2.5	0.10	0.50 (0.99)	0.46 (0.98)	
1.5	0.10	0.48 (0.99)	0.44 (0.98)	
2.0	0.0075	0.49 (0.97)	0.43 (1.00)	0.11 (0.93)

Table 3-3. Effect of Ionic Strength on Adsorptive Capacity of P'asa and Bentonite (from Johns 1986a)

Ionic strength (M)	pH	Adsorptive capacity [g/g (r^2)]		
		P'asa	Bentonite	Dleesh
0.15	5.5	0.40 (0.99)	0.30 (0.98)	0.03 (0.98)
0.15	4.5	0.45 (0.97)	0.33 (0.99)	
0.10	4.5	0.45 (0.98)	0.41 (0.96)	
0.05	4.5	0.58 (0.99)	0.46 (1.00)	
0.01	4.5	0.68 (0.99)	0.65 (0.99)	
Unbuffered	2.5 (initial)	0.61 (0.99)	0.64 (0.99)	

of 20 mg/100 g (Gregory 1984) are minuscule in comparison to the amounts that are consumed at an actual meal (Whiting 1939).

The determinations of adsorptive capacity made at pH 5.5 and 0.1 M ammonium acetate are representative of binding under oral conditions. The high adsorptive capacity under these conditions accounts for the decrease in bitterness reported by native peoples.

The overall detoxification properties of the clays are highly dependent on the degree of adsorption in the stomach, where conditions will be most favorable. Determinations of adsorptive capacity at pH 2 and 0.075 M monovalent cation concentration (ammonium ion) reflect binding during the gastric residence period. Under these conditions 271 mg of the least adsorbent of the clays, dleesh, would reduce a toxic level of 50 mg tomatine in 100 g of potatoes to the recognized nontoxic level of 20 mg/100 g of glycoalkaloids. P'asa, the most effective adsorbent, would provide the same detoxification with only 60 mg of clay.

Desorption at the increased pH and ionic strength of the intestines could reduce the effective detoxification of clays. At pH 5.5 and 0.15 M ammonium acetate the adsorptive capacities of p'asa, bentonite, and dleesh are the lowest in comparison to all the conditions tested. The reduced solubility of glycoalkaloids above pH 5.5 would be expected to reduce adsorption under the normal intestinal pH of 6–8 (Lentner 1981). Previous work (White and Hem 1983) has shown that in vivo, in the presence of clays, the binding of protonated forms of basic molecules such as alkaloids to the clays drives the equilibrium between protonated and unprotonated forms toward the left (i.e., pKas are effectively higher) as follows:

RNH$^+$ ⇌ RN + H$^+$

The resulting production of more of the protonated form at higher pHs in the presence of clay would facilitate clay-tomatine binding in vivo. Even if clay adsorption of glycoalkaloids is reduced in the basic environment of the intestines, their insolubility under these conditions should minimize adsorption by the intestine and decrease the resultant toxic effects.

Culinary practices of Andeans and Indians of the American Southwest should facilitate glycoalkaloid adsorption. Aymara meals are simple: potatoes are usually eaten as the sole or dominant constituent of a meal. Salt is used sparingly by many Aymara (see chapter 6). Under these conditions the competitive effect of other compounds is reduced and clay-glycoalkaloid binding is enhanced. The ingestion of clays as an aqueous slurry by both Andeans and the Hopi (Whiting 1939) facilitates clay-glycoalkaloid contact. The glycoalkaloids, which are concentrated in the outer cortex of the potato (Jadhav et al. 1981), readily contact surface-adhering clay particles. The procedure of dipping potatoes in a clay slurry makes bitter potatoes more palatable by preventing soluble glycoalkaloids from interacting with the oral mucosa and taste receptors. The Navajo mixed dleesh with water also (Hough 1910), although they more often boiled potatoes with clay (Bailey 1940). In the American Southwest clays were more often eaten with every bite of potato.

THE EVOLUTIONARY SIGNIFICANCE OF GEOPHAGY BY HUMANS

As weedy annuals occurring in abundance on sites of human disturbance, wild potatoes meet the classic criteria for a domesticable plant (de Wet and Harlan 1975) (see chapter 4). However, toxic glycoalkaloids, while not eliminating the possibility of consumption, would limit the use of potato tubers to casual occurrence or an emergency food. Geophagous practices associated with gathering potatoes from the wild (Laufer 1930) and with cultivation of wild potatoes (cf. Pennington 1963; Whiting 1939) and primitive domesticates (Weiss 1953) link this detoxification phenomenon with the process of human-directed evolution leading to plant domestication. Association of the detoxification function of geophagy with stages in the domestication of the world's premier vegetatively propagated cultigen has important implications as well for the domestication of many crops where wild relatives contain toxic levels of allelochemicals.

The clay-potato association provides insight into the general phenomenon of geophagy. Potato clays are used habitually with a resource which otherwise is only modestly exploitable because of the presence of nonlethal but quantitative toxins. The use of clays in this situation corresponds to geophagous practices of primates that include at least some leaves in their diet (see chapter 2).

If early human foraging was analogous to the feeding behavior of modern nonhuman primates (see chapter 7), then food procurement by humans must have been subject to the constraints imposed by plant defensive compounds. Geophagy is a behavior with antecedents that are certainly prehominoid, and higher primates, including humans, have apparently maintained it as a mechanism for dealing with naturally occurring toxins.

The conditioned aversion response is discussed in chapter 2. The well-studied model linking toxicosis and conditioned responses, mediated through gastrointestinal upset, can be extended to include geophagy as part of an integrated "psychological" response to environmental toxins (cf. Mitchell et al. 1977). This model may be extended to include geophagy as an adaptive behavioral response for generalist primates as well as rats.

Widespread medicinal use of clays to treat diarrhea (Vermeer and Ferrell 1985) and other gastrointestinal ailments may relate to the detoxification function of geophagy either directly or indirectly via the "psychological" response to gastrointestinal upset. Kaolin has been shown to adsorb diarrhea-causing enterotoxins produced by *Pseudomonas aeruginosa* (Said et al. 1980).

Periodic deficiencies in food availability are a fundamental problem facing humans and other primates (Stahl 1984). Wild potatoes are typically exploited by humans during famines or periods of seasonal scarcity (Laufer 1930; Whiting 1939). Geophagy accompanying the use of famine foods is a widely documented behavior of traditional agriculturalists around the world (Laufer 1930; Solien 1954). In these cases geophagy can be a response to gastrointestinal stress in the form of hunger (Solien 1954), or poisoning by ingestion of barely tolerable wild plants or foods contaminated by bacterial enterotoxins (cf. Said et al. 1980) and mycotoxins (cf. Carson and Smith 1983). The most appropriate responses to this stress are maintained and transmitted through human culture and institutions, and maintenance of geophagy as a cultural trait may be a relic of its historical significance (Anell and Lagercrantz 1958; Hunter 1973; Laufer 1930).

The use of clays as minor but regular additives to food by many human groups is perhaps a relict of the greater importance of clay consumption in the past. Perhaps the largely symbolic nature of clay use in some cases serves as a device for cultural memory of knowledge essential for survival in time of famine. That clay is classified as food means that the adjustment to famine requires only a shift in proportion of food items, not the consumption of an item that is regarded as inedible (Fallon and Rozin 1983).

Crises in food production are an inherent component of many models of the development of agriculture (Rindos 1984). The transition of geophagy from a general response to toxin-related stress to a more specialized detoxification technique can be interpreted as an important step allowing expanded resource exploitation and the domestication of particular plants. Wild potatoes were presumably a minor dietary constituent or a famine resource under primitive conditions of exploitation, but eventually they came to provide a subsistence base for complex prehistoric civilization in the central Andes. By analogy, wild plants were probably more available to early hominids than has previously been supposed (Leopold and Ardrey 1972).

The case of clays used with potatoes is exceptional in the directness with which it demonstrates a detoxification function for geophagy. The ethnographic data from two separate places are unusually explicit in indicating that detoxification is the intent of those who eat clay with potatoes. Glycoalkaloids are probably ideal candidates for adsorption by clays, both in the practice of geophagy by humans and in in vitro studies. They are large-molecular-weight molecules that, as well as having a large surface area and numerous polar functional groups, are cations under oral and stomach pHs.

It is tempting to extrapolate from the potato example and explain geophagy solely as a detoxification phenomenon, just as in the past researchers wanted to explain clay eating on solely nutritional grounds. Acceptance of the adaptive significance of the detoxification of secondary compounds by clays should not be unqualified. Further research must test and compare the conclusions from the specific potato case with other examples where clay and plants are consumed together. As shown here and elsewhere, clays are variable in their adsorptive properties. Ferrell et al. (1985) suggest that kaolinite clays may be most useful as detoxification agents because their lower cation-exchange capacity means they are less likely to cause mineral deficiencies. However, for the detoxification of alkaloids, the high cation-exchange capacity of montmorillonites is an advantage. The relative costs and benefits of clay consumption need further investigation. The ongoing difficulty that scientists have in explaining geophagy indicates that it is a complex issue, and the medicinal, nutritional, and detoxification functions of geophagy are perhaps all interconnected parts of this multidimensional phenomenon.

4 Domestication as a Solution for Dealing with Plant Toxins

Until approximately ten thousand years ago our species subsisted in the same way as the rest of the animal kingdom—that is, as foragers. Agriculture represented a significant departure in human history. By means of animal husbandry and the cultivation of plants, humans gained greater control over their basic subsistence resources and, on a localized scale at least, increased food supplies. Fundamental to this process was the selection for plant types with favorable levels of allelochemicals. The plant-processing techniques described in chapter 3 demonstrate a cultural-evolutionary mechanism by which humans adjusted to plant chemicals; but by selecting for particular types of plants, humans began to control the evolutionary process itself.

Why this change from foraging to agriculture should have taken place is a topic of considerable debate. Studies of the handful of remaining human foraging groups scattered around the globe and of other peoples who have engaged in similar activities until recent times provide a perspective on this stage of human evolution. It is generally agreed that foraging is an efficient means of food procurement, and that in nutritional terms many of the earliest agriculturalists were likely worse off than their predecessors (see chapter 7). Explanations for the origins of agriculture have considered cultural (Braidwood 1960), social (Bender 1978), climatic, demographic, and other factors (Rindos 1984).

Evidence accumulated from archaeological, botanical, and ethnographic sources supports the notion that agriculture was a gradual process (Harris 1989) rather than a revolution. Agriculture has roots in the post-Pleistocene pattern of exploitation and cultivation of seed and root crops that preceded agriculture (Harris 1977). Cereal exploitation was practiced by many protohistorical gatherer/hunter groups and was probably important as early as 14,500 B.P. Small-scale cultivation of tuber-producing plants using vegetative techniques may be ancient and widespread among tropical gatherer/hunters.

An essential aspect of the origins of agriculture was the domestication of plants and animals. Parallels to early stages in the processes leading to agriculture are observed today in the ongoing domestication of plants by peoples in traditional and modern societies. Although domestication is an evolutionary process affected by the conscious and unconscious actions of humans, it can be viewed with the general principles of modern evolutionary theory. Those organisms with genetic characteristics more favorable to environmental conditions will survive in greater numbers and increase the frequency of their genes. Variability in genetic traits as well as processes of genetic recombination give rise to a diversity of phenotypes upon which natural selection can operate. By most definitions, deliberate efforts by scientists to breed new crop varieties can be said to represent artificial selection. However, for plants and animals in the process of domestication, human selective influences are intimately connected to selection resulting from biotic and abiotic factors. Domestication, then, should be considered in its biological context, that is, within the realm of natural selection.

Strong human-directed selection for certain genetic traits may have far-reaching effects on the survivability of particular phenotypes and may demand complex adaptive changes in plants. In turn, the likelihood of evolution occurring under domestication is mediated by biological constraints related to the genetic makeup of the population or species. The amount of variability in a species will be determined by the evolutionary history of the species, as well as by factors such as cytogenetics and breeding behavior. Ongoing human activities that modify environments demand physiologically adaptive changes in plants, and the range of physiological variability in a plant will determine its ability to adapt to changed environmental conditions. In accordance with such issues this chapter concentrates on domestication as a process in plant evolution. Humans are viewed from this perspective as only one selective force among many determining plant chemistry. The dynamics of human selection decisions in relation to sensory perception are examined more specifically in chapter 5.

CHARACTERISTICS OF DOMESTICATED PLANTS

The special characteristics of domesticated plants that are related to human selection have been discussed by Schwanitz (1966) and

Hawkes (1983). Of these, only five are specifically important to this discussion:

1. *Gigantism.* Humans select for increased size in parts of plants such as fruit, tubers, roots, and seeds that provide sources of food.

2. *Wide range of morphological variability.* Those parts of the plant of interest to humans show great variability. For example, Hawkes (1983) comments that "one finds a tremendous range of tuber shape, color, and patterning in South American potatoes and little in the flowers and leaves, contrasted with the almost uniform tubers of all wild species of potatoes." Much of the variation may be a result of concern for aesthetic as much as useful qualities of the plant.

3. *Wide range of physiological adaptation.* The introduction of do-mesticated plants into new environmental conditions, and gene ex-change with related species that are adapted to particular conditions, have increased the adaptive capacity of crops. The specific adaptation of potatoes to cold and arid conditions of the Andean altiplano is central to the discussion below. The potato, a crop of global impor-tance, clearly is adaptable to a wide range of ecological conditions.

4. *Suppression of protective mechanisms.* The selection of culti-gens with reduced chemical defenses is the focus of this chapter. Thorns and spines have also been suppressed in many crop plants.

5. *Reduction in cross-pollination.* Among wild species of plants cross-fertilization (allogamy) and dispersal by seeds are the principal means of reproduction. Self-pollination (autogamy) and vegetative propagation are widespread in crop plants as mechanisms for "fixing" desirable genotypes.

The differences between wild and domesticated plants reflect hu-man intervention and provide insight into human interactions with plants. Because the evolution of domesticates is a recent phenome-non, it is possible to know the starting point of the process—i.e., the wild progenitors—and thus have a point of comparison for the changes that have taken place. Botanical studies have pinpointed the wild relatives of many crop plants as well as intermediates in the domestication process (Simmonds 1976).

In addition, studies of plant phylogeny and history can provide a record of human affairs. For example, they have been used to draw conclusions regarding the history of human migration and the his-torical contacts between cultural groups (Stone 1984). Maize, for instance, has a complicated dispersal history from Mexico to South

America and back which is reflected in the genetic nature of various cultivars (Bird 1984); and the root name for sweet potato, *cumar*, provides evidence of its human-directed dispersal from South America to Polynesia (Yen 1974). Comparison among chemical constituents of wild and cultivated plants provides insights into another kind of human interaction with plants, that with plant chemicals.

Although the emphasis of our discussion will be on chemical changes in plants during domestication, it is impossible to completely isolate human aspects from this interactive process. Domestication reflects a balance between plants, environmental forces, and human intervention—the three poles of the model presented in Figure 1-1. These elements will be incorporated in general terms into this discussion.

First, evidence that humans have selected for and against allelochemicals will be supported with a summary of the worldwide occurrence of this phenomenon. Subsequently, two case studies of chemical selection in Andean tubers will be presented in relation to the three poles of the model. The first example considers interactions of plant biology and human cultural values, while the second focuses primarily on the interaction of plants (potatoes) with humans as biological agents. Experimental botanical evidence that chemical selection has taken place in potatoes will be presented, followed by evidence for the ongoing exploitation of potatoes by Bolivian Aymara that provides insight into the role of human activities in determining the direction of potato evolution.

Environmental forces associated with the conditions of high altitude play a role in both of the present examples. In the first case they act primarily to affect human behavior, while in the latter case they directly determine the existence and survivability of specific plant genotypes.

Domesticates form a very small subset of the world's flora. For a plant to be domesticated it must first have some valuable characteristic that attracts people and likely results in its use before domestication. Second, plant species vary in the amount of genetic variability they possess and the degree to which they can adapt to new conditions. The dramatic changes seen during the domestication of maize, for instance, may have been a product of the exceptional variability of its wild progenitors. The third factor necessary for domestication is a fortuitous combination of ecological factors, plant characteristics, and human goals and behavior. Consideration of the dynamic interaction of biotic and abiotic environmental factors, human activities, and human cultural factors is essential for understanding the domestication process.

CHANGES IN CHEMICAL DEFENSES

Many allelochemicals function to protect plants against predation by specific insects, mammals, or microorganisms. Some compounds have specifically defined roles, e.g., the effects of phytoecdysteroids on insects (Kubo et al. 1983) or phytoalexins as a specific response to fungal infection in potatoes (Tjamos and Kuc 1982) and other crops. However, many other of the chemicals discussed in chapter 2 have general toxicity against biological organisms, including humans.

Human selection for reduced levels of those plant defensive chemicals which affect them adversely will often conflict with the interests of the plant. In fact, as plants are forced into growing in monocultures, pest populations often increase. In present-day industrialized agriculture chemical pesticides are often used to replace the natural defensive chemicals lost during domestication. Humans have accepted the elevated risks of crop failure that are associated with reducing chemical defenses. That they have thus altered the balance between herbivores and plant defenses supports the notion that selection for chemical change has been of paramount importance to humans in the domestication process.

CHEMICAL SELECTION AS A WORLDWIDE PHENOMENON

Chemical variability within crop plants is a global phenomenon, and Table 4-1 presents a sample of plants that appear to reflect selection by humans resulting in changes in secondary chemistry. The table includes plants that are used for both food and drugs.

The following discussion generally ignores selection for primary nutrients such as protein quality or caloric content. Undoubtedly changes in these plant constituents have been important in the domestication of plants (Harborne 1988). Humans domesticating fruits have tended to select for increased sucrose content in relation to hexose sugars such as glucose and fructose (H. G. Baker, personal communication). Like many mammals, humans have enzymes that digest disaccharides such as sucrose. These enzymes are lacking in most birds and insects. Sucrose is more intensely sweet to humans (Moskowitz 1974) and rats than simple sugars such as glucose (Feigin et al. 1987), and, consistent with the preference for sweetness charac-

Table 4-1. Examples of Plants Exhibiting Changes in Secondary Chemistry as a Result of Human Selection

Plant	Location[1]	Plant part[2]	Chemical	Purpose of change	Reference
Basellaceae					
Basella alba L. "Indian spinach"	Asia	L	betalains	dye	Burkill 1985 Hegnauer 1964
Brassicaceae					
Raphanus sativus L. "radish"	Asia	R	glucosinolates	flavor	Purseglove 1968 Hegnauer 1964
Chenopodiaceae					
Chenopodium quinoa Willd. "quinoa"	SA	S	saponins	reduce bitterness	Gade 1975 Hegnauer 1964
Cucurbitaceae					
Citrullus lanatus Mansf. "watermelon"	Africa	F	cucurbitacins	reduce bitterness	Hedrick 1919 Hegnauer 1963
Cucumis metulliferus E. Mey. "horned cucumber"	Africa	F	cucurbitacins	reduce bitterness	Burkill 1985
C. sativus L. "cucumber"	India	F	cucurbitacins	reduce bitterness	Hegnauer 1975
Cucurbita spp. "pumpkins"	NW	F	cucurbitacins	reduce bitterness	Whitaker and Bemis 1975
Lagenaria siceraria Standl. "bottle gourd"	India	F	cucurbitacins	reduce bitterness	Purseglove 1968
Luffa acutangula Roxb. "loofa"	Asia	F	cucurbitacins	reduce bitterness	Burkill 1985
Momordica charantia L. "bitter gourd"	Tropics	F	cucurbitacins	reduce bitterness	Burkill 1985

Table 4-1. Continued

Plant	Location[1]	Plant part[2]	Chemical	Purpose of change	Reference
Dioscoreaceae					
Dioscorea bulbifera L. "air potato"	Africa, Asia	T	saponins, alkaloid: dioscorine	detoxification	Burkill 1985; Takeda 1972; Willaman and Li 1970
D. dumetorum Pax. "bitter yam"	Africa	T	alkaloids, phenanthrenes	detoxification	Burkill 1985 Willaman and Li 1970; El-Olemy and Reisch 1979
D. sansibarensis Pax.	Africa	T	alkaloid	detoxification	Willaman and Li 1970
Erythroxylaceae					
Erythroxylum novogranatense Hieron. "coca"	Peru	L	methyl salicylate	flavor	Plowman 1979
Euphorbiaceae					
Manihot esculenta Crantz. "casava"	SA	R	cyanogenic glycosides	insect resistance	deWet and Harlan 1975 Hegnauer 1966
Fabaceae					
Lupinus albus L.	Med.	S	quinolizidine alkaloids, saponins	reduce bitterness	Hegnauer 1975; Mears and Mabry 1971; Hudson 1979
Phaseolus lunatus L. "lima bean"	WI	S	cyanogenic glycosides	detoxification	Purseglove 1968
Lamiaceae					
Micromeria viminea Urb.	WI	L	monoterpenes	pharmacological	Asprey and Thornton 1955

Table 4-1. Continued

Plant	Location[1]	Plant part[2]	Chemical	Purpose of change	Reference
Malvaceae *Malvaviscus arboreus* Cav. var. *mexicana* Schlecht.	Mexico	L		pharmacological	Alcorn 1982
Papaveraceae *Papaver somniferum* L. "opium poppy"	AM	P	morphine	pharmacological	Hegnauer 1975
Poaceae *Sorghum bicolor* (L.) Moench "sorghum"	Africa	S	tannins	bird resistance	Purseglove 1968; Subra-manian et al. 1983
Rosaceae *Prunus amygdalus* Stokes "almond"	Asia	S	cyanogenic glycoside	detoxification	Hedrick 1919
Rubiaceae *Coffea arabica* L. "Arabian coffee"	Asia, Africa	S	caffeine	pharmacological	Carvalho et al. 1965
Solanaceae *Capsicum* spp. "chili pepper"	NW	F	capsaicin	flavor	Hegnauer 1975; Tetenyi 1971

Table 4-1. Continued

Plant	Location[1]	Plant part[2]	Chemical	Purpose of change	Reference
Datura candida Safford	SA	L, Fl	tropane alkaloids	pharmacological	Bristol 1969; Bristol et al. 1969
Nicotiana spp. "tobacco"	NW	L	nicotine	pharmacological	Hegnauer 1975
Solanum melongena L. "eggplant"	India	F	glycoalkaloids, saponins	detoxification	Choudhury 1976; Bajaj et al. 1979
Solanum section *Petota* "potato"	SA	T	glycoalkaloids	detoxification	
Sterculiaceae					
Theobroma cacao L. "cacao"	C. A.	S	phenolics	reduce astringency	Schultes 1984
Taccaceae					
Tacca leontonpetalodes (L.) Kuntze. (*T. involucrata, T. pinnatifida*)	Asia, Africa	T	taccalin	reduce bitterness	Irvine 1952; Hedrick 1919; Scheuer et al. 1963
Tropaeolaceae					
Tropaeolum tuberosum Ruiz & Pav.	Peru	T	glucosinolates	flavor	Johns and Towers 1981

1. SA = South America; NW = New World; WI = West Indies; AM = Asia Minor; Med. = Mediterranean.
2. Plant part: F = fruit; Fl = flower; L = leaf; P = plant; R = root; S = seed; T = tuber.

teristic of our species, we maximize this quality in domesticated fruits by selecting for sucrose content.

Lignin reduces the digestibility of structural carbohydrates in plants (Van Soest 1982), and in domesticating vegetables humans have apparently selected for plant parts with low amounts of lignin.

Humans have selected for changes in secondary chemistry in seeds, tubers, fruits, and vegetative portions of plants. For example, within the populations of sorghum, the world's fifth most important seed crop, there has been selection for reduced levels of tannic acid. Tannic acid confers resistance to bird attack, however, and in many parts of Africa sorghum types high in tannins are purposely grown, and selection may be to increase these chemical defenses. In addition, seeds high in tannins are preferred for making beer.

The eggplant, *Solanum melongena,* is an example of a fruit that has been selected for reduced bitterness (Choudhury 1976). While cultivated varieties contain relatively low levels of glycoalkaloids (Bajaj et al. 1979) and saponins, it is likely that either or both of these classes of compounds are responsible for the bitterness of wild forms.

Members of the pumpkin family (Cucurbitaceae) have been subjects of selection for reduced levels of cucurbitacins wherever they grow. This is a common occurrence in species of *Cucurbita* that are cultivated for their fruit in the New World (Whitaker and Bemis 1975) and among both fruit and leaf crops of this family in Africa (Burkill 1985).

The poppy, *Papaver somniferum,* is a plant that has been selected both for its seeds and for its medicinal properties. Opium-producing varieties have been selected for high levels of morphine, while poppy-seed-producing varieties are low in this alkaloid.

Except for pumpkins, all of these examples are from the Eastern Hemisphere. Chemical selection phenomena in the New World show the same general patterns.

CHEMICAL SELECTION IN NEW WORLD DOMESTICATES

Domestication in the New World reflects a geographical and cultural continuum in the manner in which plants have been treated. In the most common cases, selection is concerned with limiting toxicity and improving palatability. For example, the genus *Capsicum* contains a large number of neotropical cultivars, many of which have undergone selection for levels of pungent capsaicin (Hegnauer 1975; Tetenyi 1971). However, varieties of *Manihot esculenta* differing in cyanogenic glycoside content probably reflect selection by man, per-haps unconsciously, to increase the resistance of the tubers to patho-gen attack and so allow the production of larger tubers (de Wet and Harlan 1975). Cassava has become an important staple in many parts of the world's tropics. In West Africa it has taken on particular impor-tance because it is immune to the migratory African locust. In areas that are infested with destructive populations of wild pigs, baboons, and porcupines, cassava is the one crop that resists serious predation (Purseglove 1968).

Chemical selection, like domestication itself, should not be viewed as a process that took place at some point in the past. Davis and Bye (1982), for instance, suggest that selection for sweeter fruits is occurring in ongoing domestication of *Jaltomata procumbens* (Cav.) Gentry by Indians in Mexico and Central America.

The previous examples that demonstrate human awareness and active behavior in avoiding toxins, or alternatively the selection of high levels of pest-resistant chemicals, are consistent in evolutionary terms with the general chemical role of allelochemicals in herbivore-plant interactions. Interesting and perhaps less straightforward do-mestication phenomena similar to those cited above reflect concern for the qualitative alteration of chemical constituents in relation to the medicinal value of plants. Chemical selection in medicinal plants is particularly interesting because these are often plants that are protected and encouraged rather than domesticated. As such, they may represent examples of early stages in the selection and domestication process.

Selection is likely to be directed specifically to chemical constitu-ents that are the basis for medicinal properties. For example, the Teenek Indians of Mexico differentiate, and propagate vegetatively, a dooryard variety of a common plant (*Malvaviscus arboreus* Cav. var. *mexicanus* Schlecht.) that is more effective medicinally than wild plants of the same species (Alcorn 1984). Studies on the halluci-nogens used by Amazonian Indians point to varieties of these plants that are recognized as different and are preferred by individual tribes and individual healers (Chagnon et al. 1970, 1971; McKenna et al. 1984; Schultes 1986). The Sibundoy of southern Colombia recognize at least nine cultivars of *Datura candida* (Persoon) Safford (Bristol 1969), largely on morphological grounds. To these are attributed different medicinal properties, with more toxic varieties being pre-ferred for medicinal and psychotropic use. Bristol et al. (1969) ana-

lyzed seven cultivars of *D. candida* from the Sibundoy valley for total
alkaloid content and for the identity of the tropane alkaloids present.
They concluded that "the Sibundoy natives' estimates of the relative
toxicity of their drug plants correlate well with the demonstrated
alkaloid content in five of the seven cultivars examined."

Erythroxylum novogranatense (Morris) Hieron., the most advanced cultigen of coca, is also the one highest in methyl salicylate
(Hegnauer 1966; Plowman 1979). Although this compound imparts a
pleasant flavor to the leaves that is preferred by many Andeans (Fig.
4-1), its presence does not directly reflect human selection for medicinal constituents. These plants do not show any parallel changes
in cocaine content that might reflect human selection for this alkaloid (Holmstedt et al. 1977).

Plants that are propagated vegetatively, such as *Datura* spp., *Malvaviscus arboreus*, and *Manihot esculenta*, provide particularly good
opportunities for showing chemical selection. Vegetative clones in
which particular genotypes are fixed (see below) can document human choice of favorable plant properties. Chemical selection in vegetatively propagated Andean tuber crops is particularly relevant to the
present investigation.

In the Andes several root and tuber crops form a complex of common domestication history, cultivation, and use. Selection for
changes in chemical composition appears to be a characteristic of
these vegetatively propagated crops. Hawkes (1989) provides an overview of the biosystematics of the group. For four of these crops—
olluco, oca, añu, and potato—some evidence exists to suggest that
humans have selected for changes in secondary chemicals during
their domestication. According to Brücher (1977), *Ullucus tuberosus*
Loz. (Basellaceae), or olluco (*papa lisa*), produces tubers with less
bitterness than its purported progenitor *U. aborigineus* Brücher.
Sperling, however, classifies the wild tubers as subspecies of *U. tuberosus* and disputes that they have any greater degree of bitterness
(personal communication). Any bitterness in olluco is probably due
to saponins (Hegnauer 1964; Johns unpublished data). Cultivators
broadly classify *Oxalis tuberosa* Molina (Oxalidaceae), or oca, as
either for processing or for consumption, and unpalatable varieties
are processed like tunta (see Fig. 3-2) by leaching the tubers in running water for a period of weeks and freeze-drying them to produce a
product called *caya* (Fig. 3-1) (Hodge 1951). More palatable tubers are
eaten after little or no processing. Various observers (Hodge 1951)

FIG. 4-1. Coca chewers.

have attributed the recognized differences in oca to levels of oxalic acid, although a thorough evaluation of its secondary chemistry has not been undertaken. Oxalic acid is moderately soluble in water and would be removed in the caya process. Oca tubers are usually placed in the direct sun for several days before consumption to improve their edibility. However, it is not clear that oxalate would decompose with this treatment.

Tropaeolum tuberosum R. & P. (Tropaeolaceae) (añu, *mashua, isaño, cubio, apiña-mama*), the third of the complex of Andean plants cultivated for their edible tubers, has undergone a change in the nature of its glucosinolates during domestication (Johns and Towers 1981). This change relates to taste preference, but also to cultural concerns related to medicinal uses of this plant and its relation to another sympatric cultigen with similar secondary chemistry, *Lepidium meyenii* Walp., the *maca* (Johns 1981). Folk beliefs associating añu with maca and the possible motivation for humans in selecting for qualitative changes in glucosinolate chemistry are discussed below.

Wild species of potatoes, *Solanum* subgenus *Petota*, are generally high in potentially toxic glycoalkaloids, and reduction in glycoalkaloid content has been indicated as an essential step in the domestication process (Simmonds 1976; Harborne 1988). Elaborate mechanisms of processing and preparation aimed at detoxifying potatoes (see chapter 3) underline Andeans' cognizance of secondary compounds. Potatoes are the most important of the Andean tubers and could be among the earliest domesticated plants in the New World. Consideration of the importance of toxic phytochemicals to the domestication of the potato provides insight into the chemical ecology of domestication of New World vegetatively propagated crops in general.

Invariably the compounds discussed in relation to the above domestication examples are detectable by taste and/or smell. Both the wild bean, *Phaseolus aborigineus,* and the domesticated *P. vulgaris* L. have populations with or without phytohemagglutinins (Brücher et al. 1969), compounds that are not detectable by taste. Although proteinaceous toxins such as phytohemagglutinins and proteinase inhibitors may respond to artificial selection by crop breeders, there are no data to support the idea that they were selected against during domestication, perhaps because plants containing them cannot be evaluated through the senses or because they are usually readily detoxified by heat.

PLANT SYSTEMATICS AND HUMAN BIOLOGY

Biosystematics is a fundamental part of the study of the evolution of plants and animals. Plant systematics concerns itself with the relationships of organisms and attempts to find evolutionary explanations for these relationships. Domestication results from human-directed selection of particular traits from among the diversity typically seen in natural plant populations. The relationships between domesticated and wild populations can be investigated using biosystematic methods, and where humans have acted as agents in the evolutionary process, the explanations of these relationships can provide insight into problems of direct relevance to humans and human biology.

BREEDING SYSTEMS

Physiological, morphological, and other traits of plants that determine their adaptation to particular environments affect the likelihood of domestication occurring. Plant breeding systems, for instance, influence the likelihood of a species being domesticated, but will also determine the morphological and chemical patterns that are likely to be seen in the plant under domestication. Outbreeding populations of wild plants maintain considerable variability. However, in crop plants it is desirable to fix and isolate desired genotypes (Zohary 1984). Self-pollination (autogamy) and vegetative propagation are widespread mechanisms for doing this. Wild relatives of most cereals and legume crops are self-pollinated and as such are "pre-adapted" to domestication (Zohary 1984). Vegetatively propagated crops, on the other hand, are primarily outcrossers, and in this case human exploitation of this alternative mode of reproduction provides a way of circumventing the loss of particularly favorable genotypes. Most wild potatoes are self-incompatible, that is, they have genetic factors that prevent self-fertilization. They are readily pollinated by insects, and new genetic recombinants occur readily. Vegetative propagation of tubers, then, is the means by which humans "fix" desirable potato types.

Autogamous and vegetatively propagated species are typically made up of discrete genotypes, while outbreeding species are likely to show more continuous genetic variability. The dynamics of maintaining desired chemical genotypes in most domesticates should thus result in populations with discrete differences in chemical composition.

INHERITANCE AND BIOSYNTHESIS OF SECONDARY CHEMICALS

Phenotypic traits of plants are products of complex combinations of gene expression, and secondary compounds are produced by biosynthetic pathways comprising varying numbers of enzyme-controlled steps. Each enzyme is a product of inherited genes, and the nature and quantity of individual secondary compounds are the result of different patterns of inheritance affecting the biosynthetic pathways. The way in which genes determining the synthesis of a particular compound are inherited will determine the ease with which humans can select for chemical changes. For example, consider the inheritance of three allelochemicals discussed in this book: cucurbitacins, cyanogenic glycosides, and glycoalkaloids.

Fruit bitterness in cucurbits appears to be controlled by three dominant, complementary genes, and recessive mutation of any one of these genes results in the production of nonbitter fruit (Borchers and Taylor 1988). The large number of cucurbits reflecting chemical selection for nonbitter fruit is likely a partial result of the ease of fixing new chemical genotypes in a population.

Like the production of cucurbitacins, cyanogenesis in plants such as cassava is also polymorphic. Cyanogenic plants must synthesize first the cyanogenic glycosides, and second the β-glucosidase enzyme that releases hydrogen cyanide from the glucoside (Fig. 2-2). The genetic control of this system varies from species to species (Nartey 1978). In the almond, *Prunus amygdalus* Stokes, nonbitterness is inherited as a dominant trait in a simple 3:1 ratio (Heppner 1926), and sweet almonds can be selected with ease. Alternatively, in the case of white clover (*Trifolium repens* L.), plants showing cyanogenic activity possess at least one functional dominant allele at each of two loci (*Ac* and *Li*) (Hughes et al. 1985). Homozygotes for the nonfunctional alleles at the *Li* locus (*li li*) could still be toxic when ingested by humans, and to effectively make plants containing this type of system nontoxic would require selection for the genotype *ac ac*. Plants with the genotype *ac ac Li* — are potentially toxic but lack the bitterness to indicate that they possess this property.

The genetics of cyanogenesis in cassava is not as well known (Nartey 1978), but it appears that, as in white clover, high HCN production is dominant over low HCN production, and inheritance is controlled by more than one gene.

While glycoalkaloid content is also a highly heritable trait, production of this class of compounds is controlled in a polygenic fashion (Sinden et al. 1984). Evolution in plants containing glycoalkaloids would be expected to be slower than in the previous examples. Humans can bring about a gradual reduction in the amounts of glycoalkaloids by selecting for those types with lowest toxicity. Crossing between selected plants will gradually increase the proportion of recombinants with genes for low glycoalkaloid levels, leading to the production of nontoxic varieties. In potatoes genes control the synthesis of specific glycoalkaloids, and the expression of glycoalkaloids is complicated and poorly understood for any particular hybrid plant (Sinden et al. 1984).

The dynamics of recombination is an important factor in maintaining selected genotypes, and crosses made between selected varieties of crop plants and types high in secondary chemicals will generally reverse the selection process. Humans can speed up selection by isolating desirable varieties. Human behavior and human activities thus become very important in the process. Factors such as farming practices and seed selection are considered below in relation to the cultivation and domestication of *S. × ajanhuiri* and related potatoes.

HYBRIDIZATION

Human activities, particularly the introduction of plants into new geographical areas, often bring about the formation of hybrids between plant species. Studies of hybrids between genotypes with different adaptive norms can play a central role in understanding the domestication history of many plants. Hybridization between wild and domesticated plants is a widespread phenomenon wherever crop plants remain in their centers of genetic diversity (Small 1984). In theory, when hybrids backcross repeatedly to one parental species, a population arises which resembles this parent but shows characteristics of the opposite parent (Grant 1981). This process, known as introgression, is best documented through studies of the relations of cultivated plants and their wild relatives. Introgression of genes from wild plants into cultigens may enable crops to adapt to new ecological conditions. Introgression of genes from cultigens into their wild relatives produces the widespread phenomenon of conspecific weeds accompanying crop plants. The evolutionary relationship among weeds, wild plants, and crop species has been extensively reviewed (Harlan 1975; de Wet and Harlan 1975; Pickersgill 1977, 1982; Small 1984).

HUMAN SELECTION

Human selection typically results in dramatic changes in the appearance of domesticated plants (Pickersgill 1977). Selection for characteristics of economic importance results in greater changes in one or a few characters, while others remain constant over the history of the domestication process. Such highly directed selection may make cultigens unrecognizable in relation to their wild progenitors and may overshadow the overall affinities of particular plants. It is for this reason that crop plants are notoriously difficult to study taxonomically. For example, *Brassica oleracea* L. is a species with many varieties, each a result of human selection, including cabbage, cauliflower, broccoli, brussel sprouts, and kohlrabi. The unusual morphological structures that characterize each of these vegetables reflect human interest in the particular plant part. Similar demonstration of dramatic chemical change in comparison to the constancy of other traits should be indicative of specific human interest in the chemistry of plants.

BIOSYSTEMATIC INVESTIGATIONS

Many sources of data are used in the field of biosystematics. Morphological and anatomical investigations are the basis of classical taxonomy. Today, evidence from cytological, ecological, geological, and chemical studies is routinely applied along with increasingly sophisticated techniques for examining plant structures.

Scientific decisions on the relationships among plants are made (often intuitively) on the basis of the congruity of a comprehensive body of data from multiple sources; or often deliberately on the basis of critical pieces of data that allow comparison between plants. Computers have made it feasible to deal with large bodies of data by statistical methods, and the emerging field of numerical taxonomy assists in making decisions on relationships among plants.

The detection of hybrids presents a particular problem which can be addressed through numerical studies. Organisms that have diverged during evolution will have an affinity for an intermediate hybrid that can be mistaken for a phylogenetic relationship (Wagner 1980). In the following discussion a numerical approach is taken to establish the hybrid relationships among *Solanum × ajanhuiri* and related potatoes. The assessment of these relationships forms the basis for a determination of the role of humans in selecting for chemical change in this group.

CHEMOTAXONOMY AND CHEMICAL ECOLOGY

If chemical changes have taken place in plants under human influence, they should be evident in the evolutionary sequence represented by domesticated, weed, and wild forms. The use of chemical characters to study the phylogenetic relationships of plants is called chemotaxonomy. Patterns of chemical composition provide insight into the evolutionary relationships of plants at all taxonomic levels. Specific types of allelochemicals are often characteristic of particular families and have sometimes been used successfully to solve problems concerning the taxonomic placement of families and genera. In the case studies discussed below, differences in individual chemicals are used to determine differences at both intraspecific and interspecific levels.

Chemotaxonomic studies have employed a wide range of classes of chemical compounds as taxonomic markers (Harborne and Turner 1984). Flavonoids are the most widely used, but other important examples of chemicals used in biosystematic studies include terpenoids, glucosinolates, alkaloids, cyanogenic glycosides, nonprotein amino acids, and phenolics. Macromolecules such as proteins, nucleic acids, and polysaccharides are increasingly used to address systematic problems.

The development of chemotaxonomy as a field has closely paralleled that of chemical ecology. Both areas of interest employ the same techniques for analyzing plant chemicals and have benefited in the same ways from rapid advances in analytical technology. Many phytochemical researchers are not clearly engaged in one line of research as opposed to the other, and the same people often wear both hats. Much of the data generated in phytochemistry is relevant to both fields. The distinction between chemotaxonomy and chemical ecology is primarily in the biological questions being asked. Data from one field may answer questions in the other. In this volume, for example, the chemotaxonomic record of the phylogeny of domesticates is used to answer chemical-ecological questions about the interaction of humans with plant chemicals.

The most valuable compounds for showing human selection are obviously those with biological activities, including those with recognizable taste and smell. If chemical selection has been important during domestication, it is the undesirable or highly favored compounds that are likely to show dramatic discontinuities in the chemotaxonomic data. Differences both in the concentration of active compounds and in the presence of specific markers may reflect human intervention.

TWO CASE STUDIES

CASE I. AÑU AND MACA: INTERACTION OF BIOLOGICAL AND CULTURAL ASPECTS OF CHEMICAL SELECTION

Chemotaxonomic studies of añu and maca have provided data on the phylogeny of the former, and on the associations between chemistry and cultural beliefs concerning both. Añu releases isothiocyanates, the compounds responsible for the distinctive taste and smell of cruciferous vegetables. The obligate cultigen, *Tropaeolum tuberosum* subspecies *tuberosum*, releases only *p*-methoxybenzyl isothiocyanate, while the wild subspecies, *silvestre*, is characterized by benzyl, 2-propyl, and 2-butyl isothiocyanates (Johns and Towers 1981). The edible root of maca is consumed as both food and medicine. Maca releases benzyl and *p*-methoxybenzyl isothiocyanates (Johns 1981).

The relationship between maca and añu provides an excellent example of the interrelations of biological and cultural factors in human interactions with secondary chemicals. Similarities in folk beliefs associating añu and maca with effects on human reproduction are striking (Johns 1981, 1986b). This association of two isothiocyanate-releasing "root" crops with fertility suggests that secondary chemistry is in some way related to these folk beliefs. A negative effect of *T. tuberosum* on male reproductive processes in rats (Johns et al. 1982) supports the belief of Andean peoples that the plant has an antiaphrodisiac effect on males. In addition both añu and maca are regarded as having positive fertility effects on females. Both of these plants are characterized by aromatic isothiocyanates, and, at least in the perception of Andean peoples, there is a relationship between aromatic isothiocyanates and human reproduction. Although the use of isothiocyanates to affect human reproduction in general has a Western scientific basis, the specific emphasis on aromatic isothiocyanates seems culturally determined.

The significance of this association is emphasized by the replacement of the three nonaromatic constituents of wild *T. tuberosum* subsp. *silvestre* with *p*-methoxybenzyl isothiocyanate in the cultivated subspecies *tuberosum*. Although conceivably this could occur without human intervention, in light of selection on tuber characteristics leading to the domestication of the añu, it seems likely that cultural selection for the particular chemistry of the cultigen has been important.

Human reproductive rates are indeed lower at high altitudes (Clegg 1978); whether or not añu and/or maca have positive effects in ameliorating this situation, the stresses of the high-altitude environment provide a strong motivating force for the use of these plants and the beliefs associated with them.

A basic male-female duality which underlies Andean social interaction and extends to interrelations with the natural world may influence the association made between maca and añu. This duality is particularly prevalent in plant taxonomy (Johns 1986b). Maca and añu are both "female" in having a horizontal growth form and growing close to the earth. The añu has as its antithesis the oca, a plant with an erect form which produces tubers similar to those of the añu but notably phallic in shape and in their role in Andean ritual and jest.

The relation between biological characteristics and cultural significance of these plants becomes somewhat circular; however, the experimental results supporting antireproductive activity in males are a logical starting point for drawing associations. Antiandrogenic activity adversely affecting libido, or general well-being, could lead to a conditioned taste aversion. The female-benefiting designation could derive from the conceptually opposite effect on males and could be reinforced by its life form. A present-day tonic use of maca may be purely a cultural artifact with no physiological basis. The maca could have been associated with the añu as an enhancer of female fertility only on the basis of taste and smell and because of the similarity in life form. Whatever the case, the empirical basis for some of the uses of these plants, the motivation provided by environmental effects on human fertility, and the associations made through taste and smell stand out as driving forces behind these medicinal uses and the domestication of these plants.

CASE 2. CHEMICAL SELECTION AND POTATO EVOLUTION

The potato has had an impact on human affairs for at least six thousand years (MacNeish and Patterson 1975). Today it is the world's most important tuber crop (Food and Agriculture Organization 1983), ranking only behind wheat, maize, and rice in its contribution to global food production. Potatoes' high yield of calories and protein contributed to the subsistence base for the industrial revolution and the accelerating population growth in eighteenth- and nineteenth-century Europe (Salaman 1985). Europeans' reliance on this food is demonstrated by the infamous Irish Potato Famine, which was precipitated by crop failure in 1845 and 1846 and resulted in mass starvation and migration.

While these events are well known, the potato was not introduced into Europe until about 1570, and in fact its origins, and the majority of its history, go back to a much different time and place. Although problems of potential toxicity have always accompanied potato use, the real battles between humans and the defensive chemicals in potatoes have taken place in South America.

The center of genetic diversity of the potato is the central Andes Mountains of Peru and Bolivia between 2,000 and 4,100 meters above sea level. Eight species and several thousand varieties of cultivated potatoes are known from this area (Huamán 1979). Over 110 of the approximately 154 wild species of tuber-bearing *Solanum* are found in South America (Hawkes 1978).

In spite of the availability of indigenous species, a wild ancestor of the potato is not obvious. Undoubtedly the earliest domesticate was a wild diploid[1] ($2x = 24$), and several candidates have been proposed (Simmonds 1976). Among diploid cultigens the highly variable *Solanum stenotomum* Juz. et Buk. is considered ancestral. Possible progenitors of *S. stenotomum* include *S. canasense* Rydb., *S. leptophyes* Bitt., *S. multiinterruptum* Bitt., *S. sparsipilum* (Bitt.) Juz. & Buk., and *S. soukupii* Hawkes, none of which have been studied for glycoalkaloids. Wild species of potatoes have been exploited for food, although the toxic levels of glycoalkaloids in many species limit their usefulness (Osman et al. 1978).

All other species of cultivated potatoes have been derived in some way from the diploid *S. stenotomum*. *Solanum tuberosum* L. is a tetraploid ($4x = 48$) believed to be derived from a hybrid of *S. stenotomum* and *S. sparsipilum* through a process of amphidiploidy (a spontaneous doubling of the chromosomes of the diploid hybrid) (Hawkes 1972b). *Solanum tuberosum* subsp. *andigena* (Juz. et Buk.) Hawkes is the most widespread cultigen in the Andes; subspecies *tuberosum* is the potato of global importance.

Potential for the formation of hybrids comes from the weak reproductive barriers between most wild South American species. Prior to human intervention, species barriers were maintained largely by geographical and ecological isolation mechanisms (Hawkes 1972a). With human introduction of cultivated varieties into new areas and the disturbance of the landscape to create fields, hybridization would be expected.

Solanum × *ajanhuiri* is a hybrid between *S. stenotomum* and *S. megistacrolobum* Bitt. (Huamán 1975; Huamán et al. 1982). Interploidy-level hybridization produced the allotriploid *S.* × *juzepczukii* Buk. ($3x = 36$) from wild *S. acaule* Bitt. ($4x = 48$) and *S. stenotomum* (Hawkes 1962b; Schmiediche et al. 1980), and the allopentaploid *S.* × *curtilobum* Juz. et Buk. ($5x = 60$) from *S.* × *juzepczukii* and *S. tuberosum* subsp. *andigena* (Hawkes 1962b; Schmiediche et al. 1980). The role of wild potatoes in potato evolution is clearly evident

---

1. Diploid potatoes contain two pairs of each of twelve chromosomes ($2x = 24$).

from the number of economically important hybrids they have formed with *S. stenotomum*. However, the dynamics and extent of the interaction between wild and cultivated potatoes is largely obscured by time.

Nontoxic species of potato do exist, and it is possible that the progenitor of the potato was, in fact, a species naturally low in glycoalkaloids. *S. cardiophyllum* Lindl., a weed in maize fields in San Luis Potosí and Aguascalientes, Mexico, may be analogous to South American progenitors of the potato. It is harvested and sold in local markets, requires no processing before it is eaten (Galindo Alonso 1982), and is low in glycoalkaloids (Johns, unpublished data). Perhaps more intriguing from an evolutionary perspective is the nature of *S. maglia* from Chile. This wild species was extensively exploited by the Araucanian people and may have some relation to the domestication of the potato (Ugent et al. 1987). Although Ugent et al. (1987) suggest that this species is nontoxic, others say that populations of *S. maglia* are mainly bitter (Correll 1962). Clearly, our present level of knowledge about the evolutionary relationships of potatoes, not to mention our ignorance of the chemistry of various species, makes it impossible to say with any certainty whether domesticates were selected from a nontoxic population or species, or whether chemical change did, as I suggest, come about through a gradual process.

Rather than try to trace six thousand years of domestication history, I have focused on a secondary stage of domestication: the adaptation of potatoes to high altitude. As potato cultivation was extended above 3,800 meters in the Andes, genes for physiological adaptation to cold and dry conditions were obtained through hybridization of *S. stenotomum* with the wild endemic potato species *S. acaule* and *S. megistacrolobum*. The hybrids formed with these wild potatoes, which are high in glycoalkaloids, are themselves potentially toxic. While this apparent reintroduction of high levels of glycoalkaloids into the domesticated potato poses a problem for human cultivators, it provides an excellent case study for reconstructing human interactions with glycoalkaloids during early stages of the domestication process.

Solanum × *ajanhuiri* would appear to provide considerable potential for demonstrating an ongoing selection process in relation to glycoalkaloids. It is diploid and is likely to backcross with its parents. Unlike the self-compatible tetraploids, diploid potatoes are self-incompatible and outcrossing, and therefore offer more oppor-

tunities for the generation of new diversity and hybrids. *Solanum* ×
ajanhuiri is reported by Huamán (1975) to encompass both bitter and
nonbitter varieties. Its potential variability in glycoalkaloid content
contributed to my choice of *S.* × *ajanhuiri* as a candidate to show
chemical selection.

Solanum × *ajanhuiri* has been reported to grow sympatrically
with its purported progenitors—wild *S. megistacrolobum* and cultivated *S. stenotomum* (Huamán 1975)—at high altitudes in Bolivia
and Peru. There is a good basis for assuming that *S.* × *ajanhuiri* is
repeatedly re-created in field situations. Noncultivated potato populations in western Bolivia that have been referred to as *S. megistacrolobum* are probably hybrids themselves, and as such are examples of *S.* × *ajanhuiri*. Aymara cultivators are probably responsible
for limiting diversity and maintaining recognizable taxonomic entities among the cultivars.

A spectrum of genetic variability maintained by true seed production and spontaneous plant propagation would theoretically provide
a range of tuber variation, particularly with regard to glycoalkaloid
content and flavor. In hybrids between wild and cultivated potatoes,
the α-solanine content shows continuous variation consistent with a
polygenic form of inheritance. Variation in the glycoalkaloid content
of the *S. megistacrolobum–S.* × *ajanhuiri–S. stenotomum* complex
provides a model system for examining human selection for potatoes
on the basis of secondary chemical constituents. The following biosystematic study demonstrates that such selection has occurred.

The *Solanum* × *ajanhuiri* Complex. Table 4-2 summarizes the taxa
that are considered in this discussion. A group of five cultigens called
yari (in Aymara) are the reputed F_1 hybrids between wild *S. megistacrolobum* and the primitive domesticate *S. stenotomum* (Huamán
et al. 1982); two clones called *ajawiri* are viewed as representing the
backcross of yari to *S. stenotomum*. While such a scenario seems
simplistic in comparison to our understanding of the origin of other
crops such as cereals, the vegetative propagation of potatoes may be
responsible for this scenario.

Weediness and association with human disturbances is a common
but not universal characteristic of *S. megistacrolobum*. Populations
of this species that are associated with fields of cultivated potatoes
may be best categorized as weeds, whereas populations not associated with cultivated potatoes may preserve the original nature of this
wild species (cf. Small 1984). The populations found in potato fields

Table 4-2. Taxa of Solanum × ajanhuiri *Complex (from Johns et al. 1987)*

Taxon	Ploidy	Description
S. acaule	$4n$	Wild; widespread at high altitudes in Peru, Bolivia, and NW Argentina
S. stenotomum	$2n$	Domesticate of central Peru and central Bolivia
S. megistacrolobum	$2n$	Wild; includes var. *toralapanum*, found on the eastern slopes of the Bolivian Andes, and var. *megistacrolobum* from S Peru and NW Argentina
S. × *ajanhuiri*	$2n$	
(a)		Domesticated clones on altiplano of Bolivia and S Peru
Yari		*S. megistacrolobum* × *S. stenotomum*
Ajawiri		*S. megistacrolobum* × *S. stenotomum* × *S. stenotomum*
(b)		Weed; populations of noncultivated potatoes sympatric with cultivated *S.* × *ajanhuiri*; generally classified as *S. megistacrolobum*
Sisu	$3n$	Domesticate; *S. acaule* × *S. ajanhuiri* clones sympatric with *S.* × *ajanhuiri*
S. × *juzepczukii*	$3n$	Domesticate; *S. acaule* × *S. stenotomum*

that are morphologically intermediate between *S. megistacrolobum* and cultigens of *S.* × *ajanhuiri* are referred to here as "weed *S.* × *ajanhuiri*."

Solanum megistacrolobum var. *toralapanum* Ochoa (Ochoa 1984), which is not found on the altiplano, was used as a reference point from which to test the relative affinity of altiplano populations of weed *S.* × *ajanhuiri* with cultivated populations of this species. This variety may be a relic of a more widely distributed taxon and therefore may preserve many of the primitive characteristics of *S. megistacrolobum*. Notably, the growth form and leaf shape of weed forms of *S.* × *ajanhuiri* are intermediate between var. *toralapanum* and cultivated populations of *S.* × *ajanhuiri* and *S. stenotomum*. Significantly, var. *toralapanum* is unique in *S. megistacrolobum* in containing the commertetraose sugar moiety (in the glycoalkaloids commersonine and dehydrocommersonine) characteristic of other wild species of series *Megistacrolobum* (see below).

During the investigation we realized that the origin of triploid

Table 4-3. Morphological Characters Distinguishing Sisu from
Solanum × juzepczukii (*from Johns et al. 1987*)

Characters	Sisu	Solanum × juzepczukii
Decurrence of primary lateral leaflets	0.3–0.5	0.1
Terminal leaflet length / Lateral leaflet length	1.4–2.2	1.1–1.3
Number of secondary leaflets	0–1	6
Length of peduncles (mm)	to 6	1–4
Corolla diameter (mm)	to 18	to 25
Habit	rosette	semirosette

potato cultigens may be linked to the domestication of *S. × ajan-huiri. Sisu* is a group of poorly understood clones of frost-resistant potato cultigens that have been placed with *S. × juzepczukii* (Ochoa 1958). Ochoa reported that sisu clones are triploids ($3x = 36$); those I observed in the field were, from all indications, sterile.

Sisu cultigens are distinctive in their unique morphological (Table 4-3) and agronomic traits and their nonbitter tubers. They are particularly well adapted to the driest and coldest conditions under which potatoes are grown in western Bolivia or elsewhere in the Andes. However, unlike *S. × juzepczukii*, they are eaten without any detoxification. Sisu clones are sympatric with *S. × ajanhuiri* and resemble it in the field. I developed the hypothesis that sisu clones are in fact hybrids between the wild tetraploid *S. acaule* and cultigens of *S. × ajanhuiri* (Johns et al. 1987).

Systematic studies of the taxa shown in Table 4-2 were carried out in order to demonstrate the relationships among these members of the *S. × ajanhuiri* complex. Once a likely evolutionary history of the group was determined, differences in chemical content were examined for evidence of selection by humans. While the majority of traits were expected to follow a continuous pattern, in accordance with the general principles of domestication discussed previously, those on which humans direct selection pressure were expected to show greater discontinuities.

Numerical Studies of Potato Affinities. Sixteen populations of weed *S. × ajanhuiri* and *S. megistacrolobum* var. *toralapanum* were collected in the Department of La Paz (Fig. 4-2), where *S. × ajanhuiri*

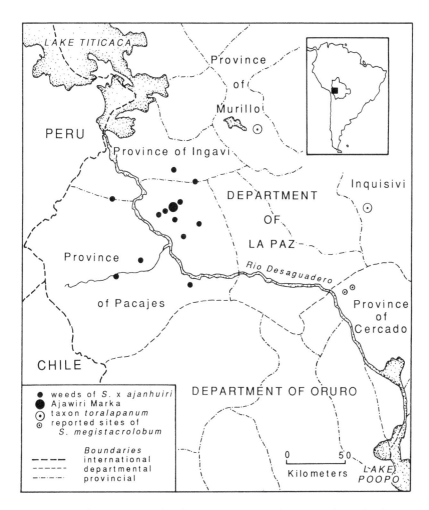

FIG. 4-2. Collection sites of *Solanum megistacrolobum* and weeds of *Solanum × ajanhuiri* used in this study.

might have originated (Huamán 1975). A large, highly variable population of weed *S. × ajanhuiri* from Ajawiri Marka, Caquiaviri, was divided into eight subpopulations. *Solanum acaule, S. × ajanhuiri* cultigens, *S. stenotomum, S. × juzepczukii,* and clones of sisu were also collected in the Departments of La Paz and Oruro. Forty-six morphological and eight chemical characters were examined initially. Details of the numerical taxonomic analyses using Similarity-Graph Clustering (SIMGRA) and Principal Components (PCA) analyses are described elsewhere (Johns 1985; Johns et al. 1987).

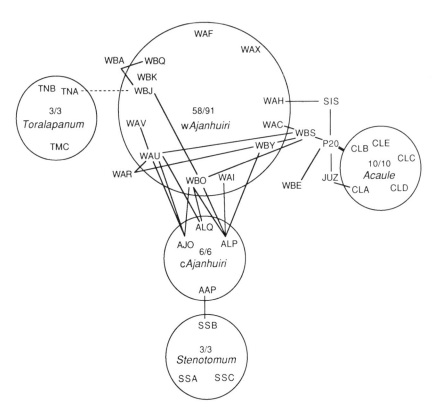

FIG. 4-3. In the SIMGRA analysis shown here, all interconnections between closely related members of a cluster are not represented. When all
members were connected to at least three others, a circle was drawn
around the group (from Johns et al. 1987).

Relationships of wild and cultivated diploids. Both SIMGRA (Fig. 4-3)
and principal components (Fig. 4-4) analysis of the data for cultigens,
weed, and wild diploids suggest the existence of the same groups:
weed *S.* × *ajanhuiri*, cultivated *S.* × *ajanhuiri* including yari and
ajawiri, *S. stenotomum*, and *S. megistacrolobum* var. *toralapanum*
groups.

Relationships of S. acaule *and the* S. × ajanhuiri *complex.* The SIMGRA
and the principal components analysis of the data for *S. acaule* and
crop and weed diploids (Fig. 4-5) provides insight into the nature of
three putative triploids (SIS, JUZ, P20). The SIMGRA (Fig. 4-3) shows
P20 to be a bridge population with connections to *S. acaule*, to weed
S. × *ajanhuiri*, and to the other intermediates: sisu (SIS) and *S.* ×
juzepczukii (JUZ). Sisu is connected to weed *S.* × *ajanhuiri* and P20,

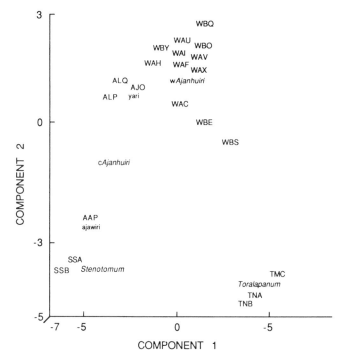

FIG. 4-4. Principal components analysis (from Johns et al. 1987). The first and second principal components are plotted on the y and x coordinates, respectively. The first principal component separates *S. stenotomum, S. × ajanhuiri*, and *S. megistacrolobum* var. *toralapanum* in an order that reflects a sequence from domesticated to wild. The second principal component includes large portions of the leaf and total glycoalkaloid data that are not represented in the first principal component. The second component better separates the groups by distinguishing the *S. × ajanhuiri* complex from its putative parental taxa.

reflecting its likely intermediacy between *S. × ajanhuiri* and P20. Sisu is related as well to yari, *S. acaule*, and *S. × juzepczukii*. *Solanum × juzepczukii* shows similar affinities, but it connects to ajawiri before it connects to yari. In the PCA the diploids—*S. stenotomum, ajawiri, yari*, and weed *S. × ajanhuiri*—are separated as in Figure 4-4. Populations of *Solanum acaule* have similar values on the second principal component to those of weed *S. × ajanhuiri*. This relationship between noncultivated and cultivated plants is paralleled in the three intermediate clones studied. *Solanum × juzepczukii* lies closer to *S. stenotomum*, whereas sisu has more affinity to weed *S. × ajanhuiri*. P20 shows only slightly more of this tendency than sisu.

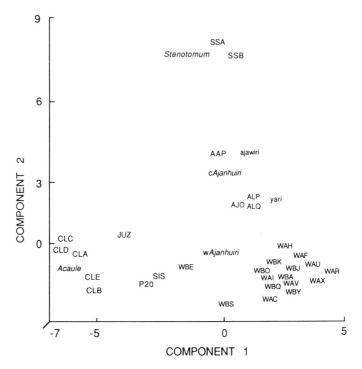

FIG. 4-5. The first principal component clearly separates the tetraploid, diploid, and intermediate taxa from one another. The second principal component separates wild, weed, and cultivated taxa (from Johns et al. 1987).

Chemotaxonomic Data. Data from glycoalkaloid analyses (Johns and Osman 1986; Osman et al. 1986) add additional information regarding relationships of taxa and insights into the issue of chemical selection. The structural relationships of the glycoalkaloids identified in this study are shown in Figure 4-6. Total glycoalkaloid determinations (TGA) and glycoalkaloid identities of *S. stenotomum* and *S.* × *ajanhuiri* are recorded in Table 4-4. These results are generally consistent with data from previous studies (Osman et al. 1978) in that the major glycoalkaloids of *S. stenotomum* and ajawiri clones of *S.* × *ajanhuiri* are α-solanine and α-chaconine (Fig. 4-6; 2 and 3).

Weeds and the primitive yari cultivars of *S.* × *ajanhuiri*, which Huamán (1975) had suggested are bitter, surprisingly have glycoalkaloid levels considerably less than ajawiri, *S. stenotomum*, and *S. megistacrolobum*. Glycoalkaloid levels in yari cultigens and weeds were actually so minimal as to make it impossible to conclusively identify the constituents present or to have confidence in the ac-

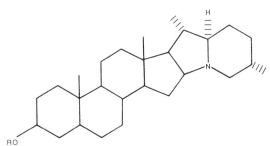

1 R=H; solanidine
2 R=O-Glc-(1→3)-[Rha-(1→2)]-Gal; α-solanine
3 R=O-diRha-(1→2, 1→4)-Glc; α-chaconine
4 R=O-diGlc-(1→2, 1→3)-Glc-(1→4)-Gal; dehydrocommersonine

5 R=H; demissidine
6 R=O-Xyl-(1→3)-[Glc-(1→2)]-Glc-(1→4)-Gal; demissine
7 R=O-diGlc-(1→2, 1→3)-Glc-(1→4)-Gal; commersonine

8 R=H; tomatidenol
9 R=O-Glc-(1→3)-[Rha-(1→2)]-Gal; α-solamarine
10 R=O-diRha-(1→2, 1→4)-Glc; β-solamarine

11 R=H; tomatidine
12 R=O-Xyl-(1→3)-[Glc-(1→2)]-Glc-(1→4)-Gal; tomatine
13 R=O-diGlc-(1→2, 1→3)-Glc-(1→4)-Gal; sisunine

FIG. 4-6. Structural relationships of glycoalkaloids (from Johns and Osman 1986).

Table 4-4. Glycoalkaloids of Cultigens of Solanum × ajanhuiri *and* S. stenotomum *(from Johns and Osman 1986)*

Species and clone identity	TGA mg/100 g	Aglycones as % of total				Glyco-alkaloids[a]
		1	5	8	11	
S. × ajanhuiri						
Ajawiri (two clones)	10.5–13.0	100	0	0	0	2, 3
Yari (six collections of three clones: Y1, Y2, Y5 [2])	<4.3					9, 10[b]
S. stenotomum (three clones)	11.3–20.8	100	0	0	0	2, 3
S. acaule × ajanhuiri Sisu (three clones)	12.0–24.8	0	55	0	45	7, 13

All analyses were made on tubers collected in the Departments of La Paz and Oruro, Bolivia.
[a]See Fig. 4-6 for alkaloid codes.
[b]Best guess on the basis of co-chromatography with other standards.

curacy of the total glycoalkaloid determinations. Total glycoalkaloids measured ranged from 4.3 to 2.2 mg/100 g. On the basis of retention times on thin-layer chromatography, the two spots characteristic of yari and weed S. × ajanhuiri are probably α- and β-solamarines (Fig. 4-6; 9 and 10).

Glycoalkaloid determinations of S. × juzepczukii, sisu, and related wild material are recorded in Table 4-5. The total glycoalkaloid content of the confirmed S. × juzepczukii clones studied here—*luq'i* and *kaysalla*—is 37.5 and 63.7 mg/100 g, respectively, compared to a range of 11.7–46.8 mg/100 g reported previously for the species. The clones of S. × juzepczukii studied here contained the aglycones demissidine, tomatidine, solanidine, and possibly tomatidenol (Fig. 4-6; 5, 11, 1, and 8, respectively), which is consistent with previous reports from this species (Schmiediche et al. 1980). However, in the identity of individual glycoalkaloids, clones of S. × juzepczukii varied among themselves and from the specimens examined by Osman et al. (1978). Variability in glycoalkaloids in S. × juzepczukii, sisu, P20, and *ankanchi* (an unusual altiplano potato probably related to S. × juzepczukii) are shown in Figure 4-7. With the exception of sisu, these clones appear to have an additive glycoalkaloid profile that is often seen in hybrids where the different genomes of two different parents are expressed.

Table 4-5. Glycoalkaloids of S × juzepczukii, *Sisu, and Related Potato Collections*

Species and Aymara name	Accession number and/or province	TGA mg/100 g	Glycoalkaloids[a] or aglycones (% of total aglycones)
	Cultivated		
S. × *juzepczukii*			
jank'o luq'i	Pacajes	37.5	dem(54), tom(41), sol(5.4)
laram kaysalla	HJA-1225, Carangas	63.7	dem(68), tom(17), sol(12)
		11.7–46.8[c]	demissine(40), α-solanine(38), α-chaconine(14), α,β-solamarines(8)
Sisu			
larem sisu	Pacajes	12.0	commersonine(55), sisunine(45)[b]
jank'o sisu	Pacajes	17.1	"
sisu	HJA-1519, Pacajes	24.8	"
Ankanchi			
wila ankanchi	HJA-1472, Sajama	2.9	
larem ankanchi	HJA-1547, Pacajes	3.7	
chanu suiti ankanchi	HJA-1470, Sajama	1.9	
	pooled sample		tom(29), dem(28), sol(22), tmd(21)
	Noncultivated		
Population 20 (P20)	Pacajes	54.3	tom(59), dem(33), sol(4), tmd(5)
S. *acaule*	Pacajes	—	demissine(70), tomatine(30)
		35–126[c]	demissine, tomatine (% variable)

[a]Variation in glycoalkaloids in luq'i, kaysalla, ankanchi, and sisu and P20 is recorded in Figure 4-7.

[b]Tomatidine aglycone, linked with commertetraose.

dem = demissidine; tom = tomatidine; sol = solanidine; tmd = tomatidenol.

[c]Reported in Osman et al. 1978.

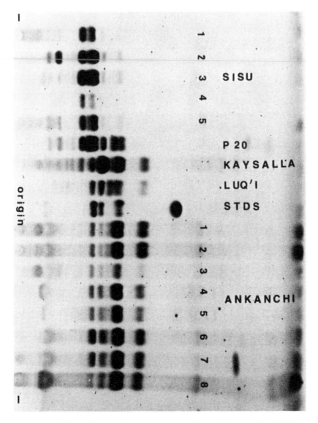

FIG. 4-7. Thin-layer chromatography of glycoalkaloids.

Although sisu has only two glycoalkaloids, it is unique nonetheless. Its clones are characterized by commersonine (Fig. 4-6; 7), a glycoalkaloid that has been reported previously from wild potatoes (Osman et al. 1978; Sinden et al. 1980), and a novel glycoalkaloid called sisunine (Fig. 4-6; 13) (Osman et al. 1986) (Fig. 4-8), which has the tomatidine aglycone and the commertetraose sugar moiety (Fig. 4-6; 11 and 14).

The presence of the tomatidine and demissidine aglycones in *S.* × *juzepczukii* and sisu is not surprising since their wild parent, *S. acaule,* is characterized by tomatine and demissine. The only specimen of *S. acaule* studied from the Province of Pacajes, Bolivia, confirmed the patterns of demissine and tomatine for this species described by Osman et al. (1978). Sisu clones are somewhat striking for their low levels of total glycoalkaloids, which fall in the lower end of the range reported for *S.* × *juzepczukii.*

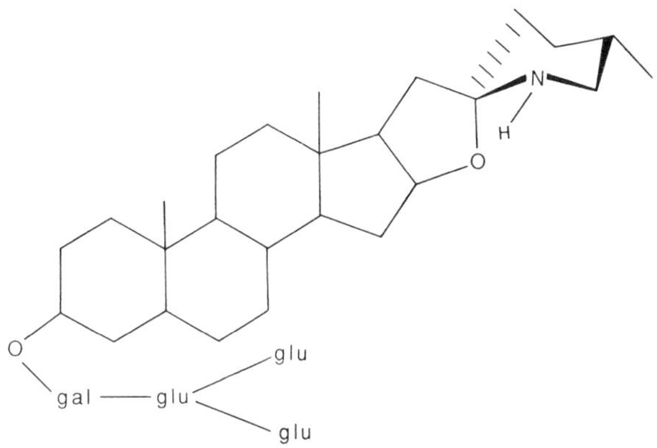

FIG. 4-8. Structure of sisunine.

Chemotaxonomic evidence from wild species of Solanum series Megistacrolobum. Because of the likely long history of hybridization of wild and cultivated potatoes in the Andes, the wild *S. megistacrolobum* ancestor of *S.* × *ajanhuiri* is difficult to determine. On morphological grounds *S. megistacrolobum* has been described as a highly variable species. The glycoalkaloid analyses reported in Table 4-6 confirm this observation. Several wild species within *S.* series *Megistacrolobum* were examined for patterns that might shed light on the nature of *S. megistacrolobum* prior to human involvement in its evolution, and on the relationship of *S.* × *ajanhuiri* to wild potatoes. Most clones of *S. megistacrolobum* examined were characterized by tomatine, with some accessions also containing demissine as either the major or minor constituent. One clone of *S. megistacrolobum* from Argentina and *Solanum megistacrolobum* var. *toralapanum* stand out by being characterized by the commertetraose sugar plus additional glycoalkaloids.

Solanum megistacrolobum var. *toralapanum* had the highest TGA levels of any of the material studied, ranging from 57 to 125 mg/100 g. Other wild species in *Solanum* series *Megistacrolobum* ranged in TGA between 20 and 72 mg/100 g.

Glycoalkaloid patterns in the series *Megistacrolobum* (Table 4-7) are closer across species lines than within *S. megistacrolobum* alone (Table 4-6) and show distinct geographical patterns. Here again the commertetraose sugar is a key character. *S. boliviense* Dun. from Bolivia and *S. sanctae-rosae* Hawkes from Argentina are similar to the *S. megistacrolobum* containing commersonine from Argentina and to *S. megistacrolobum* var. *toralapanum* in containing this entity.

Table 4-6. *Glycoalkaloids of Solanum megistacrolobum, Including Taxon toralapanum and Weeds of* S. × ajanhuiri *(from Johns and Osman 1986)*

Taxon and accession	Collection location	TGA mg/100 g	Aglycones as % of total				Glyco-alkaloids[g]
			1	5	8	11	
S. megistacrolobum							
P1210034[c]	Pot-BOL						12
P1265578[c]	Tar-BOL						12
P1265873[d]	Pot-BOL						12
P1265874[a]	Pot-BOL	2.4					
P1275149[a]	ARG	1.2					
P1283133[c]	ARG						12
P1458346[a]	ARG	26.0					12
P1458347[b]	ARG						12
P1458350[b]	ARG						12
PIOKA4520[c]	ARG		0	0	0	100	12
P1275148[d]	ARG						6, 12
PIOKA6758[c]	ARG		0	22	0	78	12, 6
P1233124[e]	ARG						6 TLC spots
P1320303[e]	Vic-ARG						7, 6, 12
Taxon *toralapanum*							
P1458397[b]	Tar-BOL	57	0	100	0	0	7, 6

Table 4-6. Continued

Taxon and accession	Collection location	TGA mg/100 g	Aglycones as % of total				Glyco-alkaloids[g]
			1	5	8	11	
Johns 83-92	Lpz-BOL, Huaynacota, Prov. Inquisivi	125	75	22	0	0	4, 6
Johns 83-102	Lpz-BOL, Palca, Prov. Murillo	88	29	71	0	0	6, 4
Weed S. × *ajanhuiri*							
7 collections	Lpz-BOL, Prov. Pacajes and Ingavi	<4.0					9, 10[f]
Johns 83-38	Lpz-BOL, Ulloma, Prov. Pacajes						7, 12
Johns 83-87	Lpz-BOL, General Campero, Prov. Pacajes	12.0					

[a] Tubers grown at *Mattaei* Botanical Gardens.
[b] Tubers from Potato Introduction Station.
[c] Vegetative material grown at *Mattaei* Botanical Gardens.
[d] Rhizomes grown at *Mattaei* Botanical Gardens.
[e] Leaf morphology similar to taxon *toralapanum*.
[f] Best guess on basis of co-chromatography with other standards.
Department codes: Pot = Potosí; Tar = Tarija; Vic = San Victoria; Lpz = La Paz. Country codes: BOL = Bolivia; ARG = Argentina.
[g] See Fig. 4-6 for alkaloid codes.

Table 4-7. Glycoalkaloid Characterization of Accessions of Series megistacrolobum *Obtained from Potato Introduction Station, Sturgeon Bay, Wisconsin (from Johns and Osman 1986)*

Species and accession	Collection location	TGA mg/100 g	Aglycones[a] as % of total 1	5	8	11	Glyco-alkaloids[a]
S. boliviense							
PI310974	BOL	72	11	89	0	0	7, 6[b]
PI310975	BOL	61	10	90	0	0	7, 6[b]
S. raphanifolium							
PI310951	Cuz-PER	37	100	0	0	0	3, 2[b]
PI310999	PER	28	100	0	0	0	3, 2[b]
S. sanctae-rosae							
PI205397	ARG	20					6, 7, 12[b]
PI218221	Tac-ARG	25	8	75	0	17	6, 7, 12[b]
S. sogarandinum							
PI230510	Lib-PER	28	0	0	100	0	9, 10

[a]See Fig. 4-6 for alkaloid codes.
[b]Traces of glycoalkaloids with aglycone 2.
Country codes: BOL = Bolivia; PER = Peru; ARG = Argentina. Department Codes: Cuz = Cuzco; Tac = Tacuman; Lib = Libertad.

The Relationships of S. × *ajanhuiri*. Both numerical analyses support the hypothesis that *S. × ajanhuiri* originated as a hybrid between *S. megistacrolobum* and *S. stenotomum*. The nature of glycoalkaloid constituents (Johns and Osman 1986) plus Figures 4-3 and 4-4 support the assessment of Huamán (Huamán 1975; Huamán et al. 1982, 1983) that yari is more closely related to *S. megistacrolobum* than is ajawiri, and that ajawiri is more closely related to *S. stenotomum*. The weed populations of *S. × ajanhuiri* show more chemical affinity with cultigens of *S. × ajanhuiri* than *S. megistacrolobum* from southern Bolivia and Argentina, and with *S. megistacrolobum* var. *toralapanum* from farther east (Johns and Osman 1986).

The assessment that the w*ajanhuiri* group—altiplano populations with affinities to *S. megistacrolobum*—are conspecific weeds of *S. × ajanhuiri* (c*ajanhuiri*) is supported by the similarities in their morphological and chemical characteristics and their differences from *S. megistacrolobum* var. *toralapanum*. It is clear from the relationships of weed and cultivated *S. × ajanhuiri* that the cultivated types—yari and ajawiri—are not simply F_1 and backcross hybrids between *S.*

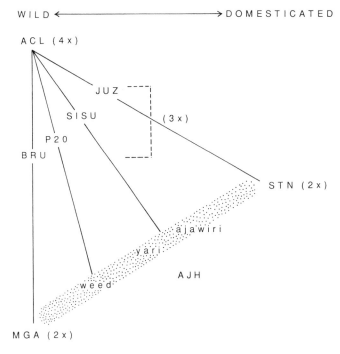

WILD ←————————————→ D O M E S T I C A T E D

FIG. 4-9. Diagrammatic summary of hybridization relationships involving *S. acaule, S. stenotomum,* and *S. megistacrolobum* (from Johns et al. 1987). ACL = *S. acaule;* AJH = *S.* × *ajanhuiri* hybrid swarm; BRU = *S.* × *bruecheri;* JUZ = *S.* × *juzepczukii;* MGA = *S. megistacrolobum;* SISU = *S. acaule* × *ajanhuiri;* STN = *S. stenotomum;* P20 = population 20.

stenotomum and *S. megistacrolobum* as proposed by Huamán (1975; Huamán et al. 1982). More likely the specific crosses responsible for the modern yari and ajawiri clones were between *S. stenotomum* and weeds of *S. megistacrolobum* that were already engaged in introgressive hybridization with *S. stenotomum.* In Figure 4-9 this process is represented on the axis between MGA and STN as a hybrid swarm situation where repeated crosses are hypothesized to continue to take place between the two parents and their progeny. Only human intervention maintains discrete cultivated entities.

Since *S. megistacrolobum* is high in glycoalkaloids and *S. stenotomum* is low, glycoalkaloid levels would be expected to follow a continuum along the lower axis of Figure 4-9 from the wild to the cultivated diploid parent. As we have already seen, they do not.

***Solanum* × *ajanhuiri* and Heteroploidal Hybridization.** The hybridization of the wild tetraploid *S. acaule* with *S. stenotomum* to produce triploid cultigens of *S.* × *juzepczukii* is well known (Hawkes 1962b). Clones of *S.* × *juzepczukii* vary in their concentrations of total glycoalkaloids and degree of frost resistance (Schmiediche et al. 1980). Schmiediche et al. (1980) suggested that *S.* × *ajanhuiri* might have contributed to frost resistance of some members of the diploid gene pool (presumably through introgressive hybridization). The hybrid nature of sisu suggests that cultivated diploids, through the *S.* × *ajanhuiri* hybrid swarm, have contributed directly to both the frost resistance and low total glycoalkaloid properties of some triploid potato cultigens.

The results of the numerical analyses (Figs. 4-3 and 4-5) support the argument that sisu (SIS) is intermediate between *S. acaule* and diploid cultigens. This conclusion is strongly supported by the chemotaxonomic data. Sisu is unique among cultivated potatoes in containing commersonine and the novel glycoalkaloid sisunine (Osman et al. 1986). McCollum and Sinden (1979) showed that aglycones and sugar moieties are inherited independently and that hybrids may contain new combinations of these characters. Sisunine, a hybrid compound reflecting the hybrid nature of sisu, is composed of the aglycone tomatidine, characteristic of *S. acaule,* and the commertetraose sugar moiety common in many species of series *Megistacrolobum.*

The heteroploidal hybridization that characterizes the relations of *S. acaule* with diploid members of the *S.* × *ajanhuiri* complex is summarized graphically in Figure 4-9. *Solanum acaule* hybridizes with diploids on any part of the spectrum from wild *S. megistacrolobum* to cultivated *S. stenotomum. Solanum* × *bruecheri* Correll has been thought to be derived from the well-documented *Solanum acaule* × *S. megistacrolobum* hybrids (Hawkes and Hjerting 1969; Okada and Clausen 1982). There is strong evidence that P20 is a hybrid between *S. acaule* and weed *S.* × *ajanhuiri.*

This study demonstrates that sisu has a different genetic composition and origin than all clones of *S.* × *juzepczukii* that have been studied. It is distinguishable morphologically from *S.* × *juzepczukii,* although in the field their similar ecological context and their shared features strongly link the two. In sisu, *S.* × *juzepczukii,* and P20, characteristics of *S. acaule* dominate, producing "lopsided" hybrids (Wagner 1983) where the hybrid resembles the parent of the higher ploidy level. This effect may overaccentuate the genetic similarities of sisu and *S.* × *juzepczukii* and underestimate their evolutionary divergence. Significantly, most of the characters by which they differ (Table 4-3) are those that ally the similarities of sisu to *S. megistacrolobum* and *S.* × *ajanhuiri.*

Solanum megistacrolobum is considered more drought resistant than *S. acaule* (Hawkes and Hjerting 1969). This trait has probably contributed to the success of *S.* × *ajanhuiri,* particularly yari, in frigid and arid areas of western Bolivia (cf. Huamán et al. 1980), and also may have facilitated the adaptation of sisu to arid conditions via its *S.* × *ajanhuiri* parent. Sisu's superior frost resistance is a contribution of *S. acaule.* Its nontoxicity compared with *S.* × *juzepczukii* is likely related to the extremely low concentrations of total glycoalkaloids (<4 mg/100 g fresh weight) in its yari progenitors.

The formation of sisu seems to be a culminating step in altiplano adaptation in combining the virtues of both the heteroploidal and introgressive hybridization processes. As a hybrid between *S. acaule* and *S.* × *ajanhuiri,* sisu embodies these two trends. Most important, through selection by both environmental and human forces, sisu has the benefits of frost resistance, drought resistance, and low glycoalkaloid levels. It facilitates human adaptation in environments where its triple qualities are sought. In turn, it is propagated widely by humans.

Implications of the Biosystematic Data. The unexpectedly low glycoalkaloid concentrations in yari (and indirectly in sisu) are the best evidence of human selection for nontoxicity in the origins of *S.* × *ajanhuiri.* Weeds and the primitive yari cultigens are unique in having levels of total glycoalkaloids considerably less than either *S. megistacrolobum* or the more advanced cultigens ajawiri and *S. stenotomum.* The total glycoalkaloid contents of the components of this domestication complex, then, are contrary to the expected continuum decreasing from *S. megistacrolobum* to weeds to yari to ajawiri to *S. stenotomum.* Glycoalkaloid levels and the actual discontinuous pattern are summarized in Figure 4-10. The route to the domestication of *S.* × *ajanhuiri* from *S. megistacrolobum* must have brought about a reduction in levels of glycoalkaloids of some fifteen- to twentyfold. Because characteristics of crop-weed-wild complexes that show the greatest discontinuity have typically been subject to directed human selection, the glycoalkaloids in the *S.* × *ajanhuiri* complex are among the tuber characteristics that show evidence of human selection in the domestication process.

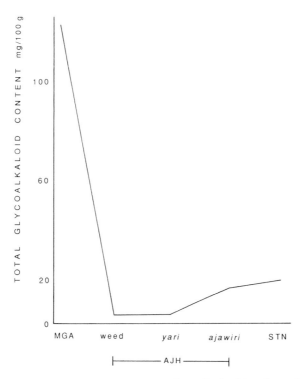

FIG. 4-10. Discontinuities in glycoalkaloid levels in the *S.* × *ajanhuiri* hybrid swarm (from Johns 1989).

Dynamics of Domestication. Chemical change depends on the same biological processes as any other aspect of domestication. In order for selection to occur, variation of chemical genotypes must exist. New genotypes which are the raw material for selection, and which also fix the characteristics selected, must be produced through genetic recombination and become established as part of the gene pool. Stages in this process involve cross-pollination, seed dispersal, seed germination, and survival and maturation of seedlings to the point where selection and another round of genetic recombination can occur. In crops such as the potato, natural biological factors in plant biology and ecology and various anthropogenic manipulations of the environment, as well as direct human interventions, will affect the outcome of the process.

Classic studies on the dynamics of domestication have focused on seed crops such as maize (*Zea mays*) (Wilkes 1977), wheat (*Triticum* spp.), and sorghum (Harlan et al. 1973) in their centers of genetic diversity. But potatoes present a different situation. In this vege-tatively propagated crop, sexual reproduction is poorly documented, and our understanding of the evolutionary process responsible for the wealth of variation in the cultivated gene pool is primarily extrapo-lated from other evidence. The few field studies directed to the problem have emphasized the potential, if not the reality, for genetic recombination in potatoes through hybridization in field situations with different clones of cultigens, and between cultivated and wild plants.

Wild species often grow in and around planted fields. Ugent's studies in Mexico on *S. × edinense* (Ugent 1967) give the most specific evidence of hybridization. He inferred from morphological evidence that the weed potato *Solanum × edinense* Berth. originated in Mexico as a hybrid between the domesticate *S. tuberosum* subsp. *andigena* and the wild *S. demissum* Lindl. (native to Mexico). He suggested that introgression of genes from *S. demissum* into pota-toes cultivated in Mexico since their postconquest introduction in-creased variability in the latter. Genes from *S. demissum* have also been an important source of resistance to potato late blight (Ugent 1968). The frost-resistant hybrid domesticates *S. × juzepczukii* and *S. × curtilobum* from the Andes of Peru and Bolivia similarly may represent examples of hybridization between wild and cultivated potatoes (Ugent 1970b). *Solanum × edinense* (3x), *S. × juzepczukii* (3x), and *S. × curtilobum* (5x) are all odd polyploids.

Diploid potatoes such as *S. × ajanhuiri* have undoubtedly played a greater role in the introgression of wild germ plasm into cultivated species than have sterile odd polyploids. Introgression between the diploid wild species *S. chacoense* Bitter and *S. microdontum* Bitter in Argentina (Hawkes 1962a), and the probable stabilization of a wild hybrid species, *S. raphanifolium* Cardenas & Hawkes, in Peru (Ugent 1970a) have been documented. Wild potatoes include a number of species that are highly cross-compatible with the diploid cultigens, and backcrosses of hybrids to wild and cultivated parents are to be expected.

Field studies of the evolution of Andean potatoes. In the Andean gene center, traditional agricultural practices involve the mixed planting of from as few as five to more than forty (Brush et al. 1981) clones of several species and ploidy levels of potatoes. Recent studies (Brush et al. 1981; Jackson et al. 1980) have underlined the suitability of An-dean potato fields for genetic recombination and have made preliminary attempts at elucidation of both cultural and biological processes that lead to ongoing human-mediated evolution. Farming practices on the Bolivian altiplano are among the most traditional in the Andes. The few examples of naturally occurring *S. stenotomum* × *ajanhuiri* reported by Huamán (1975) suggest that traditional farming practices do result in wild-cultigen crosses.

In addition to introducing cultigens and altering the environment through cultivation, humans may specifically affect the ongoing evolution of potatoes by actively selecting seed for the next season on the basis of yield, cooking time, storability, and frost and disease resistance, as well as flavor and reduced toxicity (Huamán 1975). The behavior of Andean farmers in relation to newly evolved forms has not been studied. If hybrids are formed in potato fields, it seems likely that farmers would deal with them according to recognizable properties and select for or against them depending on their desirability as defined by cultural values.

Folk wisdom in the Department of La Paz, Bolivia, credits the town of Caquiaviri (Ajawiri in the Aymara language) as the place of origin of ajawiri varieties of *S.* × *ajanhuiri,* and Caquiaviri was thus picked as the center of this study. A protected valley adjacent to the town and locally known as Ajawiri Marka was identified as a favorable location where hybridization between wild and cultivated potatoes takes place. Much of the present investigation focuses on this location and its possible significance in the evolution of altiplano potatoes. I hypothesized that ongoing introgressive hybridization was most likely to occur in the valley of Ajawiri Marka where abundant wild and domesticated plants grow in close association, and I present statistical evidence here to support this hypothesis.

The hybrid swarm comprising both wild and cultivated forms of *S.* × *ajanhuiri* provided a case study for considering the environmental and human factors that determine the introduction of new diversity into the potato gene pool. Johns and Keen (1986a) considered the role of insect pollinators and human cultivation practices in providing a situation conducive to the production of new genetic diversity, as well as the more deliberate role of humans in directing the introduction of new genetic types into the potato gene pool. The following discussion is based on that paper.

During the period from October 1982 to June 1983 we made specific observations on the ways that local farming practices could affect the genetic nature of cultivated and wild tuber-bearing *So-*

lanum. Informal survey methods with Aymara-speaking subsistence farmers were carried out from planting until after harvest.

Insects as pollination vectors. We evaluated the likelihood of insects acting as pollination vectors effecting crossing among wild and cultivated diploid potatoes in the field through observing and collecting insect visitors to potato flowers. All insect collections were carried out in the valley of Ajawiri Marka in cultivated fields of mixed diploids (*S. × ajanhuiri* and *S. stenotomum*) and tetraploids (*S. tuberosum* subsp. *andigena* and *S. acaule*).

During our nine collection periods three species of bees (*Bombus robustus* Smith, *Leioproctus* spp., and *Anthophora incerta* [Spin.]) and two fly species (*Scaeva melanostoma* [Macquart] and Tachinidae) were captured on potato flowers. Rabinowitz et al. (personal communication) have recently provided additional evidence of the importance of bees in the pollination of Andean potatoes.

Bumblebees (*Bombus* spp.) are common pollinators of potatoes (Camadro and Peloquin 1982; Glendinning 1976). Studies of the pollination biology of the genus *Solanum* indicate that a group of vibratile pollinating bees that includes *Bombus* spp. are important in releasing pollen from the poricidal anthers of *Solanum* species (Buchmann et al. 1977; Symon 1979). The bee genera *Bombus* and *Anthophora* recorded as visitors to altiplano potatoes are among the vibratile pollinators previously associated with the genus *Solanum*, and undoubtedly they play a role in genetic recombination.

Human activities and potato evolution. We adapted field-mapping methods employed previously (Brush et al. 1981) to show the diversity and intrafield relationships of varieties of cultivated plants in order to illustrate the possibilities of wild-cultigen hybridization in the Bolivian altiplano south of Lake Titicaca. Potato fields in this area typically show the same heterogeneity of varieties that has been recorded elsewhere in the Andes (Brush et al. 1981).

The data and several statistical comparisons derived from the mapping of five potato fields are summarized in Tables 4-8 and 4-9. The comparisons are useful as a basis for understanding the potential dynamics of various field mixtures and for predicting the likelihood of wild-cultigen hybridization.

Although all the fields contained mixtures of species, the relative proportions of species were highly variable. The fields contained varying amounts of weed *S. × ajanhuiri*, as reflected in the number of plants per square meter and the percentage of weed plants. Although

Table 4-8. Cultivated Plants Recorded in Field-mapping Experiments (from Johns and Keen 1986a)

Study number	Total plants (weed and cultivated)	Total plants/m²	Cultivated *Solanum*		Cultivated plants/m²
1	279	2.3	AJH[a]	10	0.5
			ADG[b]	16	
			STN[d]	38	
			Total	64	
2	639	6.9	AJH	3	3.5
			ADG	286	
			STN	32	
			Total	321	
3	2,515	4.4	AJH	105	3.9
			ADG	1,325	
			STN	806	
			Total	2,236	
3A	2,930	3.9	as field #3		2.9
4	869	4.0	AJH		3.0
			Ajawiri	307	
			Yari	10	
				317	
			ADG	159	
			JUZ[c]	4	
			STN	177	
			Total	657	
5	564	3.8	AJH	4	3.6
			ADG	20	
			STN	540	
			Total	564	

[a]AJH = *S.* × *ajanhuiri*.
[b]ADG = *S. tuberosum* subsp. *andigena*.
[c]JUZ = *S.* × *juzepczukii* (Johns 83-4. MICH).
[d]STN = *S. stenotomum*.

fields were chosen for the presence of weed *S.* × *ajanhuiri*, three fields also contained the wild tetraploid *S. acaule*. The presence of *S. acaule* × *ajanhuiri* (P20), the weed population that the numerical analyses suggest is probably a hybrid between *S. acaule* and weed *S.* × *ajanhuiri*, indicates the importance that *S. acaule* plays in hybridizing with diploids. The spontaneous occurrence of this weed is an interesting event probably similar to the one that produced the im-

Table 4-9. Solanum *Weeds Recorded in Field-mapping Experiments and Their Relationships to Cultivated Diploid Neighbors (from Johns and Keen 1986a)*

Study number	Weed *Solanum*		Weeds/ m^2	% weed	wAJH[a]/ cAJH[b] & STN	wAJH nearest neighbors		% wAJH nearest neighbors as 2n cultigens
1	AJH	215	1.8	77.1	4.5	wAJH	195	7.0
	Total	215				ADG[d]	5	
						STN[f]	9	
2	ACL[c]	112	3.5	49.8	5.9	wAJH	160	2.9
	AJH	206				ADG	39	
	Total	318				STN	6	
3	ACL	13	0.5	11.1	0.1	wAJH	75	28.6
	AJH	125				ADG	13	
	P20[e]	141				STN	23	
	Total	279				cAJH	13	
						ACL	1	
3A	ACL	13	0.9	23.7	0.6	wAJH	490	10.2
	AJH	540				ADG	13	
	P20	141				STN	23	
	Total	694				cAJH	13	
						ACL	1	
4	ACL	19	1.0	24.0	0.4	wAJH	14	14.0
	AJH	193				ADG	13	
	Total	212				STN	11	
5	AJH	26	0.2	4.4	0.05	wAJH	14	42.0
	Total	26				ADG	1	
						STN	10	
						cAJH	1	

[a]wAJH = weed *S.* × *ajanhuiri.*
[b]cAJH = crop *S.* × *ajanhuiri.*
[c]ACL = *S. acaule.*
[d]ADG = *S. tuberosum* subsp. *andigena.*
[e]P20 = *S. acaule* × *ajanhuiri.*
[f]STN = *S. stenotomum.*

portant frost-resistant cultigen *S.* × *juzepczukii.* Figure 4-11 is a map of a portion of field 3 that contained weed *S.* × *ajanhuiri* as well as P20.

From the point of view of the process of introgression and the production of genetic diversity, it is the relationship of diploid weed

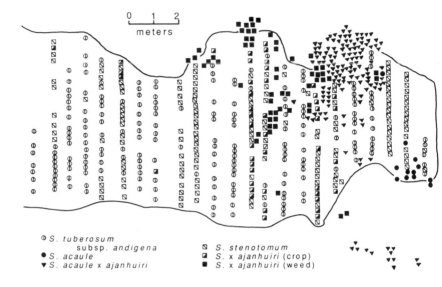

FIG. 4-11. Map showing the juxtaposition of weeds and cultigens of potato in a portion of field 3, Ajawiri Marka (from Johns and Keen 1986a).

S. × *ajanhuiri* to the cultivated diploids that is most significant. The ratio (wAJH)/(STN, cAJH) is highly variable. This statistic has been supplemented with data on the nearest diploid neighbors of all weed *S.* × *ajanhuiri* plants in a plot. The distribution of plants is likely to be as important in encouraging crossing as the total number of plants. The data do not take into consideration the overlap in field flowering times of particular entities.

***Variability among populations of* S. × ajanhuiri.** If weedy forms of *S.* × *ajanhuiri* from Ajawiri Marka are subject to ongoing hybridization with cultivated potatoes, they may contain greater diversity than those from other populations. They would therefore be expected to show greater variance in the values of certain characters. We supported this hypothesis by comparing the variances in fifty-two morphological and chemical characters of eleven populations from Ajawiri Marka to those of eleven populations from other areas in the Department of La Paz using a one-way analysis of variance (ANOVA).

Among the fifty-two characters, the variance was significantly greater (equality of variance: $p < .05$) in eight characters for the Ajawiri Marka populations. By comparison, populations from the whole collection area showed significantly greater variance in only six characters. This difference is even more convincing given that

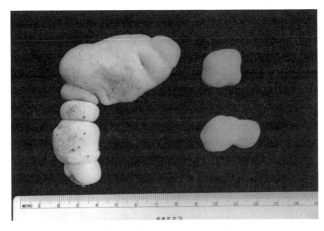

FIG. 4-12. Tuber of weed *S.* × *ajanhuiri* from Ajawiri Marka (*left*) compared with two smaller tubers typical of most populations of *S. megistacrolobum* and weed *S.* × *ajanhuiri* (from Johns and Keen 1986a).

Ajawiri Marka occupies a couple of hectares while the rest of the collections were made over approximately 2,700 km².

It is not clear from these data whether the higher variance in Ajawiri Marka reflects only introgression from cultivated plants. The milder environmental conditions of this protected location, in comparison with other areas of the altiplano, might allow the survival of a wider range of genetic variants within the populations of *S.* × *ajanhuiri*. In contrast, the more exposed conditions encountered elsewhere might be expected to impose strong selective pressure and limit the diversity of genetic types present in this species. Certainly the presence in Ajawiri Marka of plants with striking cultigen-like characters suggests that hybridization does occur here. Several populations produced large tubers (Fig. 4-12) resembling those of yari, the primitive cultivated form of *S.* × *ajanhuiri*.

***Genetic recombination and seed dispersal in* Solanum × ajanhuiri.** None of the approaches described above provides direct evidence for the natural evolution of the potato in the Bolivian altiplano. In a detailed ecological study of potato hybridization in Cuzco, Peru, that was carried out subsequent to the study in Bolivia, Rabinowitz came to the same conclusion that hybridization and evolutionary change in this crop are difficult to demonstrate. In this last work before her premature death, Deborah Rabinowitz laid a solid methodological groundwork for the continuation of studies in this area.

The following discussion considers circumstantial evidence from experiments and observations that supports the likely occurrence of the process of ongoing evolution in potatoes. The obvious indication that sexual reproduction takes place in diploid potatoes is the frequent occurrence of fruits on *S. stenotomum* and both weed and cultivated types of *S. × ajanhuiri*. Synchrony of flowering time is crucial for ensuring outcrossing in these taxa, and the production of wild-cultigen hybrids is facilitated by the long flowering period of the noncultivated taxa.

The statistical comparisons generated from the five field mappings provide a focus for discussing wild-cultigen hybrids in relation to pollination biology. Although plants of weed and cultivated diploids may occur in close proximity in the field, cross-pollination will occur only if insect pollinators move directly between the species. Levin (1978) has observed that "by definition, the frequency of hybridization will be an inverse function of flower specialization and constancy, since both constrain the wanderings of pollinators." If the similarity of foraging cues offered to insects by flowers of weed *S. × ajanhuiri* and those of cultivated *S. stenotomum/ajanhuiri* is high, the likelihood of cross-pollination will be enhanced. This is certainly likely to be the case with cultivated and weed forms of *S. × ajanhuiri*.

The frequency of hybridization is also likely to depend on the juxtaposition of plants in the field. Bees tend to favor nearer flowers, particularly when the flower resource is abundant (Heinrich 1983). Hybridization will therefore be favored by small distances between individuals of different potato species in the field. Wild-cultigen hybridization in the field would be greatest, therefore, when weed *S. × ajanhuiri* plants have cultivated diploids as nearest neighbors. Large numbers of weed *S. × ajanhuiri* in a field are less likely to facilitate hybridization with cultivated diploids if the distribution encourages crosses between plants of weed *S. × ajanhuiri*. Fields 1 and 2, while highest in weed *S. × ajanhuiri*, were thus less likely to produce hybrids between weed *S. × ajanhuiri* and cultivated diploids than field 5, where 42 percent of weed *S. × ajanhuiri* plants had a cultivated diploid as a nearest neighbor.

Seed dispersal beyond field boundaries is probably unnecessary for ensuring propagation of new potato genotypes. Cultivated fields in general do provide an appropriate environment for seedling survival. In the altiplano the cycle of rotation consists commonly of three years of cultivation followed by a fallow period of four to twenty years (commonly seven to ten). The potato is the initial crop, customarily followed by a year of quinoa (*Chenopodium quinoa*) and a year or two of barley (*Hordeum vulgare*) or cañahua (*Chenopodium canihua*). Potato berries may fall to the ground when mature, leaving seed for subsequent germination. The cultivation cycle maintains a disturbed habitat suitable for seed germination.

Domesticated grazing animals—sheep, camelids, cattle, and donkeys—may provide the most effective dispersal of wild and cultivated potato seeds. After potato plants have died down, either at senescence or because of frost kill, animals are allowed into fields to graze on weeds (non-*Solanum*) and potato plants. Farmers say that all the grazing animals will eat potato plants, and when forage is scarce, as it was in 1983, they are less particular. In arid areas of scant forage production, potato plants are gathered at the time of tuber harvest and are stored dry for forage (Fig. 4-13). Such plants still carry fruits.

Although the viability of seeds of herbaceous weeds passed through the digestive tracts of common farm animals has been investigated, no studies have reported on the survivability of seeds of potatoes or related *Solanum* species (Johns and Keen 1986a). Seed passage in Andean camelids specifically has not been studied. That potato seed can be dispersed by livestock is supported by Ugent's (1981) observation that seeds of *S. acaule* are found in animal droppings in the Andes. We discovered seedlings of *S. acaule* emerging from sheep fecal pellets that were excavated in a field near Caquiaviri (Fig. 4-13). Manure use in the Province of Pacajes follows the general patterns outlined for the Peruvian altiplano by Winterhalder et al. (1974). Sheep and camelid manure is almost the only fertilizer added to altiplano fields and is usually applied in some form at planting time. Human-assisted distribution of manure provides a possibility of seed propagation of potatoes even in the first year of the rotation.

Assuming that seedlings do germinate under favorable conditions, farming practices can encourage survival. Potatoes are mounded only at planting time and are usually not weeded. Even when fields are weeded, spontaneous crop plants are seen as valuable and are invariably encouraged.

Introduction of new genotypes into cultivation. The best opportunity for a tuber arising from true seed to be harvested is along with volunteer potatoes (ground-keepers; Aymara, *kipa-papa*) growing from tubers not harvested the previous year. Ground-keepers are often the first potatoes of the season, harvested in February and March. Alterna-

FIG. 4-13. Domestic animals may act as agents of potato seed dispersal. A. A woman gathers dried potato plants to be used as fodder (near Huamburuta, Prov. Pacajes, lat. 17°24′ long. 60°04′). B. Seedlings of *S. acaule* arising from true potato seed dispersed by sheep (*center* plants) and from tubers (*outside* plants) (Johns and Keen 1986a).

tively, they are harvested in May with quinoa, cañahua, and barley. Considerable opportunity exists during seed potato selection for the passive introduction of new types into cultivation. Initial separation of tubers for consumption or seed is based largely on tuber size and is done at harvest time. Most farmers make efforts only to separate frost-susceptible (nonbitter) from frost-resistant (bitter) types. Tschopik (1946) reported a similar practice for the Aymara of southern Peru. Within these groups, varieties are stored mixed. Selection of specific varieties generally takes place at planting time.

Most farmers are knowledgeable about potato varieties and skilled at distinguishing among them. Although the average number of varieties grown by one family is approximately ten, farmers recognize many other varieties. Individual farmers (men and women) grow and recognize as many as thirty-two varieties. However, disagreements between individuals are not uncommon. In many situations classi-

FIG. 4-13. B.

fication of varieties is imprecise and may facilitate the introduction of new genetic types.

The importance of language as a mediator of human actions was introduced in chapter 1. The usefulness of folk taxonomies in considering human interactions with the environment is discussed again in chapter 5, where the ways the Aymara classify potatoes are discussed. The Aymara have an elaborate nomenclature for potatoes which reflects the importance of this crop to them. They classify potatoes in a way similar to that outlined for the Quechua of central Peru by Brush (1980; Brush et al. 1981). An Aymara potato taxonomy using hierarchical categories similar to Brush's is recorded in Table 4-10. In many cases differentiations are not made below the folk variety level, for although tubers may be recognized as distinct, the differences are sometimes unimportant in relation to human treatment of them. This degree of imprecision is illuminated by general observations on folk taxonomy by Waddy (1982):

> If the identifiable characteristics of two or more scientific taxa are minimal and there is little or no difference in their cultural significance then they may be perceived as one entity even though the differences between them may be recognized. Similarly, if any animal or plant is only rarely encountered it may be included with another scientific taxon, and thus again be perceived as one entity.

The latter observation is particularly relevant to understanding how diversity might be introduced into cultivated potatoes. I have observed that when subsistence farmers are asked to classify a large group of tubers, although they may acknowledge some clones as unfamiliar, they place slightly different clones in existing taxonomic

Table 4-10. Schematic Diagram of the Folk Taxonomy of Potatoes among the Aymara of the Bolivian Altiplano (from Johns and Keen 1986a)

Folk level	Taxon		
Genus		*ch'oke* (*amca*)	
Species	*ch'oke* (*amca*)	*luq'i ch'oke*	*apharu*
Variety	A B C D E F	L M N O	*yari-yari apharu* . . . X Y
Subvariety	$A_1A_2A_3$ F_1F_2	L_1L_2 O_1O_2	

The designations of folk levels follow the classification by Brush (1980) of potato names from central Peru.

Folk species: *ch'oke*: cultivated potatoes that are edible after boiling. *amca* is an archaic word that retains some importance. *luq'i*: frost-resistant potatoes which are recognized as bitter. *apharu*: wild potatoes; although it is recognized that some of these are edible, they are very seldom eaten.

categories. Sorting of seed tubers at planting time is often done hurriedly, and fading colors and aging tubers are more difficult to distinguish. There is also considerable phenotypic variation in tuber morphology, and tubers may be misidentified. Resulting mistakes are apparent in fields of growing plants. Farmers acknowledge that errors happen, and they are not particularly concerned about them. The imprecision inherent in various aspects of the tuber-seed selection procedure indicates that ample opportunity exists for the introduction of true seed-derived tubers. Unfamiliar types could easily be overlooked, at least until they were propagated in sufficient numbers to be noticed. Even if they were recognized as novel, they would not deliberately be discarded until judged undesirable. Tubers that are similar to their parents would undoubtedly be classified in recognized taxa and planted along with them. New types eventually acquire names of their own, or perhaps are referred to by an existing name at the folk variety or subvariety level (Table 4-10).

In general, altiplano farmers are disposed to experiment with new varieties. Today, most new varieties are deliberately introduced by a farmer who purchases or barters for them in a weekly *feria* (market) or in a major center (e.g., La Paz). Deliberate introductions are referred to as *pruebas* (Spanish, trials). New varieties are grown and evaluated for survivability, yield, and culinary quality under local conditions, and after a year or two they are either accepted or discarded.

People are very keen observers of the potatoes they consume, and they evaluate them at this time. The Aymara have various nomenclatural categories concerned with potato quality, particularly relating to culinary properties, frost resistance, and precociousness (cf. Hawkes 1947). Among nonbitter potatoes (*ch'oke*), the three most widely mentioned categories of culinary quality are *khati* (*p'ujsa*, De Lucca D. 1983), potatoes best cooked in their skins; *monda*, potatoes best peeled and cooked in soups; and *k'ene*, floury potatoes. The two former categories are essentially opposites that divide up most nonbitter potatoes. These names are not part of the hierarchical taste taxonomy; they are covert categories dealing with potatoes at the variety level. Brush (1980) points to the importance of these types of categories in the selection decisions that likely affect potato evolution. The importance of cognitive factors and preferences as a basis for the human selection that determines the direction of potato evolution are discussed in greater detail in chapter 5.

Deliberate production of new varieties. Over the course of informal discussions with farmers I obtained two anecdotes of deliberate human propagation of botanical seed and selection of new potato genotypes. The first account was unsolicited and occurred in response to a farmer's (male, forty-five years old) interest in my collection of potato berries. When he was a young man the informant would collect the fruits (*teke-teke*) from cultivated potatoes and the next year plant the seeds like cañahua (scattered followed by plowing). When the plants came up, he transplanted them to typical rows. The first year they produced small tubers, but the tubers from these clones were big in the second year of propagation. He saw these tubers as identical to their cultivated parents.

Subsequent questioning of numerous farmers produced only one similar story. These anecdotal accounts are exceptional because of the number of persons expressing ignorance of the possibility of producing new varieties through breeding. Most people understand little more about the biology of potato reproduction than that the seeds of *Solanum* berries are capable of producing a plant.

However, the anecdotal accounts of potato seed propagation suggest that direct human production of new potato clones, perhaps only on a localized scale, may have been more important in the past. Certain towns throughout the Andes retain their traditional importance as sources of seed potatoes (Brush et al. 1981). The town of Caquiaviri, the reputed home of the ajawiri variety, may in the past have been a center of both the production and distribution of new potato types.

Natural disaster and the procurement of new tuber seed. Throughout history periodic droughts have necessitated the replenishment of potato stocks. Natural catastrophes may represent ecological bottlenecks, and it is possible that it is only in such critical situations that extreme behavior on the part of humans results in the introduction of new types. The drought of 1982–83 represented the most severe example of such phenomena in this century. Although we were not present in the area in 1983–84 to see how farmers responded to the previous year's failure, we were able to gain some insight into the unusualness of the situation.

Harvests in the Caquiaviri area were approximately 25 percent of normal. Most farmers saved their meager crop primarily as seed for the next year. Some had already purchased examples of desirable clones they had lost. The western sections of the Departments of Oruro and La Paz were especially hard hit, and the worst-case situation—farmers having no harvest at all—was not uncommon. Those people who were not totally demoralized by this catastrophe hoped to buy seed potatoes in Oruro or La Paz. Modern market and travel networks and foreign aid doubtless disguise the more localized response that would have been seen in the past for obtaining food and seed during bad years. Chuño (freeze-dried potatoes) is an important commodity in traditional Andean barter (Tschopik 1946); tunta (chuño blanco) was probably more widely used in the past for survival and as a trading commodity in renewing food and seed supplies.

Although Caquiaviri was perhaps once a center of seed tuber production, its real importance may have been as a source of tubers in years of crop failure. As I said above, clones of weedy potatoes in this valley show striking similarity to clones of cultivated *S. × ajanhuiri* and *S. × juzepczukii*. Large tubers were recognized by all persons we showed them to as being yari, and most informants were dubious that we had in fact collected them from wild plants. A possible derivation of the name ajawiri from ajayari, "domesticated wild potato" (see Huamán et al. 1980), emphasizes the role of natural evolutionary processes in its origin and argues against the deliberate production of seed potatoes. *Marka* means place; Ajawiri Marka is the place where the domesticated wild potato grows. While the rest of the area was ravaged by drought and frost in 1983, yields in this protected valley, while significantly reduced, were greater than anywhere nearby. The names by which weed *S.* × *ajanhuiri* is known in the Province of Pacajes—*yari-yari* and *pampa-yari*—strongly link these forms and the primitive yari cultivars of the species *S.* × *ajanhuiri* in the minds of local farmers. When supplies of the cultivated potatoes were depleted, weed potatoes would be obvious sources of tubers for food and seed. Under severe ecological conditions in the past, Ajawiri Marka likely represented a reservoir of seed potato from which at least ajawiri originated.

It is possible that deliberate attention to, and intervention in, potato evolution is an agricultural practice that historically has been carried out by a few people. It is more likely that the evolutionary process has always been passive in its essential aspects. Circumstantial biological evidence, including what I outlined above, indicates that potato domestication continues today without direct human intervention. Years with crop failures may represent exceptional but important times when interactions with new genotypes may have been more conscious.

The evolution of chemical change in potatoes. The preceding paragraph suggests that while humans are capable of making judgments on glycoalkaloid content and selecting for nontoxic levels, the kind of selection decisions leading to domestication of favorable types take place only under exceptional circumstances. The biological processes that encourage genetic recombination and variability in potato populations do appear to take place. In fact, glycoalkaloid content among plants from natural populations of wild potato species can vary dramatically. Levels of glycoalkaloids in *S.* × *bertaultii* Hawkes, for example, range from 10 to 600 mg/100 g (Tingey and Sinden 1982).

The chemical analyses of weed *S.* × *ajanhuiri* populations from Ajawiri Marka and elsewhere reveal that they are very low in glycoalkaloid content. The fact that almost no glycoalkaloids are found in any of these populations is difficult to explain from the perspective of current knowledge of glycoalkaloid genetics (Sinden et al. 1984). Particular segregates of interspecific hybridizations may have particularly low levels of various individual glycoalkaloids and of total glycoalkaloids, or reduced levels of glycoalkaloids may result from a gene or genes that suppress the production of specific glycoalkaloids (McCollum and Sinden 1979). In *S.* × *ajanhuiri,* solanidine and tomatidine glycoalkaloids may be suppressed, leaving only the tomatidenol-based glycoalkaloids, which are found in small amounts in many potato cultivars.

What is most unusual about *S. × ajanhuiri* is the universally low levels of glycoalkaloids in noncultivated populations. Events that led to reduced levels of glycoalkaloids in *S. × ajanhuiri* likely took place at an early stage in the domestication of this species, i.e., certainly before backcrosses of low glycoalkaloid types with *S. stenotomum* to form ajawiri clones. It is, however, somewhat unclear why and how suppressed glycoalkaloid expression became fixed in *S. × ajanhuiri.* The propagation of cultivated clones would certainly disseminate this trait across the altiplano. The vegetative mode of propagation of potatoes isolates the crop from wild populations and facilitates the fixing of the trait.

It is even less clear how genes for low glycoalkaloid expression should predominate in the weedy populations that propagate sexually as well as vegetatively. Perhaps the large numbers of crop potatoes in the area led to genetic swamping of populations of indigenous populations of *S. megistacrolobum.* This suggests that population sizes of *S. megistacrolobum* were significantly smaller than those of the crop, or considerably less adapted to the environment. Considering that weed *S. × ajanhuiri* plants are relatively less common than the crop types today and are found only on disturbed sites, it is conceivable that *S. megistacrolobum* was a minor component of the altiplano flora prior to human intervention. It may not have been present in this area at all, and the present-day weeds may have been totally derived from the crop.

It is apparent that glycoalkaloids play little or no adaptive role in altiplano potatoes; otherwise there would be strong selection pressure on the weeds for higher levels of these compounds. Although some pathogens and insect pests are a problem in the area, many organisms may be unaffected by glycoalkaloid defenses (Tingey 1984). At lower altitudes in the Andes, where a wider range of pests occurs, glycoalkaloids may be more important as natural resistance factors.

It is also clear that glycoalkaloids confer no resistance against frost. It has been suggested that the high glycoalkaloid levels in frost-resistant potatoes such as *S. × juzepczukii* and *S. × curtilobum* indicate that the two traits are genetically linked (i.e., located on the same chromosome). More likely the connection between frost resistance and high levels of glycoalkaloids reflects the sterile nature of

S. × *juzepczukii* and *S.* × *curtilobum* and the lack of opportunity for selection to occur.

Among frost-resistant potatoes, *S.* × *ajanhuiri* represents a unique opportunity for selecting for types with reduced glycoalkaloids. Although it is clear that genetic recombination takes place in this diploid hybrid, the ongoing role of humans in selecting for glycoalkaloids is difficult to delineate. But considering the uniqueness of the low glycoalkaloid levels in *S.* × *ajanhuiri*, especially in comparison to *S. megistacrolobum* and *S. stenotomum*, it is clear that the fixation of this trait reflects human intervention at some point in time.

5   Human Perception, Cognition, and Behavior in Relation to Plant Chemicals

The decisions that humans make in response to encounters with environmental chemicals depend on their perceptions and interpretations of the significance of sensory stimuli. While organisms vary in their adaptations for dealing with environmental chemicals, in general, genetic and environmental factors that govern perception, preferences for particular stimulatory chemicals, motivation, and learning processes have much in common throughout the animal kingdom. The increased intelligence of humans improves our capacity for memory and for processing information. Intelligence affects not just the extent to which technological and cultural means allow us to manipulate the chemical environment but also the way in which we make basic decisions about it.

The use of language accompanies intelligence, and our ability to retain, retrieve, and communicate knowledge about plant chemicals is clearly an important cultural adaptation. The structure of language is a reflection and determinant of cognitive processing of information and is fundamental to how we interact with the world. While classification schemes and conceptualization of abstract ideas are intrinsically interesting for linguists and cognitive psychologists, for human ecologists they are more important as aspects of the way we come to terms with the complexities of the natural environment. The manner in which humans perceive, classify, and conceptualize chemical stimuli affects the decisions made by both individuals and human populations regarding plant chemicals and the plants that contain them. For example, selection decisions made during domestication of potatoes and other crops will proceed via these processes.

This chapter takes the view that taste sensitivity and taste classification are part of the biological and cultural adaptations, respectively, of human populations to the nutritional and toxicological characteristics of their environment. First, we consider taste perception and strategies for communicating information on taste stimuli from a cross-cultural perspective; then we will examine specific characteristics of the perception and the language of the Aymara of Bolivia that relate to their adaptive experiences and to their treatment of their staple food, the potato.

FOLK BIOLOGICAL CLASSIFICATION: UTILITARIAN ASPECTS

Ethnobiological studies of the way in which different cultures perceive and structure the variability of their natural environment have concentrated on the visual classification of plants and animals. Much of the effort in this field has been on classification for its own sake. In determining general principles that govern the form of folk taxonomies, work by Berlin and others (Berlin et al. 1973, 1974; Brown 1984) has emphasized the universality of human cognitive processes. While these approaches have direct relevance for understanding perceptual processes, they say less about the nature of the biological world itself and human interactions with it. Hunn has emphasized the relation of classification categories to natural discontinuities and has taken an adaptionist perspective in looking at the practical significance of folk biological classification (Hunn 1977, 1982; Hays 1982).

Most folk taxonomic studies tend to avoid dealing with utilitarian factors as something peripheral to the underlying principles of folk classification. In fact, however, utilitarian concerns, as reflected in so-called special categories, have a fundamental role in most folk taxonomies. Taxonomies that are elaborated in relation to highly specific aspects of adaptation to economic and ecological conditions do not fit well into the universal categories. However, these are the most interesting aspects of taxonomies for a cultural-ecological study. While universal aspects of taste and smell classifications likely reflect physiological perception and biological adaptations of our species to natural discontinuities in the chemical universe, cross-cultural differences are potential indicators of the impact of experience on sensory perception and preferences.

Among sensory classifications, attention to color has dominated other aspects of sensory experience (Berlin and Kay 1969; Conklin 1980). Although taste terms have been considered as plant descriptors (Berlin et al. 1974), cross-cultural classification by taste per se and its significance since the turn of the century have been largely ignored (Myers 1904; Chamberlain 1903). More recently, O'Mahony has examined taste descriptors in English other than the "primary" concepts and has provided some comparative evidence of taste description strategies in Malay (O'Mahony and Muhiudeen 1977), Cantonese (O'Mahony et al. 1980), Spanish (O'Mahony and Manzanoa 1980), and Japanese (O'Mahony and Ishii 1987).

CROSS-CULTURAL PERSPECTIVES ON TASTE

PERCEPTION

Comparative studies on human taste perception have shown general uniformity among Western populations. A few anomalous taste patterns have been explained as genetically determined traits relating to differences in dietary history of human populations. For example, a high incidence of PTC (phenylthiocarbamate, phenylthiourea) taste sensitivity among Ecuadorian Andeans was associated with the prevalence of naturally occurring and bitter goitrogens in their diet (Greene 1974). Among inhabitants of Yucatan, Mexico, diminished sensitivity to the bitter taste of PTC was associated with the high acceptance of niacin-rich but bitter coffee (Davis 1978), although the biological basis for any such link is very tenuous.

Dietary history with or without a genetic component was considered responsible for a preference for sour and bitter tastes by laborers in the Karnataka region of India (Moskowitz et al. 1975). Greater preferences for salt and sucrose in Chinese subjects compared to U.S. subjects of European descent were difficult to explain (Bertino and Chan 1986). There is little evidence of an important genetic component in salt and sucrose preference (Bertino and Chan 1986). In general, the dietary experience of individuals may be more important than genetic differences among populations in determining most cross-cultural differences in patterns of sensory perception and preference.

Dietary experience, perhaps simply familiarity, is clearly important in the learned acceptance of usually distasteful foods such as chili pepper (Rozin and Schiller 1980). Learned preferences for substances such as chili, or common stimuli like salt and sugar, may have a physiological and/or cognitive basis. The role of learning as a determinant in taste perception is discussed in chapter 2.

CLASSIFICATION

Languages must structure and communicate the collective experience of sensory stimuli in such a way that members of a society can respond appropriately to the chemical characteristics of their environment. Taste classifications do reflect physiological perception, but it is clear that they are also strongly culturally dependent (O'Mahony and Ishii 1987). The physiological classification of four primary tastes—sweet, salt, sour, and bitter—recognized by most sensory psychologists (Bartoshuk 1978) closely parallels the taxonomies of taste in most modern Western languages. In reality, electrophysiological studies show that receptors are not discrete in the stimuli to which they respond, and while humans may have a greater number of certain receptors that reflect our dietary adaptations, perception represents more of a continuum than a series of separate sensory types. Sour and bitter stimuli are often confused among English speakers (O'Mahony et al. 1979), as are sour and salty (cf. Seetle et al. 1986). Similarly, salt and sour stimuli are confused among Torres Straits Islanders (Myers 1904), thus underlining the lack of clear correspondence between taste terms and "primary" sensory stimuli.

Classification schemes do differ in the ways they codify the complexity of chemical stimuli, and alternate strategies for dealing with this complexity should be considered within their cultural context. Examination of the strategies for dealing with chemical stimuli in different languages can contribute to the understanding of sensory perception and to overcoming any bias imposed by the Western viewpoint.

Classification systems in the Western world as recently as the late eighteenth century recognized as many as ten taste qualities (Myers 1904; Bartoshuk 1978). In traditional Chinese scholarly thought five tastes—sweet, salt, sour, acrid, and bitter—were linked symbolically to philosophical views of the universe (Needham 1956). The taste taxonomy of some cultures may be expanded to include concepts such as astringents, irritants (such as capsaicin from chili peppers), numbing agents, and emollients. These types of substances affect oral perception through general effects on cellular components or the trigeminal nerve.

In some languages taste vocabulary is reportedly reduced to as few as two terms—usually pleasant (sweet and salt) and unpleasant (sour and bitter) (Myers 1904). Myers (1904) gave the example of the Baganda (Bantu), who were reported to have two taste words, *kuwoma* and *kawa*, the former applied to sweet, salt, and other agreeable flavors, and the latter to unpleasant flavors. *Kuwoma* was used for the "pleasant" tastes of sugar, salted meat, and certain sour fruits, and *kawa* was applied, for example, to brackish water and quinine. This Baganda nomenclature was based on a more or less anecdotal report lacking systematic methodology, and in order to gather some insight on this system I constructed a taste and smell taxonomy in Maragoli (Abalogoli), a Bantu (Luhyia) language spoken in western Kenya, based on interviews in English with two bilingual speakers whose first language is Maragoli (see Table 5-1).

While Maragoli has complex concepts of taste, including words for specialized foods not included in the formal classification scheme, it follows the same basic dichotomy described for Baganda, although without making such a strong judgment between liked and disliked tastes. *Kinolu* (north Maragoli) (or *kenoru*, south Maragoli) and *ki-lulu* describe sweet and sour/bitter substances, respectively, but it is important to note that to say a food or medicine is kilulu does not imply that it is disliked. The real dichotomy between good and bad tastes is made using the words *vulahi* and *vundamanu* (north) (*vuvi*, south). All other Maragoli taste and smell concepts relate to specific known substances. Interestingly, in the Maragoli language the verb *kufunya* (to smell) is the word used to describe the state of tasting.

In the taste conception of the Ndut Serer of Senegal a similar dichotomy exists with the terms *hay* and *sɔs* (Dupire 1987). While hay is a primary sensory term encompassing salt, bitter, and hot pepper into a single concept, sɔs, which translates as "pleasant," contrasts by meaning without flavor or insipid. While sɔs overlaps to some degree with *sen*, sugar taste, its importance is as the dichotomous term for hay in taste, tactile, and thermal senses.

The primary senses in Ndut language are represented by hay, sen, and *kɔb* (sour). The latter term encompasses tastes of substances such as green fruits, vinegar, alcohol, and spoiled meat.

Perhaps most interesting in Ndut taste taxonomy is the grouping of salt and bitter into the same concept. While confounding stimuli may have a physiological basis, the importance of the hay/sɔs dichotomy in the Ndut view of the world in more philosophical ways suggests that experience, not physiological perception, is the important determinant of these concepts. The hay/sɔs dichotomy appears to have a basis in everyday experience. Ndut staple foods such as couscous are bland; items that are hay—whether they be bitter, salty, or spicy—all serve to season food and therefore relieve dietary monotony.

The context in which these terms are used emphasizes that cul-

Table 5-1. Taste and Smell Taxonomy of Maragoli (Luhyia), Kenya

Maragoli	English translation(s)
Kuzazama	To taste, transitive
Kufunya	To smell (to taste), intransitive
Kive (keveye) nu mwayu	It has a taste (*mwayu* usually means good)
	It has the right amount of taste
Dichotomy of taste qualities	
Kifunya vulahi[a]	It tastes good; it smells good; it tastes sweet
Kifunya vundamanu (north)	{ It tastes bad (something inedible)
Kifunya vuvi (south)	{ It smells bad
Duality of taste descriptors	
Kinolu (north)	Sweet, adj., e.g., sugarcane, honey
Kenoru (south)	
Ni kinulu	It is sweet
Kilulu	Sour, bitter, adj., e.g., sour milk, fermented porridge, lemons, medicine, pepper, chili pepper, some bitter vegetables, *munyu mukeleka* (alkali plant ash added to food)
Ni kilulu	It is bitter (sour, etc.)
Kifunya kuli . . .	It tastes like . . .
	It smells like . . .
Kifunya kuli ijumbi	It tastes like salt
Kuzura	Insipid, having lost its taste
Lichinyu	Taste of tubers and of soil (a good taste)
Vilungi	Taste of onion, coriander
Limisi	Nauseous (smell)
Kifunya matunda	It smells aromatic (like fruit)

[a] *Vulahi* and *vundamanu* (*vuvi*) are adverbs.

tural and ecological factors must be examined in any attempt to understand the correspondence of any taste taxonomy with physiological perception. This issue will become important as we attempt to understand the evolutionary significance of the unique features of the Aymara taste taxonomy described below.

The Maragoli strategy of likening odor and taste to known stimuli is common to many languages. Consider the terms for taste and smell in the language of the Tewa Indians of New Mexico (Robbins et al. 1916) reproduced in Table 5-2. This is one of the most detailed taste nomenclatures found in the anthropological literature. Of particular interest is the Tewa odor classification, which includes a

Table 5-2. Tewa Taste and Smell Terms (Robbins et al. 1916)

Tewa	English translation
Tfā̱	To taste, intransitive
'Ā̱	To be sweet, sweet, sweetness
'Ā̱kikinǎtfǎ̱	It tastes insipid (*'a,* sweet; *kiki,* like; *na,* it; *tfa,* to taste)
Tsiṉ̄	To be sticky (also said of taste)
'Oj̄ohe, 'Oj̄e	To be sour, sour, sourness; prickling or puckering taste
P'ahaṉ	To be burnt (also said of taste)
'I'ǣ̱	To be bitter, bitterness
Saē̱	To be hot or burning to the taste, like chili pepper; hot or burning taste; prickling or puckering taste
Suwà	To be warm, warm, warmth
Tsǎnwae	To be hot, hot heat
'Ōkaiλ̄	To be cool, cool, coolness
'Asae̱	To taste salty, or alkaline
dīhewo'O	Nauseating taste ("it makes me sick")
Su̱	To smell, intransitive. This verb appears in all terms denoting kinds of odor. Thus:
Nǎsuke̱	It smells strong (*na,* it; *su,* to smell; *ke,* to be strong)
Heiā̱'an nǎsu̱	It tastes faintly (*heia'an,* slight)
Nǎ'asu̱	It smells sweet (*'a,* sweet)
Nǎsīsu̱	It stinks (*si,* giving the meaning to stink)

Nouns with the postfix *wagi̱,* "like," are very common with *su,* "to smell." Thus:

sawagi̱ nǎsu̱	"It smells like tobacco" (*sa,* tobacco)

number of specific odor qualities and the provision for saying that something smells like something else, e.g., *sagawi nasu,* "it smells like tobacco."

The differences between English and systems such as Maragoli suggest an inverse relationship between the number of basic terms and the degree to which modifiers or descriptive phrases of taste are employed. Such a relationship would be similar to the pattern seen in color nomenclatures. Berlin and Kay (1969) concluded from the examination of ninety-eight languages that

> first, there exist universally for humans eleven basic perceptual color categories which serve as the psychological referents of the eleven or fewer basic color terms in any language. Second, in the history of a given language, encoding of perceptual categories into basic color terms follows a fixed partial order. . . . Third, the overall temporal order is properly considered an evolutionary one; color lexicons with few terms tend to occur in association with relatively simple cultures and simple tech-

nologies, while color lexicons with many terms tend to occur in association with complex technologies.

While a fixed sequence in the evolution of color terms may exist, any assumption that one strategy for dealing with color perception is superior to another is questionable. The basic color terms can be considered "summary" terms which, while useful in complex cultures as a form of perceptual shorthand, sacrifice much of the rigor in color identification possible in a society where everyone has a common shared experience of the complexities of their environment (Bousfield 1979). Thus a two-term system employing a wealth of descriptive terms may have as much classificatory power as a system with many more color terms but fewer specific descriptors.

The construction of generalizable principles of taste and smell classification depends, first, on a clearer establishment of the degree to which chemical perception is universal, and, second, on more detailed investigation of the relationship of classification systems of chemical senses to other parameters of culture and ecology.

Modifying words may evolve as necessary to deal with the complexity of environmental stimuli. Malay speakers, for example, use more modifying descriptors for describing salt tastes than English speakers. Part of this difference is perhaps attributable to the small part that taste description appears to play in English culture compared to visual stimuli such as color (O'Mahoney and Muhiudeen 1977). It is unclear, however, why Malay speakers use so many more modifiers for salt, although one assumes salt has considerable salience for them. Certainly the relationship of taste classification to chemical characteristics of the environment and to human adaptation remains to be tested in any systematic or generalizable way.

TASTE AND THE AYMARA CLASSIFICATION AND SELECTION OF POTATOES

In keeping with the dependence of the Aymara on potatoes, their potato nomenclature reflects considerable sophistication in recognizing and distinguishing varieties, and attention to subtle differences in potato characteristics and quality (Hawkes 1947; La Barre 1947). The Aymara use over a thousand names to directly describe potatoes and have many more words that relate to potato cultivation and taxonomy, and to specialized ways of classifying potatoes (De Lucca D. 1983; Hawkes 1947). Potato nomenclature and taxonomy is a store of cultural knowledge concerning potatoes. It is the medium for communication about potatoes and an instrument in decision making, and it affects human attitudes and actions toward potatoes (Brush 1980).

The Aymara potato taxonomy has essentially the same structure (see Table 4-10) as the potato taxonomy outlined for the Quechua of central Peru (Brush 1980; Brush et al. 1981). At the folk species level, taste quality is important for differentiating bitter and nonbitter potatoes, designated as *luq'i* and *ch'oke*, respectively. Although attention to texture is important in folk variety groupings (La Barre 1947), reference to taste per se, beyond the broad separation of bitter and nonbitter, is not evident in the potato nomenclature. However, the importance of differences in potato taste is clear to any observer who has shared a typical Aymara meal made up of a number of varieties of boiled potatoes. Palatability and culinary quality based primarily on flavor and dry-matter content are important in the way Quechua- and Aymara-speaking people evaluate potatoes (Brush et al. 1981; Huamán 1975; Jackson et al. 1980).

The conclusion that humans selected for reduced quantities of glycoalkaloids in potatoes assumes a capability to recognize differences in alkaloid levels between varieties. I hypothesize that Aymara cultivators have placed direct selection pressure on the chemical constituents of potatoes through their attention to taste quality. Selection is likely to be applied at the level of the human community acting over time on the potato gene pool. Both perceptual and cognitive processes would be involved in selection for particular tastes, and each must be considered directly.

It is possible that the Aymara possess an experience-determined taste specificity. Although short-term exposure to altitude affects taste sensitivity and preference (Frisancho 1981), no studies have considered the taste perception of acclimatized populations. The cold and hypoxia of high altitude, and the impoverished plant and animal resources, make the altiplano an extreme environment demanding unique biological and cultural responses.

Quantitative data on taste perception and preferences obtained in suprathreshold taste tests (i.e., tests using concentrations of stimuli above the lowest detectable concentration) of pure compounds carried out with Aymara-speaking people who maintain a basically traditional subsistence life-style are discussed below (Johns and Keen 1985). By means of a dietary survey, data on current diet composition were obtained that help to explain the unique features of Aymara taste preferences. The ability of an Aymara community to distinguish potatoes on the basis of taste was examined using standard sensory testing procedures as well as taste panel tests with several potato varieties (Johns and Keen 1986b).

The cultural mechanisms that mediate communication about potato tastes affect the community's evaluations of potato quality. The Aymara taxonomy of taste (Johns and Keen 1985) provides a specific focus for this study. In addition the Aymara taste taxonomy may reflect taste experience and other environmental influences.

GLYCOALKALOIDS AND POTATO TASTE QUALITY

Although differences in taste quality are distinguishable to a casual taster or experimenter, it is not often clear what principal or combination of principals is determinant in a complex food item. The characteristic flavor of potatoes results from a complex interaction of free amino acids, $5'$-nucleotides, simple sugars, phenolics, and glycoalkaloids (Solms and Wyler 1979; Sinden et al. 1976). In order to test the hypothesis that humans select for reduced quantities of glycoalkaloids, I focused on the perception and preference of these compounds while generally ignoring the possible role of other factors contributing to taste and quality. The role of glycoalkaloids in determining human preference for whole potatoes was tested by analyzing potatoes used in the panel test for these constituents.

Taste tests were designed to address the capability of humans to affect the domestication of the hybrid potato *Solanum* × *ajanhuiri.* Clones of ajawiri and yari varieties of *S.* × *ajanhuiri* and five samples from weed *S.* × *ajanhuiri* populations, similar in tuber appearance to the primitive yari varieties, were included in a panel test along with tubers of other cultigens.

Preliminary analyses of clones of *S. megistacrolobum*, the wild relative of *S.* × *ajanhuiri,* indicated that tomatine was the major alkaloid in *S. megistacrolobum* (cf. Johns and Osman 1986). Since ajawiri clones of *S.* × *ajanhuiri,* at least, were known to contain α-solanine and α-chaconine (Osman et al. 1978)—the alkaloids characteristic of most cultivated potatoes—I hypothesized that preference and selection for solanidine-containing glycoalkaloids (α-solanine and α-chaconine) versus those containing tomatidine (e.g., tomatine) could be a driving force in potato domestication (Johns 1986b). Evidence from animal studies indicated that in general tomatine is no more toxic than α-solanine and α-chaconine (Nishie et al., 1975). However, in a series of potato glycoalkaloids including α-solanine and α-chaconine, tomatine showed the greatest potency in producing cardiotonic effects on the isolated frog ventricle (Nishie et al. 1976).

Although the solanidine versus tomatidine hypothesis was not supported, the study nevertheless provided some interesting results. In the experiment described here the ability of humans to discern quantitative and qualitative differences in glycoalkaloids and to distinguish solanidine- and tomatidine-based glycoalkaloids was examined using suprathreshold tests of sixfold concentrations of tomatine and of a mixture of α-solanine and α-chaconine.

TASTE PANEL TESTS OF POTATO QUALITY

A panel of fourteen residents of Caquiaviri, Bolivia, was enlisted to evaluate fourteen varieties of potatoes selected to represent a probable wide variation in glycoalkaloid content. Potatoes were obtained both locally and outside the Caquiaviri area. Included with three clones of *S. × ajanhuiri* cultigens and five clones of weed *S. × ajanhuiri* were tubers of *S. stenotomum, S. tuberosum, S. × juzepczukii,* and *S. acaule × ajanhuiri* (sisu) (see chapter 4). The varieties and their sources are recorded in Table 5-3. Details of sample cooking and the testing procedure are found in Johns and Keen (1986b). All subjects were tested for PTC sensitivity (see below) and were uniformly tasters.

The subjects responded to the stimuli in two manners. Both a scoring method and a ranking method (Larmond 1977) were used with each subject. First, individuals were asked to score the specimens for bitterness or sweetness on the following five-point scale:

5   very bitter
4   slightly bitter
3   without flavor, neither bitter nor sweet
2   flavorful, sweet
1   very flavorful, very sweet

Bitterness and sweetness are used as contrasting concepts of good and bad in the Aymara taste taxonomy (Johns and Keen 1985), much as we saw earlier for African languages, and are used in this sense here. Sweet means good in this context, not the taste of sugar. This experiment was confounded by the apparently contextual nature of

Table 5-3. Identity and Source of Potato Samples Used in Taste Panel Tests of Potato Quality (from Johns and Keen 1986a)

Sample Number	Name	Species	Source
1	Phiñu rojo	*S. stenotomum*	La Paz market
2	Huishla ppacki	*S. tuberosum* subsp. *andigena*	Corocoro market
3	Wila kunurana	Unknown	La Paz market
4	Laram ajawiri	*S.* × *ajanhuiri*	La Paz market
5	AJH:6	Weed *S.* × *ajanhuiri*	Ajawiri Marka, Caquiaviri
6	Jank'o luq'i	*S.* × *juzepczukii*	Caquiaviri
7	Jank'o yari	*S.* × *ajanhuiri*	Vichaya, Prov. Pacajes
8	Wila yari	*S.* × *ajanhuiri*	Caquiaviri
9	Jank'o sisu	*S. acaule* × *ajanhuiri*	Caquiaviri
10	Wila sisu	*S. acaule* × *ajanhuiri*	Caquiaviri
11	AJH:15	Weed *S.* × *ajanhuiri*	Ajawiri Marka, Caquiaviri
12	AJH:5B	Weed *S.* × *ajanhuiri*	Ajawiri Marka, Caquiaviri
13	AJH:5C	Weed *S.* × *ajanhuiri*	Ajawiri Marka, Caquiaviri
14	AJH:25	Weed *S.* × *ajanhuiri*	Ajawiri Marka, Caquiaviri

the Aymara terms for bitter and sweet, and ratings should be considered from this perspective.

After they had scored each potato sample, the Aymara subjects were also asked to rank samples in order of preference. Taste scorings and preference rankings were compared by estimation of 95 percent confidence intervals on mean values. The mean values of the ranking (in order of preference) and scoring (for bitterness or sweetness) test methods are recorded in Table 5-4. *Phiñu rojo* stood out as the specimen scored as least bitter (and therefore with the best flavor), and the one ranked as most preferred. The wide 95 percent confidence intervals produced few significant differences between samples, although the score and ranking methods show similar patterns. Phiñu rojo was scored as significantly different from jank'o luq'i (6), *wila sisu* (10), *huishla ppacki* (2), and two of the weed *S.* × *ajanhuiri* samples (5, 3) in both analyses.

Table 5-4. *Mean Values of Preference Ranking and Flavor Scores and Glycoalkaloid Characterization of Potato Samples Used in the Taste Panel Test*

Each subject ranked ten specimens for preference and scored them for bitterness or sweetness on a 5-point scale. The specimens are ordered and grouped according to preference rankings. Ninety-five percent confidence intervals overlap for all samples except those of groups 1 and 3. The Bonferroni adjustment for simultaneous estimation of confidence intervals (Neter and Wasserman 1974) was used. (From Johns and Keen 1986a)

	Group sample		Mean order rank (95% confidence interval)	Mean score (95% confidence interval)	TGA[a] mg/100 g	Glycoalkaloids[b]
Number	Number	Name				
1	1	Phiñu rojo	1.5 (1.0–2.0)	4.6 (4.1–5.1)	17	sol, cha
2	3	Wila kunurana	3.4 (1.8–5.0)	3.9 (2.9–4.9)	6	sol, cha
	4	Laram ajawiri	3.6 (0.5–6.7)	3.9 (2.7–5.1)	11	sol, cha
	8	Wila yari	5.6 (3.6–7.6)	3.3 (1.9–4.7)	4	α,β-slm
	7	Jank'o yari	5.7 (3.4–8.0)	3.0 (1.8–4.2)	2	α,β-slm
	12	AJH:5B	6.0 (3.3–8.7)	3.1 (1.8–4.4)	5	α,β-slm
	14	AJH:25	6.3 (4.0–8.6)	2.8 (1.9–3.6)	2	α,β-slm
	9	Jank'o sisu	6.1 (1.3–10)	2.9 (0.8–5.0)	17	com, sis
	11	AJH:15	6.4 (1.8–11)	3.1 (1.6–4.6)	7	dem, α,β-slm
3	13	AJH:5C	6.6 (3.4–9.8)	2.0 (0.8–3.2)	3	α,β-slm
	6	Jank'o luq'i	6.6 (4.8–8.4)	2.4 (1.6–3.3)	38	dem, tom, sol, cha
	10	Wila sisu	6.9 (4.2–9.6)	2.6 (1.2–4.0)	12	com, sis
	5	AJH:6	7.5 (4.4–10)	2.3 (1.0–3.6)	3	α,β-slm
	2	Huishla ppacki	7.7 (5.3–10)	2.2 (1.2–3.2)	12	sol, cha

[a]TGA = total glycoalkaloids fresh weight.

[b]Constituents comprising 95% or more of total glycoalkaloids: cha = α-chaconine; sis = sisunine; com = commersonine; slm = solamarine; dem = demissine; tom = tomatine; and sol = α-solanine.

Uncooked samples of all fourteen tuber specimens were preserved in 95 percent ethanol and saved for subsequent qualitative and quantitative analyses of glycoalkaloids. The glycoalkaloid constituents and total glycoalkaloids of the tuber specimens used in this test are also recorded in Table 5-4. The glycoalkaloids detected in individual clones are consistent with the results recorded previously for these taxa (see chapter 4). Samples 1, 2, 3, and 4 contained α-solanine and α-chaconine as major constituents. Other samples were characterized by mixtures of demissidine-, tomatidine-, tomatidenol-, and solanidine-containing glycoalkaloids.

The manner in which the panel subjects rated potatoes for flavor or preference is not correlated with total glycoalkaloids or with the pattern of individual glycoalkaloids. The specimen with the greatest amount of glycoalkaloid, jank'o luq'i, was one of the most bitter and least preferred. However, specimens that were even less preferred had glycoalkaloid constituents and total glycoalkaloid levels less than or comparable to specimens that ranked much higher. Although phiñu rojo (1) and huishla ppacki (2) ranked second and third in total glycoalkaloid content, and both were characterized by α-solanine and α-chaconine, they are the most and least preferred specimens, respectively. Jank'o luq'i, with 38 mg/100 g, was the only sample in the study containing more than 20 mg/100 g total glycoalkaloids.

TASTE PERCEPTION

Aymara-speaking subsistence farmers living in the Department of La Paz rated compounds representing sweet, salty, sour, and bitter tastes for both taste intensity and taste pleasantness. Solutions of glucose (2 M), sodium chloride (1 M), citric acid (0.06 M), and quinine sulfate (0.00075 M) were prepared by successive half dilutions with distilled water to provide six different concentrations of each (Johns and Keen 1985).

While responses to taste intensity have been extensively characterized in Western populations, the more limited studies on pleasantness responses have dealt primarily with sweet and salty stimuli. Moskowitz et al. (1975, 1976) characterize responses to sweet, salty, sour, and bitter tastes, in terms of both intensity and pleasantness, for particular populations and provide a field method appropriate for cross-cultural studies. In their study East Indian medical students responded to taste stimuli with ratings that fit polynomial functions considered equivalent in shape and slope to those of previously studied Western populations. Anomalous bitter and sour preferences among laborers from Karnataka, India, were established by comparing responses with data generated from the group of Indian medical students. I calculated predicted intensity and pleasantness values for each concentration of the four test compounds used in my study from the polynomials in Moskowitz et al. (1976). The regression lines representing predicted responses to concentration were then compared to actual responses from my study using an analysis of covariance. The relation between intensity and pleasantness for my response data and that predicted from the literature was also compared using an analysis of covariance.

The Aymara test population directly related subjective taste intensity to physical molarity for all compounds (Fig. 5-1) in a manner similar to previously studied human populations. Data for taste pleasantness ratings are normal in showing increasing preference for glucose (twenty-one of twenty-two subjects) and decreasing pleasantness for citric acid and quinine (all subjects) as concentration increases. Individual pleasantness ratings for sodium chloride generally decrease with concentration, although two of twenty-two subjects preferred sodium chloride of increased concentration (cf. Pangborn 1970).

The greater rating for 2 M glucose in comparison to 1 M glucose is anomalous in comparison to most populations, which characteristically produce an L-shaped function with a break point at 1 M (Moskowitz et al. 1974). Extraordinary patterns similar to those seen here have been associated with biological factors such as body weight and hypoglycemia (Cabanac 1979). Individuals showed considerable variation from the curve of the group means: although 55 percent of subjects demonstrated greater preference for 2 M solutions, 32 percent responded with a break point at 1 M and a decreased preference for 2 M glucose. The increased pleasantness rating for 2 M glucose is most pronounced in subjects under thirty-eight years of age (Fig. 5-2). The variability for this group at 2 M (S.D. = ± 0.02) is much lower than the total population (S.D. = ± 0.23), with all nine subjects responding in the same manner. Overall, the Aymara function for glucose pleasantness was equivalent in slope, intercept, and variance to that of the comparison population.

The Aymara population responded with lower pleasantness ratings for sodium chloride, citric acid, and quinine than other populations in similar tests. Comparison with the functions generated

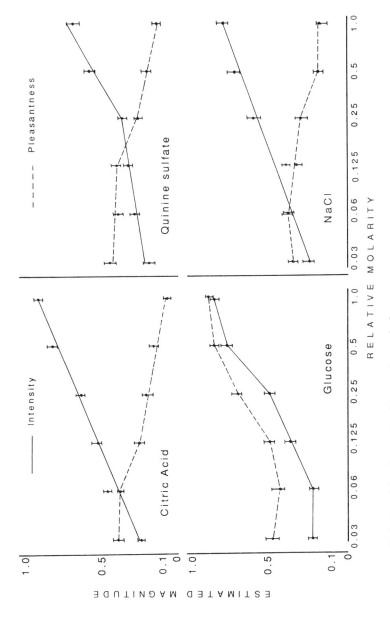

FIG. 5-1. Relationship of perceived intensity and pleasantness to concentration (from Johns and Keen 1985).

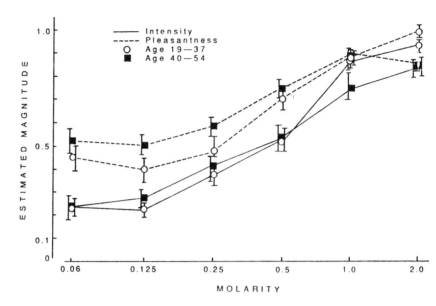

FIG. 5-2. Comparison of glucose preference between old and young Bolivian adults (from Johns and Keen 1985).

using parameters reported from the East Indian population show that the anomalies are not in perception per se. Functions generated from intensity ratings for the Aymara are not significantly different from those calculated for the comparison population. In each case functions did not differ in slopes or intercepts. Curves of pleasantness ratings for sodium chloride, citric acid, and quinine sulfate for the Aymara are all significantly lower than curves calculated for the East Indian students. The slopes of the functions of Aymara pleasantness ratings for citric acid and quinine sulfate do differ statistically from those recorded for the East Indian population. Pleasantness functions for sodium chloride did not differ significantly in slope.

Interpretation of these results is complicated by the scaling method used, which makes all ratings relative to the highest-rated compound, usually glucose. Although the methods used in this experiment parallel the methods of Moskowitz et al. (1975) as closely as possible, the comparison of data from this study with that of the previous investigators is strongly limited by inevitable differences in experimental situations. These limitations must be considered in interpreting the data. Nonetheless, a striking pattern emerges that makes sense in relation to the dietary and taste classification data: at

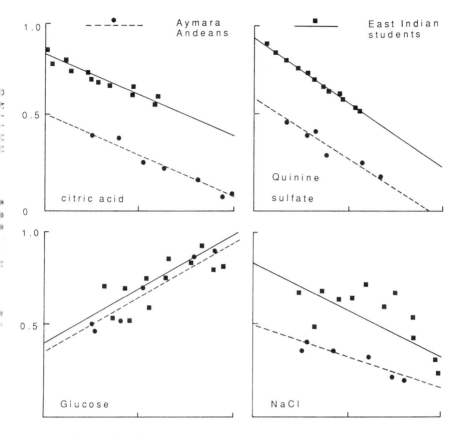

FIG. 5-3. Relationship of pleasantness adjusted for intensity for Aymara Andean subjects and East Indian students (from Johns and Keen 1985).

a constant intensity level Aymara subjects score pleasantness for salt, sour, and bitter consistently and significantly lower than the data generated from the comparison population (Fig. 5-3). At a constant sweetness level the Aymara glucose pleasantness ratings appear equivalent to the comparison population. This result for glucose pleasantness and the expected intensity curves for glucose and for the other three compounds support the argument that the anomalous results for sodium chloride, citric acid, quinine, and concentrated glucose are not due to experimental differences but represent real differences in Aymara tastes. This interpretation of the results is discussed below in relation to Aymara dietary patterns and taste classification.

My comparison of Aymara data with East Indian data reflects my inexperience with sensory perception experiments at the time the research was carried out. Better controls of the role of testing methods and of experiential, environmental, and genetic factors in determining the Aymara pattern of taste would be provided if the same tests were carried out among both altiplano agriculturalists and Aymara people with a different life-style outside the high-altitude environment. That this is easier said than done points to a fundamental problem in doing cross-cultural studies on tastes, and perhaps to why so few have been carried out.

PTC TASTE SENSITIVITY

Forty-one subjects who participated in taste perception and other experiments were tested for PTC taste sensitivity using simple PTC paper strips. PTC tasters far outnumbered nontasters. Thirty-eight of the forty-one subjects (93 percent) were tasters.

TASTE PERCEPTION OF TOMATINE AND SOLANINE-CHACONINE

Sixteen of the subjects included in the taste perception procedure were also presented with two sets of triplicate samples made up of six successive half dilutions each of tomatine and an α-solanine–α-chaconine mixture. Subjects ranked the solutions for both preference and intensity. The highest concentration of both solanine-chaconine and tomatine (0.00092 M) corresponds to 80 mg/100 ml of solanine, the standard compound to which glycoalkaloid levels in potatoes are routinely compared. (I express concentrations in mg/100 ml in this discussion to facilitate comparison with studies examining potato glycoalkaloid content.) The concentrations presented cover a range that corresponds to the range (in mg/100 g) that is found in cultivated potatoes.

The subject population responded to the test for tomatine and solanine-chaconine perception with relatively little sensitivity to these glycoalkaloids over most of the concentrations presented (Fig. 5-4). An analysis of variance showed no significant difference in intensity rating between glycoalkaloid concentrations of 40 mg/100 ml or less. Subjects did perceive 80 mg/100 ml solutions as significantly stronger than 40 mg/100 ml. Equivalent molar solutions of tomatine and the solanine-chaconine mixture were perceived as equivalent for all concentrations.

As the concentration of solanine-chaconine increased above 20

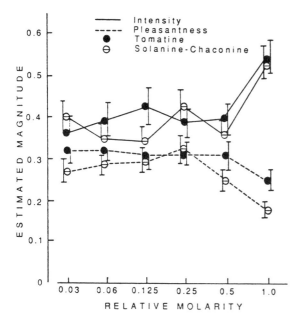

FIG. 5-4. Relationship of perceived intensity and pleasantness to concentration for the glycoalkaloids tomatine and a solanine-chaconine mixture (from Johns and Keen 1986b).

mg/100 ml, subjects showed a decreasing preference for solanine-chaconine, with significant differences in preference over the range 80–20 mg/100 ml. Subjects were unable to detect significant differences between 20 mg/100 ml and 2.5 mg/100 ml ($p = .63$). The subjects were not able to detect any significant differences in the concentration of tomatine over the full range of the test. The subject group showed no differences in response to tomatine versus solanine-chaconine at concentrations below 20 mg/100 ml. However, solanine-chaconine was rated less favorably than tomatine at both 40 mg/100 ml and 80 mg/100 ml, although the difference was significant at 80 mg/100 ml (t test, $p = .04$) but not significant at 40 mg/100 ml ($p = .16$).

TASTE TAXONOMY

Both formal and informal interviews contributed to my understanding of the Aymara taste classification (Johns and Keen 1985). Subsistence farmers in the Provinces of Ingavi and Pacajes learn Aymara as a first language, and Spanish is learned both in the home and at

Table 5-5. Aymara and Spanish Taste Taxonomies of Bilingual Aymara Speakers (from Johns and Keen 1985)

Aymara	Spanish	English equivalents
mojjsa	agradable	pleasant
mojjsa		sweet
	dulce	sweet
misq'i		sweet (of honey)
c'ara	salado	salty to taste
jajjo	agrio	unpleasant
jajjo		bitter, hot, sour
c'allc'u		sour, bitter, hot
c'asc'a		bitter, sour
	amargo	bitter
c'ara	muy salado	unpleasantly salty
ch'apaka chua	sin gusto	blandly unpleasant, insipid

school (since 1953). Aymara words were transcribed according to De Lucca D. (1983). This recently published dictionary and the postconquest dictionary of Bertonio (1612) were also searched for all Aymara terms relating to taste. Although our methods (Johns and Keen 1985) were far from exhaustive in establishing the contextual significance of Aymara taste concepts, they have produced at least a satisfactory working taxonomy of Aymara taste that defines the normal range of physiological taste experience.

The hypothesis that the Aymara would have an elaborate taste nomenclature was largely unsupported. Nevertheless, in relation to the concepts of sweetness and bitterness the taxonomy appears somewhat expanded. Aymara is essentially a spoken language, and it shows considerable regional variation (Briggs 1976). Although the field data do not correspond completely to the terms found in the dictionaries of De Lucca D. (1983) and Bertonio (1612), the significant patterns are the same; the ambiguities seen in the field are also present in both dictionaries. The discussion that follows relates to the field data from the Province of Pacajes, Department of La Paz, unless otherwise stated. The Aymara and Spanish taste taxonomies

Table 5-6. Aymara Taste Terms Recorded by De Lucca D. (1983)

Aymara	English equivalents
mojjsa	pleasant, sweet (of sugar)
malli	pleasant, sweet, smooth
misqu'i	sweet (of honey), pleasant
kawi	sweet (of freeze-dried *oca*), pleasant, smooth
c'ara	salty, very salty
jayu, jayuni	salty
jayu c'ara	very salty
jajjo, jaru	unpleasant, bitter, hot, sour
c'allc'u	unpleasant, sour, bitter
c'asc'a	sour
c'ata (Chucuito, Peru)	unpleasant, bitter, sour
kata (north altiplano)	unpleasant, bitter
chanayana	sour
chiwacu (north altiplano)	sour, unpleasant, puckery
ch'apaka	blandly unpleasant, insipid
k'ayma	blandly unpleasant, insipid
umachata, umanchata	watery
umani, umanina	watery, juicy
qhacha	rough, granular

of the bilingual Aymara speakers interviewed in this study are summarized in Table 5-5. Taste terms recorded by De Lucca D. (1983) are tabulated in Table 5-6.

Of the four classes of taste stimuli presented, only two—sweet and salt—were recognized and classified unequivocally. Sour and bitter were confounded, certainly in terms of classification, and perhaps also in terms of perception. The Aymara generally followed the usual human pattern in considering sweet pleasant and sour and bitter unpleasant. The sweet taste was described most commonly as *mojjsa*, occasionally as *misqu'i* (literally, bee's honey; also the Quechua [*misk'i*] word for sweet). Mojjsa was also expanded to mean favorable (Spanish, *agradable*) and to refer to anything with a pleasant taste that was not strongly salty, bitter, or sour. Water was sometimes described as mojjsa. De Lucca D. (1983) records two additional terms for pleasant and sweet: *malli* is a general term, while *kawi* distinguishes the sweetness of a specific food, freeze-dried oca.

A salty taste, whether perceived as pleasant or unpleasant, was invariably referred to as *c'ara* in Aymara. Occasionally subjects used the Spanish term *agrio* to describe extremely salty samples, although they generally used the term *salado*.

For many bilingual speakers agrio was the only Spanish term used to describe unpleasant tastes. Bitter substances were usually appropriately termed *amargo*, but the term *ácido*, the usual Spanish equivalent of sour, is unknown to the Aymara. On the surface this might appear to be a language problem related to the poor command of Spanish by Aymara speakers. However, closer examination of both Spanish and Aymara taxonomies indicates that the patterns of Aymara taste perception, and the lack of a one-to-one correspondence between Aymara and Spanish, are complicating factors. Agrio as the Spanish designator of unpleasant tastes is the polar opposite of the Aymara mojjsa or the Spanish *bueno* (good). This dichotomy is also made using purely Aymara concepts; mojjsa was usually opposed by *jajjo (jaru, jarro)*. Thus in this sense the Aymara used agrio and jajjo as directly translatable terms. However, the one-to-one translation breaks down when dealing with bitter and sour per se. The Aymara used three overlapping terms—jajjo, *c'allc'u*, and *c'asc'a*—none of which referred specifically to bitter or sour. Any two of these terms could be used as synonyms in a particular situation, but in general they were recognized as distinct. Jajjo was the word most commonly associated with the hot pepper taste, but it was also the term most commonly associated with bitter tastes and was occasionally associated with sourness. *C'allc'u* was the word most commonly associated with sourness, but on occasion it was associated with bitterness or hotness.

The third term, c'asc'a, appears to be associated with bitterness and sourness. An educated translator expressed c'asc'a as referring to the sensation left by any substance that leaves the mouth in a bad condition; that is, anything that damages or "burns" the mucosa. Informants used both c'allc'u and c'asc'a to describe this effect. The word *c'ata (kata)* as recorded by De Lucca D. (1983) further expands the Aymara vocabulary of ambiguous terms for unpleasant tastes.

The Aymara freely use taste terms in relation to various foodstuffs, and it is in relation to food that they are perhaps best understood. Here again there was no universal agreement among informants on how things taste, although individuals were usually unequivocal. Bitter potatoes and bitter quinoa usually were described as jajjo, while nonbitter potatoes (which are called *dulce* [sweet] in Spanish) and chuño were either c'allc'u, c'asc'a, or mojjsa, depending on variety. Sweet (nonbitter) quinoa was described as c'asc'a or c'allc'u if badly prepared. Limes were referred to by all three terms, while fruits in general were usually c'allc'u.

Probably the most important aspect of the Aymara taste taxonomy is its expanded nomenclature of unpleasant substances. Interestingly, while sour foods are placed into this schema, it is notable that the Aymara apparently lack any specific word for sour.

DIETARY SURVEY

Dietary information was compiled (Table 5-7) using the twenty-four-hour recall method (Burk and Pao 1976; Johns and Keen 1985). Frequency of consumption was more important than quantity of items consumed. Thirty-one subjects were interviewed in the villages of Caquiaviri and Chacoma; these included sixteen of the subjects who participated in the physiological taste tests. The thirty-one interviews conducted comprise 114 reported "meals" (eating or drinking episodes). Dietary components are listed in order of frequency of use. Although the contribution of individual items is not recorded, I observed that chuño, potatoes, bread, sugar, quinoa, pita, and meat (usually small portions of dried charqui) are the dietary staples. Chuño, potatoes, bread, quinoa, and pita are commonly eaten simply, with no seasoning. Additives such as carrot, onion, rice, noodles, and various condiments form minor parts of soup and stew dishes, the bulk of which are chuño, potatoes, quinoa, and/or meat. None of the additive items are locally produced; the necessity of purchasing them was given as the explanation for their sparing utilization. Partially refined sugar is usually consumed in tea, coffee, and mate.

Seasonality in dietary habits was not taken into account in the testing methodology. My experience over eight and a half months (October–June) in the test area indicates that diet is fairly constant throughout the year. Tschopik (1946) includes June in the part of the year when food was abundant for the Aymara of Chucuito, Peru, on the shores of Lake Titicaca and records minor seasonal variations through the year. In areas away from Lake Titicaca, like those I studied, the Aymara diet is more restricted; the simple diet contains few seasonally available items and depends heavily on staples stored in large quantities for lengthy periods. There is similar seasonal stability in the diet of high-altitude Quechua populations in southern Peru (Mazess and Baker 1964).

Table 5-7. *Frequency of Food Utilization by a Sample of Bolivian Farmers (from Johns and Keen 1985)*

Item and description			Number of meal reports
Sugar (partially refined)			41
[a]Chuño (includes *tunta* and *chuño negro* [*Solanum* spp.])			37
[a]Potato (*Solanum* spp.)			33
Bread (wheat flour)			19
[a]Meat (sheep, cattle, llama), as small portions of *charqui*			19
Carrot			17
Onion			16
Coffee			13
Tea			13
[a]Mate (herbal tea, mostly *Spacele* sp.)			13
[a]Quinoa (*Chenopodium quinoa* Wild.):	sweet	7	
	bitter	6	
	total		13
Pita (grain, toasted and ground):	wheat	6	
	barley	5	
	[a]cañahua	2	
	(*Chenopodium canihua* Cook)		
	total		13

Table 5-7. Continued

Item and description	Number of meal reports
Rice	12
Noodles	7
Toasted wheat	6
Cheese (sheep milk)	5
Chocolate	2
Egg	2
Flour (wheat)	2
Cabbage	1
[a] *Isaño* (tubers of *Tropaeolum tuberosum* R & P)	1
[a] Maize	1
Milk (cow)	1
[a] *Oca* (tubers of *Oxalis tuberosa* Molina)	1
Peas	1
[a] *P'asa* (edible clay)	1
Soft drink	1

Condiments used: [a] chili pepper, oregano, parsley, [a] salt

[a] Denotes traditional food.

DETERMINANTS OF AYMARA TASTE PREFERENCE AND CLASSIFICATION

The unique features of Aymara taste can best be understood by considering the experiential factors that have historically defined the way the Aymara express and communicate their taste perception. The dietary survey data aids in understanding the taste perception tests and the unique aspects of the Aymara taste taxonomy. Both physiological and familiarity explanations for differences in human taste preference (Moskowitz et al. 1975; Cabanac 1979) are considered here in relation to the Aymara taste-perception data.

The modern Aymara diet as tabulated in the twenty-four-hour recall study is striking in its use of very few food items. As subsistence farmers the Aymara consume very little that they do not produce themselves. Traditional foods—those consumed prior to the Spanish conquest—retain the primary place in the diet today, and introduced crops, e.g., barley, are prepared in traditional ways. Simplicity in dietary intake and the heavy emphasis on traditional foods facilitates an examination of the historical role of dietary experience in determining taste perception and taste taxonomy.

SENSORY CHARACTERISTICS OF TRADITIONAL DIETS

Traditional dietary items are lacking in intensely sweet flavors. That Aymara describe foodstuffs such as some potatoes as sweet might indicate a subtle distinction of only weakly sweet materials. The terms mojjsa, misq'i, and kawi differentiate specific kinds of sweetness. Although refined sugar would not have been part of the traditional diet, it has been readily accepted. That the Aymara function for glucose pleasantness fails to show a break point at a 1 M concentration indicates an unusual preference for intense sweetness among the population. Most likely this preference for sweetness reflects a physiological need associated with the adaptation to a high-altitude (3,950 meters) environment. Young adults show a greater glucose preference, perhaps a result of a caloric consumption insufficient for maintaining an active life at high altitude. A similar pattern of glucose preference was seen with Quechua-speaking peoples tested in the Department of Junin, Peru (Johns 1986b).

Dietary data reported here are consistent with a general pattern of low fat–high carbohydrate diets among high-altitude peoples (Picón-Reátegui 1978). It should be noted, however, that subsistence agriculturalists in various circumstances around the world are generally

FIG. 5-5. Altiplano salt deposit.

dependent on carbohydrate staples. But high carbohydrate consumption may promote more efficient glucose metabolism and appears adaptive in hypoxic (pO_2 61 percent of sea level in this case) environments (Picón-Reátegui 1978). High preferences for glucose among the Aymara could support the hypothesized adaptive mechanism favoring high carbohydrate intake at high altitudes. Low levels of blood glucose among Andean high-altitude residents are associated with chronic hypoxia (and are not a result of food habits) (Picón-Reátegui et al. 1970), and may be part of a physiological mechanism increasing preference for sweet tastes.

Although salt occurs naturally on the altiplano (Fig. 5-5), it is used sparingly and never in the preparation of carbohydrate staples. Water sources are often strikingly saline, although relatively fresh well and stream water (e.g., 225 mg salts/liter) is sometimes available. In contrast, salt levels in the largest river in the region, the Rio Desaquadero, measured at Vichaya in the Province of Pacajes, ranged from 1,095 mg salts/liter (December 11, 1982) to 1,400 mg/liter (June 10, 1983), depending on rainfall (Johns 1985). Paque (1980) explained the low salt appetite of Saharan Bedouins as a physiological response to the oversalinity of their environment. A low sodium chloride preference of the Aymara may be a similar adaptation for limiting salt intake. The Aymaras' negative attitudes toward salt are reflected in the magic belief that salt can be used to bring misfortune to another person (Otero 1951; La Barre 1951).

The traditional Aymara diet is strikingly lacking in any sour component. Historically only spontaneously fermented food would have had this quality. The two sour commodities available in markets today—vinegar and limes—are not part of the normal diet. Limes are used almost exclusively for medicinal purposes (Oblitas Poblete 1969). The increased acuity with which citric acid is perceived by the Aymara and the extremely low pleasantness ratings that it evokes may reflect their naïveté when they do encounter intensely sour substances.

The predominant taste characteristics of the plant items consumed by the Aymara are bland and mildly bitter. The traditional crops—potato, quinoa, and cañahua—contain varying degrees of bitter secondary compounds. The characteristic glycoalkaloids of potatoes are present in quantity in some varieties.

It is commonly assumed that all bitter potatoes are processed into tunta or chuño fresco (see chapter 3) as a means of detoxification. I was surprised to observe that bitter varieties, e.g., luq'i (*S.* × *juzep-czukii*), are often consumed boiled (*papa kati*). Farmers confirmed that this was a normal occurrence and explained that many people own land that will produce only bitter varieties, and that the water necessary for producing tunta is often unavailable. In contrast, Werge (1979) reported that in central Peru bitter potatoes are consumed without processing only in time of famine. Chuño and tunta appear to have different levels of bitterness depending on the source and method of processing.

Quinoa is characterized by saponins (see chapter 2). Both bitter and sweet (nonbitter) varieties are grown. Most Aymara express a preference for the bitter variety, although it is always washed before consumption, apparently to reduce saponins to a nontoxic level. Cañahua also has varieties that probably differ in saponin levels.

SEEKING AN ECOLOGICAL EXPLANATION

The feature of Aymara dietary experience that may most determine taste preference is the monotonous consumption of a few bland plant items prepared with minimal seasoning. This lack of variety of foodstuffs is common in other isolated high-altitude towns in the central Andes (Mazess and Baker 1964; Picón-Reátegui 1978). Dietary exposure to sour and bitter foods among East Indian (Karnataka) laborers was given as a reason for increased preference for sour and bitter (Moskowitz et al. 1975). Exposure has been shown to increase preference for distasteful substances (Rozin and Schiller 1980). Decreased pleasantness ratings for salt, sour, and bitter among the Aymara are consistent with a dietary explanation as well.

However, accepting this argument—that bitterness is not preferred by the Aymara—may seem inconsistent with the observation that moderately bitter items are important in the traditional and present-day diets. It should be emphasized that bitter flavors are never deliberately maximized in intensity in Aymara traditional foods. Both glycoalkaloids and saponins occur naturally in staple foods, and in potentially toxic levels (see chapter 2). Toxic levels of glycoalkaloids and saponins are characteristic of the potato and quinoa varieties that have the necessary frost resistance to ensure reasonable yields for Aymara subsistence. The inevitable presence of these compounds in the Aymara diet poses a constant threat of toxicity; this threat over time may have produced not a preference but an enhanced genetic or conditioned aversion to bitter compounds (cf. Rozin 1976). Aymara subjects did not rate the intensity of quinine greater than the comparison population. Therefore, unless they have lower threshold levels for quinine, there is no basis for attributing to them greater acuity to bitter tastes. Cognitively based behavior rather than extraordinary physiological sensitivity is likely a more important factor in the way the Aymara interact with plant toxins.

The prevalence of PTC tasters among the study population may indicate some genetic selection for recognition of bitter and toxic dietary constituents, such as the goitrogens which were proposed by Greene (1974) to be related to the high percentage of PTC tasters in Ecuador. Goitrogenic activity is associated with thiocyanates as well as thioureas and related compounds similar in structure to PTC. Although Greene (1974, 1980) was incorrect in attributing goitrogenic properties to the tarwi, the añu (see chapter 4) is an important Andean crop with potential goitrogenic properties. I have reported the presence of a thiourea from this plant (Johns and Towers 1981).

The Aymara taxonomy for bitter and sour tastes can be understood at least partially to reflect subtle differences in the quantity and quality of the bitter compounds commonly experienced. The importance of recognizing toxic compounds may contribute to the cultural significance of bitterness. The traditional diet is without a distinctly sour component, and the Aymara lack clear, unambiguous concepts of, and terms for, sourness both in their own language and in Spanish. The apparent confusion of sour and bitter seen here is distinct from the pattern of reducing sour and bitter to a single concept of distaste such as we saw in the African languages described above. Recognition of the variety in bitter stimuli is the primary determinant of the taste classification, while recognition of sour is secondary. The terms relating to bitterness and sourness—jajjo, c'allc'u, c'asc'a, and c'ata—can be understood to reflect the unique dietary experience of the Aymara over millennia of existence on the altiplano.

Examples of unique taste taxonomies such as those of the Aymara, Bantu, and Ndut reflect experience and adaptation to particular circumstances. Each must be considered within its cultural and environmental context, while the cultural context of more "universally" recognized taste classification schemes must also be examined more closely.

METHODOLOGY

Lessons learned from the Aymara case study should be helpful for developing an improved methodology for future studies. The inherent ambiguity of the experimental method of the Aymara study, where all pleasantness ratings are relative to glucose (the highest-rated compound), confounds glucose preference with decreased preference for sodium chloride, citric acid, and quinine. If the Aymara preference for glucose in fact exceeds that indicated by these data, the apparent aversion to sodium chloride, citric acid, and quinine could be nonexistent or considerably less. While a physiological explanation relative to need best explains the young adults' preference for concentrated glucose (and perhaps the decreased preference for salt), either a physiological explanation for increased sweet preference or an experiential explanation for decreased preference for sour and bitter might best fit the experimental data in relation to the latter compounds. In view of the demands of the resource-limited, high-altitude environment and the unique features of Aymara diet and taste taxonomy, both physiological and experiential explanations may be correct.

To the Aymara the terms jajjo, c'allc'u, c'asc'a, and c'ata are not synonyms, but they are difficult for the non-Aymara observer to decipher. These terms for bitter and sour may show as much relationship to levels of intensity of flavor or to context as to a specific taste. Further studies are needed to specifically test level of intensity as a factor in the differentiation between jajjo, c'allc'u, c'asc'a, and c'ata. The lack of universal agreement among informants with regard to these terms and their relation to foodstuffs suggests that the terms may have some contextual significance that was not discernible by the methods applied in this study. Although the limited methods of physiological testing and direct discussion did elicit a classification of the translatable terms for taste in Aymara, a more definitive understanding of Aymara taste classification would require in-depth linguistic analysis and field study of the working use of these terms.

Further research should attempt to put the taste taxonomy on a quantitative basis—that is, provide a more statistically testable record of the meaning attributed to particular terms in specific contexts. The fact that all primary informants used all three bitter-sour terms but were not in mutual agreement on their meaning could be a starting place for more detailed investigations of contextual significance and individual perception differences. Testable hypotheses combining more directly integrated studies of taste perception and cognition can be formulated from the initial synthesis of biological and linguistic data provided by this study.

GLYCOALKALOID CONTENT AND POTATO TASTE QUALITY

The background provided by the taste-perception study, the folk taxonomy of taste, and the dietary survey indicate that the Aymara are attuned to bitter and unpleasant substances in their diet. As well, the panel tests of potato taste quality provide quantitative support for observations that the Aymara and other Andeans are capable of making judgments of potato quality on the basis of taste. However, the groupings into which the test population placed samples of wild and cultivated potato tubers do not appear to be based, at least primarily, on glycoalkaloid content. Scores obtained for *S. stenotomum*, *S. × juzepczukii*, ajawiri, and yari are consistent with evaluations of bitterness obtained in a similar study by Huamán (1975), who attributed high bitterness evaluations given to yari clones to the high levels of glycoalkaloids that he supposed would have been obtained from the *S. megistacrolobum* parent. Yari clones of *S. × ajanhuiri* and related weeds actually contain almost no glycoalkaloids. Within the range of glycoalkaloids present in potatoes used in the panel test, with the exception of jank'o luq'i (that is, less than 20 mg/100 ml), some other factor, or combination of taste and flavor factors, determines potato quality.

Flavor in potatoes is determined by a complex combination of constituents, including glycoalkaloids. Factors such as dry matter and texture probably played some role in the judgments made by the Aymara in this study. Subjects tended to confound bitterness and unpleasantness in the scoring part of the panel test, and it is likely that their judgments of flavor quality involved a combination of factors of which bitter taste was only one.

In tests with a trained taste panel of North American subjects, potatoes with total glycoalkaloid levels above 14 mg/100 g were consistently rated bitter (Sinden et al. 1976). The Aymara population actually most preferred a potato variety with as much as 17 mg/100 g total glycoalkaloids. The present experiment is not directly comparable to that carried out in North America, but these two results are not greatly different. The difference suggests that the Aymara have a slightly higher threshold level of response to glycoalkaloid content.

Although the Aymara potato taxonomy (Table 4-10) makes only one distinction on the basis of taste, this is an important determinant in relation to Aymara perception of and behavior toward potatoes. Potatoes are either luq'i or ch'oke—bitter or nonbitter. Jank'o luq'i is typical of the former category and is usually processed before consumption (Johns 1986b; Werge 1979); all of the other cultivated potatoes used in this study belong in the latter category. The most bitter potato, jank'o luq'i, contained 38 mg/100 g total glycoalkaloids, and the participants' decisions to reject it were likely based on its high level of glycoalkaloids. Thus a threshold of glycoalkaloid unacceptability for the Aymara appears to be between 17 (the acceptable level established above) and 38 mg/100 g.

The differentiation of potato quality at levels somewhere between 17 and 38 mg/100 g is consistent with data from the study on the perception of solanine-chaconine in solution. Subjects determined no differences in strength at concentrations below 20 mg/100 ml. However, as they perceived an increase in concentration up to 80 mg/100 ml, they also showed a significant and increasing rejection of this mixture of compounds. Although concentrations expressed as mg/100 ml in aqueous solution or as mg/100 g in the fresh weight of potato tissue are not strictly equatable and are unlikely to be perceived as equivalent, the results of studies done with pure compounds support the conclusions from the panel tests using potato tubers that a break point for solanine-chaconine preference by the Aymara occurs roughly between 20 and 38 mg/100 g.

If the Aymara rejected glycoalkaloids at higher levels than other populations, their inferred acuity to bitter and toxic substances

(Johns and Keen 1985) might appear to be unsupported. However, recognition of glycoalkaloid contents at a moderately high level (such as 17 mg/100 g) may, in fact, be one key to a positive evaluation of potato quality by the Aymara. An ability to make decisions of acceptability at moderately high levels of glycoalkaloids does not mean that the Aymara are insensitive to these compounds. As noted above, bitter potatoes such as jank'o luq'i are often eaten unprocessed immediately after harvest. Perhaps in this resource-limited environment the Aymara have developed a higher level of acceptance for glycoalkaloids. In addition, they may have enhanced enzymatic detoxication capacities for dealing with these compounds.

The hypothesis that the Aymara would reject potatoes on the basis of tomatine versus solanine-chaconine content was clearly unsupported. Based on this hypothesis I predicted that solanine-chaconine would be preferred over tomatine. In contrast, while tomatine and solanine-chaconine are perceived as equivalent in intensity, solanine-chaconine received significantly lower preference ratings than tomatine. The relative rejection of solanine-chaconine by the Aymara could represent an experiential sensitivity to the solanidine glycoalkaloids encountered in their diet. Rejection of solanine and chaconine may represent a conditioned response to exposure to high levels of these compounds, or even genetic selection for increased sensitivity to these compounds which form an intrinsic part of their diet. However, differences in physiological perception of compounds do not necessarily have adaptive significance. While the differential acceptability of tomatine versus solanine-chaconine may have a physiological basis, it could also be a chance effect. The hypothesis that this effect seen in the Aymara population represents a conditioned or genetic difference could be tested under controlled conditions with Aymara and non-Aymara populations living under different dietary conditions.

CHEMICAL SELECTION OF POTATO VARIETIES

It is apparent from these data that glycoalkaloid concentration plays only a minimal role in the ongoing selection and domestication of potato varieties. Potatoes are evaluated for glycoalkaloids alone in terms of two character states—too high or acceptable—and not as part of some more subtle selection process. Any role that glycoalkaloids may play in combination with other compounds or factors as a determinant of flavor quality is difficult to discern.

In the case of *S. × ajanhuiri*, it appears that while the Aymara have the capacity to select for low glycoalkaloid levels, they may have done so directly only at the earliest stage of the history of this hybrid species (cf. Fig. 4-10). The theoretical mutational event that produced yari types resulted in plants below the hypothetical acceptable/non-acceptable level break point. *S.* × *ajanhuiri* cultivars have probably remained below this level ever since. While humans have had an apparently minimal role in selecting for new acceptable types, this is not to say that they have not played an important role in rejecting other types that were clearly unacceptable because of toxicity above the critical point.

Three of the five clones of weed *S.* × *ajanhuiri* included in the taste panel test were equally or more acceptable on the basis of taste than many of the clones currently grown by the Aymara. It would appear, then, that the weed population of *S.* × *ajanhuiri* contains clones that, if all other factors were equivalent, should be readily accepted into the cultivated gene pool. In view of the size achieved by some weed *S.* × *ajanhuiri* tubers used in this study, these hybrids between wild and cultivated potatoes would seem likely candidates for domestication. One sample (Fig. 4-12) produced tubers up to 11 cm in length in the field and was as acceptable from a culinary standpoint as yari and sisu, two cultivated potatoes with many primitive characteristics. Undoubtedly yari was domesticated under circumstances similar to those seen in Ajawiri Marka today. That more clones have not followed yari into cultivation may relate more to cognitive and cultural factors, and to the rarity of environmental conditions that demand unusual behavior on the part of humans, than to the genetic potential present in the wild populations.

6 Reconsidering the Model of Human Chemical Ecology

The model of human chemical ecology outlined in chapter 1 attempts to encompass holistically the multifaceted response of humans to environmental (plant) chemicals. In a sense the model is dangerously close to encompassing the whole rather than being representative of it. Thus this study has of necessity focused on a narrow component of the interaction of humans with plant chemicals: the domestication of plants and the origins of agriculture.

The domestication example was the basis for the generation of nine hypotheses designed to test the model in a meaningful way. Before reexamining these hypotheses, this chapter reconsiders the significance of the case study involving the Aymara and their interaction with glycoalkaloid constituents in the domestication of the potato crop. Probable stages in the process of selection for reduced glycoalkaloid levels in *Solanum × ajanhuiri* provide a focus for considering potato evolution in general. Second, this chapter examines each of the nine hypotheses against aspects of the potato example and against other information from the literature.

CHEMICAL SELECTION AND STEPS IN THE DOMESTICATION OF THE POTATO

Biosystematic studies of *Solanum × ajanhuiri* and its progenitors supported the hypothesis that dramatic reductions in glycoalkaloid levels have taken place during the domestication of this plant. Changes in glycoalkaloid levels between wild forms of the series *Megistacrolobum*, including *S. megistacrolobum*, and the primitive yari forms of *S. × ajanhuiri* dramatically overshadow other changes in morphology, with the exception of tuber size and color. By analogy with changes in specific characters of concern to humans in other crops, these chemical changes support the argument that selection by humans for improved flavor and reduced toxicity has been a major evolutionary force in potato domestication.

The crucial variable in the evolution of *S.* × *ajanhuiri* and related potato cultigens is time. If the origins of *S.* × *ajanhuiri* are due to recent intraspecific hybridization rather than traceable to early stages in the domestication of the potato, human interactions with the plant may be different than they would be if it has been around for centuries. Considering the highly reticulate nature of potato evolution under human influence, the evolution of *S.* × *ajanhuiri* probably reflects both modern and ancient processes.

Both yari clones and their conspecific weeds contain minimal amounts of glycoalkaloids. Subsequent backcrosses with *S. stenotomum* have increased the levels of glycoalkaloids in ajawiri clones. In the domestication of *S.* × *ajanhuiri*, selection for glycoalkaloid content undoubtedly took place at some stage or stages prior to the creation of the yari types we know today.

The relaxation of the restraint imposed by potentially toxic levels of glycoalkaloids on the exploitation of potato tubers has made this resource much more available to humans. Prehistoric civilizations in the central Andes depended to a significant degree on potatoes, and the removal of glycoalkaloids appears to have accompanied Andean cultural development. Although wild potatoes could have provided abundant food supplies at times, the toxic levels of glycoalkaloids in most wild species make it unlikely they were exploited by pre-agricultural gatherer/hunters as more than a casual resource. However, until humans became intimately associated with potato populations, domestication was equally unlikely.

Total glycoalkaloid content varies widely between potato species and among populations of a species. Undoubtedly, genetic segregation produces a considerable range of total glycoalkaloid content. Genotypes of wild species with little or no glycoalkaloids are the exception but are nonetheless known. For example, as we discussed in chapter 4, many wild potatoes, such as *S. bulbocastanum* Dun. and *S. chacoense* Bitt., contain no detectable glycoalkaloids (S. L. Sinden, personal communication), and others, such as *S.* × *bertaultii*, have levels as low as 10 mg/100 g. Perhaps early exploiters of wild potatoes could have successfully concentrated on species such as *S. cardiophyllum* Lindli from Mexico or on populations that were low in total glycoalkaloid content. However, it does not necessarily follow that the wild progenitors of the cultivated potato were low in glycoalkaloid content. Considering the widespread cultivation of potatoes in the Andes, the interfertility of cultivated and wild plants, and their tendency to hybridize, it is difficult to say which wild potatoes have not received genes through introgression from cultigens. Quite conceivably, low glycoalkaloid levels in certain modern wild species are secondarily derived and do not reflect the state at the onset of the domestication process.

Undoubtedly potatoes were an unpredictable and dangerous resource until some way was found to eliminate their potential toxicity. Detoxification methods such as leaching and freeze-drying were probably important in expanding the exploitation of an otherwise poisonous plant resource. Although technological developments in processing probably predate the initial stages of the agricultural revolution, sophisticated techniques such as freeze-drying are often species specific and would only come about through an extended period in which humans were in intimate association with an intrinsically poisonous food resource. These technological developments give little insight to the question of how early hominid foragers with little or no processing technology avoided toxicity.

I have provided evidence to support the argument that geophagy was a crucial factor in making potatoes available to preagricultural peoples living in the Andes. Geophagy has prehominid origins and was certainly available to humans before and during the domestication process. Significantly, the use of clays with potatoes as a detoxification technique persists to the present day and coexists with the use of nonbitter potatoes without clay.

Once a potato resource was widely exploited, and therefore familiar to its exploiters, the likelihood of particular genotypes being selected for desirable traits would greatly increase. Once cultivation of potatoes was practiced, the population size of favorable genotypes would be increased by human dispersal and propagation. The variability of total glycoalkaloid content seen in wild plants indicates that the raw material on which the initial chemical selection could act was available. Glycoalkaloids are highly heritable, and recombination between types with low levels would maintain and increase this trait.

Humans are the most important factor limiting the introduction of new types of potatoes. Hawkes (1975) has suggested that the strong selection that farmers place on potato clones decimates populations resulting from true seed propagation. There is little conclusive evidence that humans deliberately seek to produce or establish new genotypes that arise through sexual propagation. In general, it appears that new genotypes are introduced into cultivation in spite of humans. However, humans do consciously distribute clones that are obtained elsewhere, and once a clone is recognized as cultivable, Andeans have clear-cut criteria and mechanisms by which they can evaluate its qualities.

In an evolutionary sense, exceptions to the passive nature of human treatment of new types may be the more important events. Considering the limited number of potato clones that are actually cultivated, their stability over time, and the diffusion of clones, a few deliberate introductions could have a profound effect. The periodic occurrence of drought and famine may have acted as evolutionary bottlenecks. Certainly human behavior was exceptional in many ways during the drought of 1983. As we saw in chapter 4, it may have been in these situations that reservoirs of tuber seed, such as Ajawiri Marka, were turned to. Although new chemical types may be continuously produced, with high-altitude potatoes it is only during exceptional circumstances that humans direct selection that can bring about chemical change. However, this single example relating to one crop at one time and place should not be the basis for conclusions about chemical selection in general.

The role of humans in evolutionary processes depends on both human motivation and biological constraints. Considerable numbers of nontoxic varieties of potatoes exist on the Bolivian altiplano today, and the necessity for active selection of new clones is minimal. Because most frost-resistant potatoes (e.g., $S. \times juzepczukii$) are sterile and represented by only a few genotypes, the possibility for human selection is, in fact, very limited. Earlier in the domestication of potatoes, and other crops, the situation could have been very different. When fewer nontoxic plant resources were available, humans may have been motivated to take a much more active role in the domestication process.

The limited resources of the high-altitude environment dictate the dependence of the Aymara on specific plant foodstuffs, the most important of which is the potato. The Aymara have responded to their marginal environment with numerous physiological and behavioral adaptations. Their farming practices appear generally well adapted for stabilizing their agricultural output. Their experiences in this environment and their behavioral and technological responses to these experiences are preserved and communicated, primarily in the Aymara language.

Above all, the Aymara language reflects the importance of potatoes to Aymara life and survival. The cultural evolution of the Aymara and the evolution of the potato have been intimately related. The "coevolutionary" nature of this association is demonstrated particularly well by the example of sisu, which arose through hybridization between *S.* × *ajanhuiri* and *S. acaule.* Its combination of frost resistance, drought resistance, and low glycoalkaloid levels make it important to Aymara subsistence. What is perhaps most important in the sisu example is its demonstration that chemical selection is only one of the forces that have shaped the nature of modern potato cultigens. Human preference and choice of chemical properties are exercised in context and over time in a dynamic tension with other biological, environmental, and cultural factors.

TESTING THE MODEL OF HUMAN-PHYTOCHEMICAL INTERACTION

Specific information considered above in relation to potato domestication is essential to the following discussion of the nine hypotheses delineated in chapter 1. In some instances the potato case fails to provide an adequate test of the hypotheses. Realistically, the hypotheses extend beyond what can be addressed in an experimentally oriented field and laboratory investigation. Thus relevant information from other sources is drawn upon to clarify the model and provide direction for further investigations in this area.

1. *The phylogenetic relations of domesticated plants and their wild relatives will reflect chemical changes occurring during the domestication process.* The empirical evidence for human chemical selection in *S.* × *ajanhuiri* is provided by the phylogenetic and chemotaxonomic studies discussed above and summarized in Figures 4-9 and 4-10. The discussion in chapter 4 of chemical change in other domesticates attempted to demonstrate that this is a widespread phenomenon. None of those investigations intentionally demonstrated this phenomenon, and biosystematic studies directly correlating chemical change with phylogenetic change are generally unavailable. Nonetheless, in cases where wild relatives of crops and their chemical characteristics are well known, there are sufficient grounds to make fairly conclusive statements.

The family Cucurbitaceae provides clear examples of chemical change, largely because cucurbitacins are detectable organoleptically, because the genetics of cucurbitacins are simple, and because the differences between bitter and nonbitter types are very pronounced. Phylogenetic studies of members of the Cucurbitaceae in Mexico and Central America are receiving increasing attention from a number of botanists. *Capsicum* is another genus where extensive phylogenetic studies have been undertaken. Although chemical studies on amounts of capsaicin do not parallel broader biosystematic investigations, the extreme potency of most wild small-fruited peppers makes it clear that chemical selection has been important in the domestication of this group.

Biosystematic investigations that include chemotaxonomic studies provide the best chance of uncovering chemical selection. The comprehensive work by the late Timothy Plowman, an eminent ethnobotanist and plant systematist, on *Erythroxylum* spp. (coca) provides a clear phylogeny in relation to which data on methyl salicylate show a clear trend of human chemical selection. The chemotaxonomic study I undertook with *Tropaeolum tuberosum* supports a similar process that is complemented by other taxonomic work (Sparre 1973).

2. *Chemical change will be for either reduced levels of allelochemicals (in order to improve flavor or to lessen toxicity) or greater levels of allelochemicals (in order to increase resistance of the plant to pests or to enhance pharmacological properties).* In the case of the potato it is clear that selection has been for reduced toxicity. Glycoalkaloids, except at low levels where they may positively contribute to flavor, are considered negative dietary constituents. Since chemicals that serve defensive roles in plants are, generally speaking, as effective against humans as they are against other herbivores, human efforts have primarily been to eliminate them. Cucurbits and *Capsicum*, where levels of cucurbitacins and alkaloids, respectively, have been reduced, are good examples of this phenomenon. Controversy still surrounds the determination of wild progenitors of cultivated forms of cassava, and the direction of change in cyanogenic glycoside content in this crop is not clear. However, cassava appears to be analogous to sorghum, in which tannins confer resistance to birds and selection has been for both increased and decreased levels of chemical defenses.

Selection for flavor is evident in the *Tropaeolum tuberosum* and *Erythroxylum* spp. (coca) examples, while *Malvaviscus arboreus* and *Micromeria viminea* (Table 4-1) are examples in which selection has apparently been for enhanced pharmacological activity. It is unclear in the latter two cases how this is reflected in chemical properties of the plant.

3. *Genetic change is determined by factors in plant reproductive biology in combination with human treatment of the plant populations.* Circumstantial evidence compiled in this volume and by other investigators documents biological and cultural factors present in the center of potato genetic diversity that would facilitate the production of new genetic types. The demonstration of a conspecific weed in association with cultigens of *S.* × *ajanhuiri* suggests that hybridization between wild and cultivated species is part of the process by which variability is produced and maintained. Potentially cultivated types of weed *S.* × *ajanhuiri* with large tubers that are indistinguishable from, and of comparable culinary quality to, yari are available in Ajawiri Marka today. The mixed cultivation of many species and varieties of potatoes in Andean fields will facilitate the production of new genetic diversity. The ecologically diverse habitats maintained around cultivated fields combine a number of biotic and abiotic resources which support natural pollinators of potatoes. When new genetic recombinants arise, biological conditions exist to support their maturation. Human activities may assist the process. For example, noncultivated potatoes, whether wild plants, volunteers from tubers of cultigens, or new genotypes arising from seed, are tolerated in cultivated potato fields. Circumstances by which new genotypes may be harvested and introduced into the cultivated gene pool are discussed in chapter 4.

Field situations similar to those associated with the cultivation of traditional Andean crops have been studied in relation to other crops in various parts of the world. For example, in Mexico farmers encourage the introgression of genes from wild teosinte (*Zea mays* subsp. *mexicana* [Schrad.] Iltes) into cultivated maize (Wilkes 1977). Hybrid ears produce offspring with a higher yield, and humans consciously select them as seed stock.

The genetic diversity in a vegetatively propagated crop like the potato is comparatively low compared to that found in sexually propagated plants. Likewise, the diversity of chemical types will be minimized. In a sexually propagated, outcrossing plant a continuum of chemical variability would be expected even in circumstances where strong selection was being applied. One would hypothesize that in a vegetatively propagated crop, or in a self-compatible and inbreeding crop, the maintenance of favored genotypes would result in more discrete steps in chemical levels.

Although the *S. × ajanhuiri* complex contains very limited diversity, the differences between wild, yari, and ajawiri types do suggest steps in level of composition. Because the phylogeny of the potato is not well known, it is difficult to test chemical selection hypotheses in the crop as a whole. In comparison to the wider range of glycoalkaloid levels seen in wild species of potatoes, the concentrations of glycoalkaloids in cultivated potatoes are fixed at a level below what is unacceptable for palatability. Only in sterile hybrids such as *S. × juzepczukii* are high levels maintained. In this case, cytogenetic factors clearly limit genetic diversity.

It is difficult to design controlled experiments to test the hypothesis that plant reproductive methods will determine patterns of chemical variability. Data on the chemistry of cultivated or wild plants are insufficient to make comparisons of such variability in relation to breeding systems.

In theory, the best control for the potato case would be an annual domesticated species of the genus *Solanum* that reproduced sexually. An outcrossing species would provide the clearest contrast. *Solanum americanum*, a highly variable weed native to the New World, could provide a case for comparison. In equatorial Africa it is popular as a leafy vegetable, and both wild and cultivated forms are harvested. Africans recognize bitter and less-bitter types of this plant, which probably differ in foliar glycoalkaloid levels. Different types that I have observed in use in Kenya are all above the threshold level for palatability and have to be processed before consumption. Unfortunately, this species is self-compatible (Edmonds 1977) and as such is likely to compare rather than contrast with the potato in its pattern of chemical variability. Because of the relatively recent introduction of *S. americanum* into Africa, its simplified genetic diversity, and its exploitation and domestication over a wide geographical area, this species offers the potential for testing the hypothesis that self-compatible sexually propagated crops will facilitate the "fixing" of desirable chemical types.

4. *Humans can perceive differences in chemical concentration and will show preferences for chemicals consistent with changes in plant chemical makeup seen during domestication.* The taste preference experiments for glycoalkaloids described in chapter 5 directly tested this hypothesis, and the data support the contention that levels of glycoalkaloids are important to humans. Human ability to detect glycoalkaloids by taste appears to have a threshold between 17 and 38 mg/100 g. Both North Americans and the Aymara find potatoes acceptable when total glycoalkaloids are less than 17 mg/100 g. Selection decisions in the domestication process were probably made simply between acceptable and unacceptable levels of glycoalkaloids. The variation seen in levels of total glycoalkaloids below 20 mg/100 g is attributable to genetic segregation and is essentially random. Potato quality in this range depends on other factors besides glycoalkaloid levels.

5. *Human language concerning taste, smell, and other chemical properties will reflect human concerns relevant to the chemicals under selection.* Although the Aymara language does not contain an elaborate classification of potato tastes, its nomenclature follows the two-state manner by which glycoalkaloid levels appear to be distinguished. Ch'oke and luq'i describe palatable and unpalatable potatoes, respectively.

As we saw in chapter 5, the language and culture of the Aymara reflect their unique taste experience and their need to minimize toxic constituents in their diet. As well as affecting the choices the Aymara make for chemical constituents, language is the medium through which choices and preferences are made and experienced. Language separates human cultural evolution from biological evolution. It contains the accumulated human wisdom on plant properties. Whether the choices humans make on the basis of their cultural knowledge are conscious or subconscious is a moot point. Undoubtedly they are both.

6. *Geophagy and other processing techniques as practiced by humans have an evolutionary role in the detoxification of plant secondary compounds.* This hypothesis is impossible to test directly because it relates to an unrecorded part of history. In chapter 3 I argued that the evolutionary significance of geophagy and the mod-

ern pattern of its use with potatoes support the importance of the adsorptive properties of clay as a detoxicant during the domestication of this crop.

Unless potatoes were exploited on a more than casual basis it is unlikely that humans would become familiar enough with different genotypes to initiate the selection process. In general, I suggest that the processing techniques described in chapter 3 preceded domestication and were essential in the domestication of a range of crop plants. However, processing techniques that are employed with potatoes require the sophisticated utilization of freezing temperatures that are only present at high altitudes in the Andes. In the potato case, geophagy alone appears to be the key that enabled humans to overcome the constraints toxicity placed on the domestication process. The few reports of the use of clay with other plants, particularly yams, suggests that geophagy may have played a broad role worldwide.

7. *Toxic plants used in human diets provide nutrients that would otherwise be deficient.* Humans must deal with toxins to obtain adequate nutrition from plants. This hypothesis and the following one reflect the recurrent emphasis in studies of animal food procurement on the metabolic balance between nutrients gained and the costs of withstanding the negative effects of xenobiotics. Testing this hypothesis requires knowledge about the complete nutritional needs of a human population and about the nutritional composition of the resources available to them. Such data are obtainable, although they were not obtained in relation to the Aymara use of toxic potatoes. In general, it is clear that toxic frost-resistant potatoes such as *S. × juzepczukii* provide substantial amounts of food to Andeans living in high-altitude environments. The dietary survey in chapter 5 illustrates the essential importance of chuño blanco or tunta produced from bitter potatoes in the Aymara diet.

Worldwide, it appears that the most important plants that are detoxified before consumption are carbohydrate staples. It hardly requires a detailed nutritional assessment to say that potato, cassava, cycads, and acorns have all played this role in the subsistence of various groups. Acorns are also important sources of protein and fat.

The nutritional contributions of protein, minerals, and vitamins made by other plants that are processed for detoxification would require more detailed dietary analyses. Seed crops, including many legumes, and leafy vegetables are important sources of dietary protein, vitamins A and C, folic acid, and minerals (Faboya 1983; Gomez 1981; Keshrino 1983; Keshrino and Ketiku 1979; Ndiokwere 1984).

8. *Processing techniques used by humans are effective in removing toxicity while retaining nutrients.* Very few of the traditional processing techniques discussed in chapter 3 have been evaluated for either their detoxification efficiency or their capability to conserve nutrients. Cassava and potato processing techniques were discussed as examples in which processing efficacy has received some attention. The leaching and freeze-drying process leading to the production of chuño blanco or tunta removes approximately 93 percent of glycoalkaloids from bitter potatoes (Christiansen and Thompson 1977). However, the resultant product is greatly reduced in protein, and chuño blanco is exclusively a source of stored carbohydrate. In a marginal environment that is subject to the kind of droughts that I described in chapter 4, the capacity to store sufficient carbohydrate to outlast crop failures lasting up to three or four years is a primary concern. Bitter potatoes are sufficiently productive in this environment to supply the resources to make this possible, and potato energy, which in the form of fresh potatoes can be conserved for less than a year, can be preserved to last ten to twenty years when detoxified and dried.

While evidence presented in chapter 3 indicates that glycoalkaloids are effectively bound by clays, it is not clear what the cost of clay ingestion would be in loss of vitamins and minerals.

Traditional processing techniques applied to African leafy vegetables such as *Vernonia amygdalina* and *S. americanum* are being evaluated for their effects on nutrient content (Keshrino and Ketiku 1979). In these vegetables, water-soluble nutrients like vitamin C are lost during processing, while fat-soluble vitamins and protein are more likely retained. In fact, in diets that depend to a large extent on carbohydrate staples such as cassava, yams, or maize, leafy vegetables can be an important source of protein.

9. *High levels of secondary compounds are retained in the diet for ecological reasons.* High levels of glycoalkaloids are retained in certain species of cultivated potatoes grown in Peru and Bolivia. *Solanum × juzepczukii* and *S. × curtilobum*, specifically, are grown for ecological reasons—that is, for their adaptation to cold. However, the

corollary that glycoalkaloids are retained in these potatoes for ecological reasons does not hold. Glycoalkaloids do not confer any resistance to cold, and the high levels of glycoalkaloids found in these
species result from heredity and the cytogenetic factors discussed in
relation to the third hypothesis. In the case of *S.* × *ajanhuiri*, I
originally hypothesized that humans would select for low glycoalkaloids, while cold stress from the environment would select for higher
levels. As the *S.* × *ajanhuiri* example illustrates, where human selection from a pool of genetic recombinants is possible, the tension
between toxicity and cold stress does not exist and is resolved by
selecting for low-glycoalkaloid, high-frost-resistant genotypes.

Herbivorous insects and disease pose a greater problem to crop
production in tropical environments than they do in the Andes, and
evidence does exist to support the assertion that other toxic crop
plants besides potatoes have higher yields because of their greater
resistance to predators. It is less clear that this is why humans retain
these plants in their diets, although this is a likely assumption.
Cassava has been discussed in ecological terms at various places in
this volume. Bitter cassava is reputedly grown in parts of both South
America and Africa because of its resistance to insect and mammalian herbivores (see chapter 4), and varieties of *Sorghum* with high
tannin levels in their seeds are the most resistant to bird predation.
Varieties of the Andean seed legume *Lupinus mutabilis* with low
quinolizidine alkaloid levels are known (Williams 1984). They are
not cultivated in the Andes, however, because of their lower resistance to pests. In each of these examples processing techniques are
necessary as a means of detoxification.

The convergence of food and medicine will be discussed in the
following chapters. The pharmacological properties of food plants
are widely recognized in many cultures, and pharmacologically active phytochemicals may be deliberately included as part of the diet.
Evidence that humans have sought to maintain high levels of pharmacological compounds in food plants is circumstantial. The bitter
leafy vegetable *Vernonia amygdalina* is reputedly a remedy for various ailments among West Africans (Laekeman et al. 1983). Kenyans
say similar things about other bitter leafy vegetables, including *S.*
americanum (Kloos et al. 1986).

APPROACHES IN HUMAN CHEMICAL ECOLOGY

The chemical ecology model and the case study of potato domestication provide a methodology and a focus for considering the dynamics

of the various forces involved in human-plant interactions. Overcoming the restraints imposed by allelochemicals in plants has been intrinsic to human survival and success. Testable models exploiting the expanding base of scientific data in plant and human biology provide the potential for understanding this important evolutionary process.

The interactions of humans with their phytochemical environment occur on several levels. The primary focus of this study has been at the level of the human population. Alternatively, consideration of individual interactions with chemicals, on one extreme, and the evolutionary significance of human adaptation, on the other, provide important insights.

Individuals and populations show adaptive responses to environmental forces, while evolution introduces time and genetic change into the adaptive process. Examinations of individuals and populations both now and through time can be interconnected dimensions in the attempt to understand human interactions with plant chemicals. Studies at the level of the individual seem to have much to offer ethnobotanists who are interested in the interaction of humans with plant chemicals. Much of the ethnobotanical literature is, in fact, based on interviews with a single informant. Learned individuals are indeed often valuable repositories of a society's knowledge. A particular member of a society who has assumed the role of herbalist or medicine man might be the best source for information about the use of medicinal plants by that group. A herbalist's knowledge of chemical and pharmacological properties can point to a rational basis for the use of particular plants. However, such knowledge does not necessarily provide insight into the important question of how this knowledge was initially acquired. In fact, any idea that the key to understanding how plant properties are evaluated lies with the native healer who is intimately familiar with plants is probably a romantic illusion. It is certainly hard to accept information from single informants as having scientific validity (Johns et al. 1990).

While individuals do vary in their knowledge of plants, it is difficult to scientifically test any differences in their knowledge of, and their ability to detect, phytochemical constituents. Are learned individuals able to differentiate between concentrations of chemicals in plants or to assess the properties of a particular medicinal plant? Experiential knowledge of this sort is not easily communicated. A cultural component in hedonic responses suggests that any greater acuity that a herbalist might have would have a strong cognitive basis mediated by many factors. Any special acuity is unlikely to be physiologically based and would be difficult to test outside of the context in which it is exercised. Perhaps the scientific observer can learn through shared experience what it is that the healer is responding to in a particular plant. S/he too then could be a healer, but with the lack of any objective criteria will have made no scientific contribution. Tests could be designed to evaluate an individual's response to compounds that are believed to be responsible for certain biological activities. This is essentially what I did: pure solutions of glycoalkaloids and analyses of total glycoalkaloid levels in potatoes of supposedly different quality were used to test the ability of the Aymara to evaluate potatoes on the basis of glycoalkaloid content. Sufficient chemical data are not available to make even this simplified approach possible in very many cases.

Testing an individual's responses to a particular compound has little meaning without a comparative data base gathered from a larger population. Subtle differences between individuals in the perception of concentrations of glycoalkaloids may have been hidden by factors of accommodation and physiological state that are characteristic of human taste perception. Only at the population level was it possible to see patterns that could have biological meaning. It is possible to make comparisons between individuals in relation to compounds where response differences are less subtle and where patterns of variation are well studied. Among the Aymara, for example, individuals' preference responses for glucose fell into distinct categories and appeared to be age related.

Differences among human populations in taste physiology, dietary metabolism, and detoxication systems may be dietary adaptations to particular environments. Several examples of such interpopulational differences are discussed from a comparative physiological perspective in chapter 7.

Many aspects of the interaction of humans with their phytochemical environment cannot be understood without looking at factors and processes in human evolution that have determined our biological responses. The following chapters on diet and medicine focus on human-plant interactions from this perspective. Certainly physiological perception reflects basic nutritional and survival concerns. Behavioral responses such as conditioned aversions and geophagy reflect evolutionary history. These adaptations to ecological conditions are manifest in the ways in which humans use environmental resources for food and medicine. However, although crucial insights into interactions of modern humans with plant allelochemicals can emerge from an understanding of evolutionary trends, the evolutionary perspective, like the case of the individual, is difficult to approach directly.

Ultimately, experimental approaches in this field must center on the population and the comparison of populations. It is at the level of population that evolution occurs. My investigation focused on the Aymara of the Provinces of Pacajes and Ingavi. I studied the interaction of this one group of subsistence agriculturalists with one group of chemicals, glycoalkaloids, from one group of plants, potatoes. This extremely simplified vision of human-plant interactions provided a focus from which to consider the forces that determine evolution, and in turn the evolutionary determinants of diet and medicine that affect individuals and human populations today.

7 Plant Chemical Defenses as Determinants of the Human Diet

The well-being of an organism is determined first and foremost by its ability to obtain the chemical essentials for life—oxygen, water, and food. Energy and the components necessary for the growth and maintenance of the physical body and its biochemical processes require a continuous supply of chemicals. Our bodies are chemical machines, and our interactions with the environment on the most fundamental level are chemical interactions.

This chapter and the next extend the specific discussion in the previous chapters to a broader consideration of the evolutionarily determined patterns that govern our interactions with environmental chemicals. This chapter focuses on the role of plant allelochemicals in determining diet, while chapter 8 takes a similar approach in the consideration of the evolution of medicine. I believe that diet and medicine are intimately related in human evolution, and therefore these two chapters are closely interconnected.

It is my hypothesis that plant allelochemicals, while providing constraints that determine human dietary patterns, have also through time become essential health-promoting constituents. Because humans have sought to reduce the toxic chemicals in their diet, and as a result have lost the disease-controlling effects of some of these compounds, they compensate by ingesting them in other ways. Through the intimate and ongoing interaction of our ancestors with plant allelochemicals we have evolved physiological, behavioral, and cultural means of reducing the threats these chemicals pose, but, perhaps ironically, at the same time we may have made them an intricate part of our internal ecology and our adaptations to the environment. I believe the origins of medicine can be found in our physiological and biological adaptations to allelochemicals. Herbal medicine, and by extension modern pharmacological practices, may be an attempt by humans to replace nonnutrient chemicals that were always an important part of our diet.

If the chemical-ecological model of herbivore-plant interaction introduced in chapter 1 is relevant to human interactions with the environment, plant defensive compounds can be assumed to have played a significant role as evolutionary determinants of human dietary needs, food preferences, and procurement patterns. One cannot assume, however, that the interactions of hominids with plant allelochemicals have had any significance in the evolution of *Homo sapiens.* Perhaps the biological adaptations discussed in chapter 2 provided an adequate solution to the issues of toxicity well before the appearance of the genus *Homo.* The cultural solutions to plant toxins discussed in chapters 3–5 suggest that this is not the case. Indeed, the search for high-quality plant foods characterized by low levels of allelochemicals may have launched primates toward the type of mental development that predisposed hominids to the subsequent behavioral and intellectual evolution that characterizes the human species (Milton 1981; 1988). This chapter concentrates on our current understanding of the interactions of our ancestors with plant chemicals over the crucial periods that separate humans evolutionarily from our primate relatives.

Within a healthy organism metabolic processes function as an integrated whole, responding in concert to threats from the environment or disturbances from within. Health, which is obviously the desired condition, is theoretically a state when the organism is functioning efficiently. It is, however, a condition that is often more easily defined by negatives, and health is frequently understood to be the absence of disease. Dietary diseases are a result of deficiencies or excesses in essential nutrients. On the other hand, undesirable components of the environment, particularly parasitic diseases and chemical toxins, present threats to all organisms. Defensive processes of the immune system and detoxication enzymes allow animals, including humans, to remain in a state of "health" in the face of most threats. When these systems are overwhelmed, disease results.

Changes in environmental conditions require coordinated physiological responses in order for an organism to retain a state of homeostasis. Whether or not a disease is overcome depends on the capability of the animal to restore the equilibrium state of health; various physiological and behavioral functions assist in the process. Organisms that are "well adapted" to their environment are by definition capable of retaining and maintaining health over the course of their reproductive life span.

Human health, while dependent on the same ecological parameters that affect all animals, has peculiarities of its own. Human manipulation of the environment can have negative consequences for human health, but it can also provide novel solutions for the problems encountered. The genus *Homo* evolved as foragers in the forests and savannas of Africa, but as our ancestors moved into other regions they encountered a number of ecological zones and niches, each with unique problems for health. Foremost among these were problems of diet. Changes in the availability of nutrients present challenges for the maintenance of the organism. The nature and consequence of some of these changes will be discussed below. Likewise, as humans moved into new environments, new parasites, infectious diseases, and other threats presented possible negative consequences to our well-being.

Human cultural innovations have countered many adverse effects on health and have increased human adaptation in general. The expansion of culture greatly accelerated the process of human evolution in comparison to that of other primates and has been a powerful determinant of human affairs. Cultural practices that enabled efficient procurement and processing of animal and plant products made more high-quality food available to humans. In addition, by learning to use the chemical constituents of plants in a pharmacological manner, humans have positively affected their chemical balance in order to remove the causes, or at least the symptoms, of diseases.

While nutrient and nonnutrient chemicals may have counteracting effects on the well-being of humans and other animals, plant allelochemicals can have positive effects on our feeding and our physiological well-being. In making high-quality plant and animal foods available, evolutionary changes of the hominid line positively affected our nutritional status. However, in changing our interactions with plants, particularly with leaves and their allelochemical constituents, these changes affected the balance between nutrient status, disease-causing organisms, and the positive and negative properties of biologically active plant chemicals.

The interrelationship between nutritional status, infectious disease, and toxicity is diagrammed in Figure 7-1. In this scheme interactions A, B, C, and D are well recognized. While plant allelochemicals are certainly toxic to microorganisms, interaction E, which

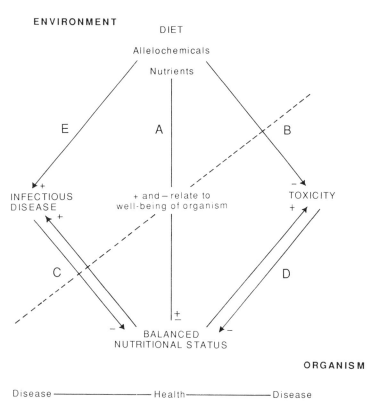

FIG. 7-1. A testable model of the evolution of human medicine.

suggests that phytochemicals in the diet limit infectious disease, is theoretical.

A synergistic relationship exists between nutrition and infectious diseases (interaction C) (Beisel 1982; Solomons 1984); the ability of humans to resist infectious diseases is very much intertwined with nutritional status, while individuals afflicted with infectious diseases may develop more serious nutritional deficiencies. Protein-energy malnutrition as well as specific deficiencies of zinc, iron, copper, selenium, vitamin B_6, and vitamin A have been linked with impaired immune response (Chandra 1988). The relationship between nutrition and disease is particularly clear in the modern world, where malnutrition among human populations in developing countries is the single most important factor contributing to vulnerability to infectious disease and the excessive mortality of children and adults (Chandra 1988).

Similarly, there is a synergistic relationship between nutritional status and the detoxication capabilities of humans (interaction D). The relationship between diet and allelochemicals in animals has been discussed as a ratio between nutrients (primarily protein) and toxins. An increasing body of literature indicates that a variety of nutritional factors can influence xenobiotic metabolism in laboratory animals (Meydani 1987; Yang and Yoo 1988). Vitamins A, C, E (Reed 1985), sulfur-containing amino acids (Reed and Meredith 1984), dietary fat (Meydani 1987), and minerals such as selenium (Chen et al. 1986), copper (Andrewartha and Caple 1980), zinc (Bray et al. 1986), and manganese (de Rosa et al. 1980) are all implicated in animals' ability to withstand the effects of drugs and environmental toxins.

Fewer studies have studied the interactions of nutrients and xenobiotics in humans. Malnourished humans do have reduced levels of detoxication enzymes and altered abilities to metabolize drugs (Krishnaswamy 1987). Consider, for example, the metabolism of cyanogenic glycosides in cassava. As we saw in chapter 2, metabolism of cyanogenic glycosides depends on the availability of a pool of sulfane derived from sulfur-containing amino acids. A diet with sufficient iodine, protein, and sulfur-containing amino acids is essential for enabling people to tolerate the potential toxic effects of cassava (Cock 1982).

In a series of experiments that manipulated protein, fat, and carbohydrate in the diets of healthy humans, Anderson et al. (1986) found that only an increase in protein had a positive impact on drug metabolism involving cytochrome P-450 mono-oxygenases. While fats in general may not have an impact on drug metabolism, specific dietary fats such as ω_3 and ω_6 fatty acids may have a positive role in modulating cytochrome P-450 metabolism. In studies unfortunately limited by the small sample sizes employed, normal individuals fed on a well-balanced diet with the addition of a pure protein supplement increased the rate of metabolism of theophylline, while a supplement of sucrose had the opposite effect (Kappas et al. 1976). Theophylline is a naturally occurring alkaloid found with caffeine in the leaves of *Thea sinensis* (tea), and it is a very important and widely used drug in the treatment of asthma.

The synergism of diet, disease, and allelochemicals would be complete if it could be demonstrated that better-nourished animals were correspondingly better able to control disease by tolerating higher levels of allelochemicals.

EVOLUTION OF THE HUMAN DIET[1]

The overwhelming evidence from various sources is that humans are by nature omnivores, opportunistic consumers of a range of animal and plant products (Isaac and Crader 1981; Milton 1987). This is the dietary pattern of most humans today; it is the dietary pattern of our closest living primate relatives; and it is undoubtedly the mode of existence practiced by all of our hominid ancestors.

While earlier researchers (Dart 1953; Ardrey 1976) argued that australopithecines—the first hominids and members of the family that includes modern human beings—were hunters engaged in a carnivorous life-style, most recent observers (Isaac and Crader 1981; Stahl 1984) have emphasized their probable reliance on plants for the bulk of their subsistence.

There is a tremendous range of variability within the omnivorous habits exhibited by human gatherer/hunters, traditional agriculturalists, and modern urban industrialists. Of particular interest are the relative proportions of animal and plant products consumed in various cultures. There is abundant evidence that many current diseases of humankind have a dietary basis which to some degree appears to relate to biochemical differences between animal and plant foods. However, the issue is not whether we are carnivores or vegetarians but rather, as omnivores, the proper balance between plant and animal foods. This is not a question that is likely to have a simple answer. The array of nutrients provided by each of the animal and plant groups of food is extremely variable, and it is the very nature of omnivory to allow flexibility for adapting to a range of environmental conditions. There are innumerable ways to obtain a nutritious balance; likewise, an exact point of biochemical homeostasis does not exist. Nonetheless, there are clearly nonadaptive dietary regimes that lead to illness, and valuable evidence of the dietary range to which we are best suited can come from evolutionary studies.

Consideration of the evolution of human dietary patterns is an area of active scientific inquiry. Recent research has focused on a number of specific questions:

---

1. Technically speaking, the diet does not evolve. Organisms evolve, and what they eat changes in response to environmental conditions. The phrase "evolution of the human diet" when it appears is a shorthand for what more correctly might be called "evolution of human dietary behavior," "evolution of human dietary patterns," or "changes in human diet during the evolution of our species."

1. What was the relative proportion of foods derived from plant and animal sources in the diet of our ancestors?

2. What were the proportions of different classes of plant and animal foods in the diet? (Plant foods include fruits, leaves, tubers, seeds, and nuts; animal foods are generally classified as large or small mammals, birds and bird eggs, fish, mollusks, and insects.)

3. Did early hominids rely on hunting or scavenging as a means of obtaining animal protein and fat? (See Binford 1984; Gordon 1987; Isaac and Crader 1981; Speth 1987.)

4. What were the responses of early humans to periodic fluctuations in food resources, and how were they affected by periods of dietary stress? (See Speth 1987 for references.)

5. How have humans adapted to major shifts in subsistence patterns such as that associated with the origin of agriculture?

Rather than attempt a complete overview of the rapidly expanding research that relates to these questions, I concentrate on a chemical-ecological approach to the problem of the evolution of the human diet. Broader discussions of the evolution of diet can be found in recent reviews (e.g., Gordon 1987). In emphasizing the role of plant chemistry in determining human dietary patterns, I will draw on other areas only as they relate to the understanding of the role of plant and animal constituents to this issue. I suggest that plants have always been a fundamental part of hominid diets, but that in the modern world the perceptual and behavioral patterns that evolved in relation to our interactions with plants and plant chemistry during a foraging life-style contribute to many of our diet-related health problems. We have efficient physiological, behavioral, and cultural mechanisms for maximizing our nutrient intake and overcoming toxicity, which should allow us to procure food from plants in a natural setting. Technological innovations have made plant as well as animal foods more available to humans, and with cultural changes in food procurement and processing (particularly the rapid change over the last ten thousand years and the accelerating change over the past hundred years), the manner in which plants and animals are consumed has in some way contributed adversely to health.

Modern humans exhibit relatively little genetic variability in diet-related physiological and morphological features, and we can assume that our dietary requirements have not changed much since the appearance of modern *Homo sapiens* over forty thousand years ago.

Current research is more concerned with the way adaptive experience determined our dietary behavior and physiology during the much longer evolution of hominids. The fossil record provides the only direct evidence of the process of dietary change that has taken place over the space of several million years. Have our nutritional needs changed appreciably from those of the first *Homo sapiens* who evolved at least a hundred thousand years ago? Have they evolved since the first appearance of *Homo habilis,* the first member of our genus (Johanson et al. 1987), some two million years ago? There is archaeological evidence that some dietary shifts occurred during these periods (Gordon 1987), but for an omnivorous genus which has relied heavily on cultural rather than biological evolution, shifts in consumption patterns may not in themselves represent significant change in biologically determined nutrition requirements. Moreover, considering the fragmentary and two-dimensional nature of the fossil record, it is usually more straightforward to ask how much we have diverged from the dietary patterns that typify the primate line to which we belong. Studies of human evolution do, in fact, tend to concentrate on comparisons between behavior, morphology, nutrition, and biochemistry of modern living primates and modern humans.

LINES OF INVESTIGATION

Considering the degree to which humans have diverged in evolutionary terms from other primates, comparative physiological and behavioral data are difficult to interpret without some independent measure of the temporal and ecological context through which differences occurred. Paleoarchaeology provides the only direct evidence from the past. Although this evidence is too incomplete to be the basis for initiating studies on diet, paleoarchaeology does provide a means for testing hypotheses on human dietary history formulated from studies of modern organisms, and for formulating further hypotheses for research with living organisms. In this volume I have concentrated on evidence concerned with detoxication enzymes, nutrition, sensory physiology, and feeding behavior of humans and other animals. The chemical-ecological approach works best with data of these types. None of these aspects of human biology, however, leaves an archaeological record that can be examined. What *can* be examined are hominid bones and teeth, and those stone and bone artifacts that provide evidence of the activities of our predecessors.

ARCHAEOLOGICAL EVIDENCE AND APPROACHES

The tools left by our ancestors reflect changes in the ways food plants and animals could have been obtained and processed, but provide frustratingly little detail of what was actually consumed. The record over the more than two million years since hominids first began to make tools is biased toward meat consumption. Bones of prey animals are easily preserved, while most parts of plants are not. Studies of the utility and wear patterns of stone artifacts and of the marks made with stone tools on processed bone supports the usefulness of these tools in butchering carcasses and processing bones, including those of large mammals (Toth 1985), but it is clear that these tools were also used for woodworking and cutting soft plant material (Keeley and Toth 1981). Stone tools that could have been used for pounding plant material —pulverizing nuts, grinding seeds, or crushing vegetables and fruit—are common in sites that were occupied by members of the genus *Homo* (Kraybill 1977). The use of stone tools for processing plant material, particularly seeds, increased dramatically prior to the onset of agriculture some ten thousand years ago. Stone tools actually are not as efficient for digging to reach underground water sources or for edible roots and tubers as are antelope bones, broken long bones from large animals, or sharpened digging sticks (Toth 1987). The discovery of pointed bones 1.5–2 million years old with wear on their ends is consistent with a use for digging up bulbs and corms (Brain, cited in Toth 1987).

Fossil remains of early hominids themselves provide a direct record of changing morphology that could have functional significance. Clues to human diet can be obtained by comparing the anatomy and morphology of the organs used in the ingestion and digestion of food by modern humans to those of fossil hominids. In none of these organs for which there is a fossil record do humans have features that reflect specialization to a particular type of diet. Tooth morphology has received the greatest amount of attention in this area, and in conjunction with studies of mastication and tooth microwear can give some indication of the dietary habits of fossil hominids (Grine 1981; Isaac 1984; Lucas et al. 1985; Walker 1981). However, the available data do not provide a clear picture of this part of the past, and interpretations vary considerably, although the emphasis seems to be toward plant consumption. In general, studies in this area allow one only to rule out the most specialized diets. In robust australopithecines, who first appeared around the same time as the genus

Homo approximately two million years ago, mastication forces were greater than in humans, although apparently designed for similar kinds of food. These creatures may have relied on plant foods (such as hard nuts, seeds, leaves, or grasses) that were resistant enough to require a considerable amount of oral processing (Grine 1981; Peters 1987), either through increased force or a longer duration of mastication (Gordon 1987). Compared with the robust australopithecines, members of the genus *Homo* show thinner molar enamel and a reduction in cheek tooth size (Grine 1981; McHenry 1984). These changes may directly reflect dietary forces.

Recent analyses of the remains of a *Homo habilis* skeleton from Tanzania (Johanson et al. 1987) reveal an individual with body size and proportions similar to those of australopithecines and may indicate somewhat arboreal habits. Did this animal have the same procurement patterns as australopithecines but rely on its greater intelligence to refine its diet through processing? While its teeth are reduced in comparison with the robust australopithecines, they may not be much changed from the australopithecine condition in general. If a more refined diet was a factor in leading to reduced dentition, the evidence for it is not obvious at that time.

Homo erectus, our immediate predecessor, certainly had features such as skeletal proportions and reduced dentition that suggest that its food procurement and eating habits were more like those of humans. It also had a larger brain. While studies of hominid fossils often raise as many questions as answers, the fact that brain size seems to have preceded changes in dentition in the genus *Homo* suggests that the capacity for technological manipulation of food was important in diet-related changes in morphology.

Changes in jaw formation and facial structure that have accompanied the emergence of modern man may reflect a shift in diet from that of Neanderthals. In the absence of any proof, nondietary explanations are just as valid for explaining these changes, and we must be careful to avoid overinterpreting these data. For example, changes in facial morphology could accompany changes in human social structure or less aggressive behavior (Sarich, personal communication).

In general, paleoarchaeology often provides only indirect evidence for understanding what humans actually ate or how plant chemicals determined food choice. Although meat and bone marrow were consumed by early hominids, we do not know if animal products were occasional or regular foods (Isaac 1984). On the other hand, we cannot say with certainty whether plants were eaten only when nothing else was available, or were a staple part of the diet.

More precise estimates of what our ancestors ate may come through techniques that employ chemical and microscopic analyses of human remains and food residues. Coprolites have provided some evidence of early dietary patterns, and have been particularly useful as a record of diets over the past ten thousand years (Kliks 1978).

Chemical analysis of food residues in ceramic vessels and on tools is possible using infrared spectroscopy (Hill and Evans 1989). Gas chromatographic analyses of coprolites have provided direct evidence of secondary chemicals present in the human diet a few centuries ago. Moore et al. (1984) have used the latter techniques to profile specific plants in the prehistoric diet of Native Americans living in Utah as far back as 4400 B.C. Unfortunately, the most critical periods, that is, the earliest, are the most difficult to study by chemical methods.

Analyses of bone chemistry appear to provide the most direct window on prehistoric dietary patterns (Ambrose and DeNiro 1986; Bumstead 1985; Price et al. 1985). The ratios of trace elements and stable isotopes that are incorporated into bone are determined by the ratios of these chemical substances in the diet during the manufacture of the bone tissue. Different classes of foods, particularly animal and plant but also different kinds of plants have different compositions.

Carbon isotope ratios ($^{13}C/^{12}C$) in bone collagen are an indication of the proportion of C_4 and C_3 plants in the diet. The so-called C_4 plants such as sorghum, millet, maize, and amaranth synthesize four-carbon compounds as the first step in photosynthesis, while C_3 plants, including barley and wheat, synthesize three-carbon molecules as the first step in this process. In addition there are differences in $^{13}C/^{12}C$ ratios between marine and terrestrial organisms.

Differences in the amounts of rare ^{15}N isotopes in an organism relate to the way nitrogen was obtained. By analyzing the amount of ^{15}N in human bone it is possible to determine the importance of nitrogen-fixing terrestrial plants (e.g., legumes), other terrestrial foods, and marine foods in the diet. In addition there is an enrichment of ^{15}N as organic nitrogen moves to higher trophic levels. The proportion of ^{15}N is therefore an indicator of the importance of meat in the diet. Although alterations in bone constituents through environmental processes may occur after preservation, analyses of bone chemistry may eventually provide information on the diet of early hominids (Ambrose and DeNiro 1986).

STUDIES RELATING TO PRIMATE FEEDING BEHAVIOR

Primatologists consider the selectivity of nonhuman primates in their food choices to be directly related to variability in the nutrient and nonnutrient composition of plants. Most nonhuman primates utilize plants as food, in some cases as their only food. Although the contributions of various plant parts vary from species to species, fruit and leaves constitute the bulk of the plant foods eaten (Harding 1981).

Most of the feeding studies that have considered plant chemistry have focused on species that utilize leaves as their primary food source. If the present-day diets of primates are any indication, leaves must have been an important food for early hominids. Glander (1982) summarizes the work of a number of investigators in concluding that "there is a dynamic and complex relationship between secondary compounds and nutrients that hinges on the type of defensive compound present, the amount of nutrient present, and the digestive system of the primate." Digestibility of the food, which relates to its fiber content, is also an important factor. While this statement suggests the complexity of the issue, it also masks the difficulties in obtaining data on all factors influencing the behavior and physiology of a particular primate species.

Phenolic content is an important factor in the difference between feeding patterns of black colobus monkeys (*Colobus santanas*) in Cameroon and red colobus (*C. badius*) and black-and-white colobus (*C. guereza*) monkeys in Uganda (McKey et al. 1978). Animals from Cameroon avoid the high levels of phenolics in leaves of abundant tree species by consuming leaves of rare species and high amounts of seeds. Although these seeds contain high levels of alkaloids, including strychnine, McKey proposed that the animals were able to consume the seeds because they were compensated with more nutrients than they would have received from leaves containing similar compounds.

This work with colobus monkeys suggests that increased nutrients can compensate for the cost of high levels of allelochemicals. It is not clear, however, that plant allelochemicals generally act as feeding deterrents to primates.

Food quality has been identified as the primary factor determining food choice in a number of primates, including howler monkeys (*Alouatta palliata*) (Milton 1979). Young leaves are preferred by these monkeys, apparently because they contain higher protein content and less fiber than mature leaves. Howlers are highly selective, at times to the extent of eating the pedicels of a fruit and discarding the fruit itself. They may be maximizing the nutrient benefits while minimizing the amount of toxin ingested (Glander 1978, 1982).

While studies that use chemical data to seek broad correlations between behavior and chemistry have pointed out logical patterns, it is very difficult to establish cause and effect. Patterns are not always clear-cut. With howlers, for example, tannins may have no deterrent effect (Milton 1979). Certainly in systems where multiple factors, including nutrients of various kinds, fiber, and various allelochemicals, are involved, recognition of causal factors for particular behaviors is difficult.

Moreover, most studies that have carried out chemical analyses related to feeding behavior have considered only total phenolic (tannins) content, total alkaloids, total protein, amino acid composition, fiber, ash, and/or water content. It is generally impractical to look at the chemistry of the dietary plants in any greater detail. The lack of attention to determining specific chemicals present can provide confusing data. Particular compounds may be very toxic to certain species while others may have little toxicity even in large quantities. Thus a correlation between phenolic content and feeding behavior is only meaningful when the specific effects of individual phenolics, as well as other potential toxins in a plant, are established for the animal in question.

Even species subsisting primarily on leaves consume other types of plant and animal foods and show elements of the omnivory that appears to be widespread among primates. More relevant analogues to understanding the evolution of the hominid diet come from studies of primates that are clearly omnivorous. The monkey *Cercopithecus aethiops*, baboons (*Papio* spp.), and common chimpanzees (*Pan troglodytes*) are the primates with the most varied diets (Harding 1981). Each of these feed to some degree on vertebrate or crustacean flesh or eggs, insects, fruit, seeds, nuts, flowers and inflorescences, and vegetative plant parts.

Baboons and chimpanzees occupy habitats in Africa which overlap those exploited by early hominids. Their ecology and feeding behavior offer useful analogues for studying the diet of early humans.

Peters and O'Brien (1981) identified possible plant food items still eaten by baboons, chimpanzees, and humans in South and East Africa and concluded that leaves and shoots and/or fruits would be the most common food items exploited by early hominids, with underground storage organs and seeds and pods also being important.

Like humans, baboons exploit underground roots and tubers. Their diets are extremely adventive and flexible in accordance with seasonal fluctuations in the availability of their preferred foods—animal matter, fruits, and seeds (Hamilton et al. 1978; Peters and O'Brien 1981). Baboons can obtain over 25 percent of their diet from leaves but often consume much less (Hamilton et al. 1978). They carefully select and process their food, and while they may be avoiding plants with high amounts of allelochemicals, there are no data to support this hypothesis. On the contrary, they persist in eating small amounts of highly toxic plants that Hamilton et al. (1978) classify as euphorics. Baboons (*Papio anubis*) in Gombe National Park, Tanzania, eat large quantities of bitter (to humans) leaves of *Pterocarpus angolensis.* They may be more tolerant than chimpanzees or humans to secondary toxins (Wrangham and Goodall 1987).

As our closest relatives, chimpanzees receive close scrutiny in many regards. Fruits make up the largest part of their diet, followed in importance by leaves, an important protein source (Hladik 1978). Shoots, stems, seeds, insects, and small mammals round out the diet (Hladik 1978).

While lacking any special physiological capabilities for dealing with allelochemicals, chimpanzees seem well suited to avoid their toxic effects. Bulk in relation to fiber content versus nutrient content may be a more important issue in their feeding than problems of toxicity. Based on alkaloid data alone, Hladik (1977a) concluded that allelochemicals in leaves are generally unimportant deterrents to chimpanzees. While many of their foods contain alkaloids, they may need only to avoid a few particularly noxious alkaloid compounds. Chimpanzees appear to avoid high levels of condensed tannins by feeding on ripe fruit rather than unripe (Wrangham and Waterman 1983), and it may be that behaviors leading to selective feeding have helped these animals to successfully minimize the potential effects of toxins.

Chimpanzees, like many primates (Galdikas 1988; Oates 1977), eat a variety of leaves and thus consume a variety of toxins in tolerable doses. Specific plant parts are used, sometimes after peeling and discarding unpalatable portions. Because they are omnivores consuming nutritionally dense foods such as fruit and animal protein, the quality of their diet is higher than that of strictly herbivorous primates, and the balance between nutrients and secondary compounds is in favor of nutrients. Clay eating by chimpanzees (Hladik and Gueguen 1974; Uehara 1982) and other primates (Oates 1978) has been hypothesized to serve the function of detoxifying plant secondary compounds.

Undoubtedly the most important adaptation of chimpanzees to dealing with plant toxins is their intelligence. Their cognitive skills give them a greater ability to deal with the spatial and temporal variability of their environment, including the chemical environment. While chimpanzees may be selective as individuals, as a species the diet is fairly flexible (Hladik 1981) in a way that has allowed them a wide geographical distribution with habitats ranging from rain forest to savanna woodlands. Omnivorous animals must achieve a balance between ingesting safe, nontoxic foods and the flexibility needed to explore for new food sources (Rozin 1982). Intelligence would seem to make this task easier.

Katharine Milton (1981, 1988) developed a very appealing hypothesis that links the evolution of brain size and mental acuity in primates with differences in predictability in the distribution of high-quality plant foods in tropical forests. She compared two forest primates in Central America—spider monkeys (*Ateles geoffroyi*) and howler monkeys. The more intelligent spider monkeys seek ripe fruit as principal dietary constituents. While this resource is generally nontoxic and of high quality in terms of digestible energy versus nondigestible fiber, it is patchily distributed and these animals are forced to forage over a large range to obtain a continuous food supply. Spider monkeys ably respond to temporal and spatial patterns in which young leaves, flowers, or fruit are available. Individuals show considerable mental acuity and recognize and remember the locations of many food species. In addition they show considerable behavioral flexibility in adding new plant species to their diets. Howler monkeys, on the other hand, feed to a greater degree on particular leaf species that are less patchily distributed and more predictable. Compared with spider monkeys they have smaller home ranges and a notably smaller cranial capacity. Howlers may depend on a cohesive social structure for remembering information about the availability of palatable and nontoxic resources rather than on the intelligence and behavioral flexibility of the individual.

The selection pressure that favors mental capabilities in forest primates appears also to operate in savanna settings. Savanna-living primates such as baboons and chimpanzees have large home ranges and considerable behavioral flexibility. Similar selection pressures undoubtedly affected savanna-living australopithecines (Milton 1987), and the trend toward greater behavioral flexibility predisposed humans to further intellectual development.

Taste Perception. Differences in the feeding behavior of primates are probably regulated by interspecific differences in physiology and sensory perception (cf. Westoby 1974). The behavior of species such as spider monkeys and macaques (Hladik 1981), which forage widely for scattered resources, must have a different basis of motivation than that of species such as howlers, which concentrate their feeding on the most abundant plants. In the latter case, internal physiological signals of satiety are probably more important than changes in sensory perception (Hladik 1981). On the other hand, animals that feed on ripe fruit would likely have an enhanced taste perception and probably a hedonic preference for soluble sugars and organic acids. Differences in taste perception could reflect differences at the receptor level or at higher cognitive levels. Studies of interspecies differences in primate taste perception are generally scarce, but data are gradually accumulating, largely from the work of Glaser and Hellekant. Differences do exist both in taste thresholds (Glaser 1980) and in the ability to taste specific compounds such as sweetness-enhancing proteins (Glaser et al. 1978).

Studies of taste thresholds for sweet, salt, sour, and bitter compounds have been carried out with more than a dozen primate species (Glaser 1980). Contrary to the pattern seen with responses to the taste-modifying proteins discussed in chapter 2, threshold differences in discrimination ability do not follow phylogenetic relationships, but rather appear to relate to the dietary history of the species.

The best experimental example correlates preference behavior in response to chemical stimuli in laboratory tests with feeding behavior in the wild of the New World monkey *Aotus trivirgatus* (Glaser and Hobi 1986). While the behavior of these animals shows that their ability to detect acetic acid is similar to other monkeys, they are able to detect citric acid at half the concentration detected by other species. Even more striking, while other species avoid concentrated citric acid, *Aotus trivirgatus* show a preference for acetic acid and citric acid in concentrations from 0.02 M down to the threshold concentration. *Aotus trivirgatus* feed on sour fruits that are avoided by other animals, and their ability to withstand high concentrations of acids appears to have opened up a particular resource.

The dietary flexibility seen in omnivorous primates such as chimpanzees may be reflected in a greater capability for hedonic response than that seen in strictly plant-exploiting species such as spider monkeys. Sensory characteristics are undoubtedly of increased importance in conjunction with the increased capability for learning that characterizes hominid evolution. Humans appear to have more acute tastes than other primates (Glaser 1980), although methods used to establish threshold test levels for humans and primates are not strictly equatable. If this acuity is real, one might speculate that it has evolved in conjunction with the capability of learning from experiences to enhance our ability to recognize particular foods.

Sensory perception is an area with great potential for understanding the evolution of the human diet. Particular dietary behaviors such as the response to variety and the capability for conditioned aversions have not been systematically studied in primates, although a few studies show general similarities in the ways humans and nonhuman primates learn from experience with dietary chemicals. When poisoned with lithium chloride, green monkeys (*Cercopithecus sabaeus*) develop strong aversions that can be provoked through visual cues (Johnson et al. 1975). In experimental situations chimpanzees acquire preferences for piquant foods (Rozin and Kennel 1983). Investigations in the area of sensory preferences and learning behavior in conjunction with electrophysiological studies of threshold sensitivities in primates have the potential to provide insights into the role of both nutrient and nonnutrient chemicals in the evolution of the human diet.

More urgently needed are studies of the feeding behavior of primates in their natural habitats. Unless greater efforts are made to preserve the habitats occupied by these creatures, such opportunities will vanish within a few years.

Gut Morphology and Digestive Physiology. Milton (1986, 1987) summarized data on the significance of primate digestive physiology and morphology to studies of the human diet. In spite of their dependence on plant foods, most primates show relatively little specialization in gut morphology when compared with mammalian herbivores of other orders. A few leaf-eating primates (Hladik 1977b; McKey

1978) ferment food in their foregut, and this may aid in the detoxification of chemicals. However, in general the nonspecialization in primate gut morphology may be connected with improved behavioral, rather than morphological or physiological, solutions to the problem of selecting healthy food.

Even though most primates lack the specialized large fermentation chambers of ruminants, some, including humans, have some capacity to digest plant cellulose through acid hydrolysis in the stomach (Andersen et al. 1988) and microbial fermentation in the colon (Cummings 1984; Milton 1986). All hominoids show the same basic gut form—that is, a simple acid stomach, a small intestine, a small cecum terminating in a true appendix, and a well-sacculated colon, which is the site of fermentation (Milton 1986). Gut proportions in modern humans differ from other hominoids in that greater volume is present in the small intestine—the primary absorptive organ. In addition, the size of the human gut relative to body mass is small in comparison with other apes. Nonetheless, chimpanzees and humans are strikingly similar in terms of their digestive kinetics and fiber-assimilation efficiency (Milton 1986; Milton and Demment 1988). Our particular gut morphology suggests that humans are adapted to exploit nutritionally dense, rapidly digested foods. However, we retain the capacity for digesting crude plant material. Most mammals are able to digest nonlignified dietary fiber, particularly pectin, some gums, hemicelluloses, and hydrated celluloses, through fermentation (Jeraci and Van Soest 1986; Van Soest 1982). Bacteria isolated from the human large intestine are able to ferment these fibers as well as cellulose from leaves (Wedekind et al. 1988).

Animals, including humans, show considerable phenotypic adaptability to diet. Individuals consuming the refined diets found in industrial societies may show phenotypic differences in gut morphology and function compared with individuals on more traditional diets (Milton 1986). People in the sub-Saharan region of Africa traditionally consume diets rich in leafy vegetables such as *Crotalaria brevidens, Gynandropsis gynandra, Solanum americanum*, and *Vernonia amygdalina*. They show transit times (the time required for food to pass through the gastrointestinal tract) two to five times as rapid as persons eating a refined Western diet, and produce four to five times more fecal matter (Burkitt et al. 1972). The traditional Ugandan diet may closely approximate the diets eaten by all humans until quite recently. Human coprolites from post-Pleistocene sites in the

American West indicate that individuals occupying these sites ate as much as 130 grams of plant fiber per day (Kliks 1978).

Clearly humans, like other primates, are able to thrive on a diet high in fiber. Like primates we can select foods from a range of plants and can tolerate a diet containing appreciable amounts of plant allelochemicals. But I anticipate myself; it is time to move from non-human primates and consider present-day human populations directly.

COMPARATIVE STUDIES AMONG MODERN HUMANS

Omnivorous Behavior. All of our hominid ancestors lived as foragers until the origins of agriculture. Present-day gatherer/hunters—human populations that live without agriculture—have been scrutinized carefully for insights into the stage of human evolution that they represent. Although gatherer/hunter populations are all omnivorous, the majority consume between 50 and 80 percent of their foods as plant products (Eaton and Konner 1985). The Inuit living in the extreme Arctic environment are unusual in that they obtain 90 percent of their food from animal sources. Although gatherer/hunters today must by necessity live in marginal habitats, the good health of these people emphasizes that they subsist within a range of acceptable nutrition. Eaton and Konner (1985) determined means of 35 percent meat and 65 percent vegetable foods from gatherer/hunter diets as a basis for calculating appropriate primitive human dietary ranges. While these are only estimates of the diet of early humans, they are undoubtedly based on the best data available for making such calculations.

Comparative Physiology and Biochemistry. Comparative physiology has revealed only a few population differences that reflect dietary adaptations. The most widely cited example is the greater tolerance to lactose in groups that drink animal milk, particularly northern European populations, compared with those that do not. Interpopulational differences may reflect genetic changes that have occurred since the initiation of dairying as part of the development of agriculture (Flatz and Rottauwe 1973). Similarly, dietary intolerance to wheat flour (celiac or wheat-eating disease) is most frequent in populations that have been recently exposed to wheat as a dietary staple (Simoons 1981). Persons with celiac disease are sensitive to the substance gliadin, which is a fraction of the gluten of cereal grains.

Another example of interpopulational differences in dietary physiology is the unusually high (3–10 percent) sucrose intolerance among Inuit and Amerindians of northern Canada (Schaefer 1986). Intolerance symptoms are prevented if lean meat is consumed prior to sucrose ingestion. The traditional Inuit diet, which is high in meat and low in sugar, may be associated with the loss of the normal hormonally mediated insulin response to sugar consumption.

Different reactions of individuals to nutrients and to xenobiotics often reflect genetic differences in the ability to synthesize particular proteins. Enzyme polymorphisms are widespread in humans (Beckman 1978), and a number of these are found in enzymes that are directly involved in our interactions with environmental toxins. While ethnic differences in gene frequency may not have evolutionary significance, there is evidence, in certain cases at least, that substantial selective forces act on polymorphisms. Other genetic differences may appear at the hormonal or organ-function level. A few of these that relate to diet and environmental toxins will be discussed below.

Lactose intolerance results from deficiencies in the enzyme lactase. Approximately 6–24 percent of Europeans are deficient in this enzyme, while 66–90 percent of Asians and Africans are lactase deficient. Sensitivity to ethanol among Asian and Amerindian populations likely relates to deficiencies in aldehyde dehydrogenase, the enzyme that metabolizes acetaldehyde formed from ethanol (Goedde and Agarwal 1986). Like sucrose intolerance, differences in alcohol metabolism among Native Americans may also be related to their history of carbohydrate consumption. Inuit and northern Amerindians are less tolerant to alcohol than populations from farther south, which may relate to the largely carnivorous diet and lack of access to fermentable carbohydrates until recent times in northern populations (Schaefer 1986). Other digestive enzymes that show polymorphisms along ethnic or geographical lines include α-amylase and pepsinogen (Beckman 1978). It is not clear whether there is any selective advantage to particular genotypes in these cases.

Differences in kidney function relative to natural selection may explain differences in hypertension patterns between Americans of African and European descent (Freis 1986). The hot African climate plus a scarcity of water in many areas may have led to selection of individuals whose kidneys best retain sodium and water (Helmer and Judson 1968).

The ability of humans to deal with toxic substances also reflects genetic differences in detoxication mechanisms (Festing 1987). In addition to variation inherited within families, numerous ethnic differences in drug metabolism have been observed (Goedde 1986), although it is generally unclear what selective forces, if any, may have brought about these differences.

Persons of Asian and European descent show differences in caffeine metabolism that may reflect enzymatic differences and/or differences in renal function (Kalow 1986). Ethnic differences in the oxidative metabolism of the antiepileptic drug phenytoin are well known (Hvidberg 1986). Eskimos, certain African populations, and possibly Japanese appear to metabolize and eliminate the drug faster than Caucasians, while Ghanaians excrete the metabolite *p*-hydroxyphenytoin at a slower rate. Ghanaians and British Caucasians show differences in the way the cytochrome P-450 mono-oxygenase system oxidizes drugs such as debrisoquine and the natural alkaloid sparteine (Woolhouse 1986) that are independent of phenytoin metabolism (Hvidberg 1986). In general, Ghanaians may have a lower capacity to oxidize certain drugs, which may place them at lower risk of developing liver tumors associated with metabolic activation of environmental carcinogens such as aflatoxins present as dietary contaminates in maize and peanuts (Woolhouse 1986).

Other polymorphisms that may reflect differential selection for dealing with natural toxins include the "silent gene" for cholinesterase, which affects response to the anesthesia succinylcholine (Schaefer 1986), and genes that result in methemoglobin reductase deficiency (Evans 1986; Ommen 1986).

Also of apparent relevance from a chemical-ecological perspective are interpopulational differences in the polymorphic ability to taste phenylthiocarbamide (PTC) and related thioureas. As we saw in chapter 2, individuals with the capacity to perceive this substance may be better able to perceive (and thus avoid) naturally occurring goitrogenic compounds containing the thiocarbamide linkage (Greene 1974). Genetic variation has also been established in sensitivity to the taste of quinine (Smith and Davies 1973) and caffeine (Hall et al. 1975) and to the smell of urinary metabolites of asparagus (Lison et al. 1980). These and other differences in perception among human populations could relate to adaptation to different diets.

Nutritional Requirements. The recommended nutritional requirements for humans are based on data from extensive physiological studies of humans and other animals. Nutritional substances have a concentration window whereby the ingestion of a food in excess of an upper limit produces toxic effects, while supply of essential nutrients below a certain threshold produces deficiency diseases. However, it can be difficult to make other than broad estimates on the optimum amounts of various food constituents that should be included in a proper diet. Nutrient requirements vary under different conditions and stages of life history. Humans' ability to adapt to nutrient intakes within broad ranges is consistent with our omnivorous nature.

Essential nutrients, particularly those which distinguish humans from other animals, can reflect our dietary history. Those nutrient requirements that appear to reflect such evolutionary adaptations are the focus of the following discussion. Recommendations on nutrient intake are subject to revision as scientific studies provide new data. Because they are formulated in modern, affluent societies, they are often incompatible with diets consumed by humans in other societies and may not accurately reflect evolutionary factors. Certainly one of the goals of studies of the evolution of diet is to help refine our knowledge of human nutritional requirements.

Essential nutrients can be divided into five categories: carbohydrates, fat, protein, vitamins, and minerals. In addition, dietary "fiber" is increasingly recognized as a necessary part of the human diet. The necessity of obtaining certain nutrients would have limited early hominids' adaptation to specific environments and life-styles (Mann 1981). The success or failure of humans in different dietary conditions to meet their nutritional needs can provide insights into the nature of the proper human diet.

Energy is the most essential nutritional need. Digestible carbohydrates that can be broken down to glucose are important energy sources. Fats and proteins provide energy contributions, although these substances also have other dietary importance. Sources of basic energy may be insufficient for gatherer/hunters at certain times of the year, and seasonal stress may be a normal factor in the human diet (Eaton and Konner 1985; Hayden 1981b; Speth 1987).

In addition to providing a substantial proportion of our energy requirements, certain fatty acids that are involved in membrane structure and the synthesis of prostaglandins and related compounds are essential components of human nutrition. The primary essential fatty acids for humans are linoleic acid and its derivative, α-linolenic acid (Bjerve et al. 1987). These compounds are commonly found in the leaves and seeds of plants and in insects. Derivatives of the essential fatty acids can be obtained from meat, fish, and selected plant foods.

The composition of fat in the diet may be an important factor affecting human health as it relates to the dietary source of the fat. Dietary factors that affect lipoprotein metabolism, such as the amount and type of fats and dietary cholesterol, have been extensively studied in relation to cardiovascular disease. Vegetables are lower in cholesterol and have a higher proportion of polyunsaturated to saturated fatty acid than meats. Eicosapentaenoic acid, a long-chain (C_{20}) polyunsaturated ω_3 fatty acid is being investigated for its apparent antiatherosclerotic properties (Eaton and Konner 1985). While this fatty acid is much more common in the flesh of wild animals and fish than in domesticated animals, the C_{18} precursor of ω_3 fatty acids, α-linolenic acid, is found in the leaves of many plants and in some seeds. Although most of our requirements for fatty acids can be met from both animal and plant foods, it is notable that humans have the capacity to meet needs for essential fatty acids from herbaceous material.

Although adequate amounts of protein can be obtained from both animal and plant foods, animal meat, eggs, and dairy products are of high quality in terms of producing a balance of the eight essential amino acids required by humans. Because many plant foods lack certain amino acids, particularly tryptophan, methionine, lysine, and/or threonine, it is necessary to consume plants containing all of these compounds within fairly short time intervals if animal foods are not part of the diet. A varied diet is mandatory, then, for humans subsisting on plant products.

Essential minerals are obtained from a variety of both animal and plant tissues, as well as from sources of inorganic minerals found in the environment. Sodium is particularly important from an evolutionary perspective. Plant materials are often low in this element, especially in mountains and inland environments where soils are naturally low in sodium. New growth, which is often preferred because of its greater digestibility, tends to be low in sodium. Humans often consume sodium more than an order of magnitude in excess of nutrient need (Beauchamp 1987), and the human taste for salt has been suggested to reflect the deficiency of this element in savanna environments where early hominids evolved. Carnivorous animals easily obtain sufficient sodium from prey (Diether 1977), and the human taste for this compound may reflect the herbivorous diet at formative stages of our evolution (Denton 1982). However, neither rats (Beauchamp and Bertino 1985) nor baboons (Barnwell et al. 1986) in controlled experiments show any dietary preference for this compound. If salt preference is a response to the lack of salt in the diets of our savanna-living ancestors, we might expect baboons to show a similar pattern.

Sodium chloride may interact with other chemicals in the diet to potentiate their effects and perhaps to provide subtle enhancements of flavors. This enhancement might interact with our intellectual capabilities, allowing us to make finer distinctions on the properties of individual foods and their dietary qualities. This idea is certainly conjectural, but in view of the perplexity caused by the human preference for excess salt, it is as worthy of consideration as the other tentative explanations of the phenomenon.

Our vitamin requirements seem to dictate that we consume both animal and plant foods. Although vitamin B_{12} is not required in large amounts, it is not found in plant foods (except fermented ones), and early hominids must have consumed some animal products to obtain it (cf. Hamilton and Busse 1978). On the other hand, vitamin C is most available in fruits and vegetables, although it is found in small amounts in meat. This nutrient is destroyed with heat and alkalinity. Thus Eskimos' consumption of raw meat may help them to avoid scurvy, one disease resulting from vitamin C deficiency (Mann 1981). Humans, several primate species, and guinea pigs are unique among mammals in their need for dietary vitamin C. Other animals synthesize this vitamin, and it has been hypothesized that the enzyme necessary for its biosynthesis was lost during primate evolution because ample supplies of vitamin C were present in the diet (Pauling 1970).

Fiber is a constituent that differentiates plant from animal foods. It is derived from the cell walls of plants, and while it is not essential as a nutrient, dietary fiber is important for the normal function of the gastrointestinal tract (Eastwood 1984). As stool bulk is increased, transit time is reduced, absorption of other substances can be affected, and the bacterial flora of the intestine may be modified.

Although there is evidence that dietary fiber has a role in protecting humans against colon cancer (Creasey 1985) and the effects of coronary heart disease (Eastwood 1984), other factors, particularly the quality of fats in the diet, may be of equal or greater importance in the etiology of these diseases. Diets high in plant fiber are characteristically low in fat, and thus these factors are difficult to separate. Nonetheless, specific fibers such as oat fiber have been shown to lower cholesterol in animals and humans (Shinnick et al. 1988).

In summary, we can restate something that has been known for a long time: the more varied the diet, the less likely humans are to develop nutritional deficiencies. However, there are many ways to obtain an adequate diet, and particular diets that are rather monotonous can be balanced nutritionally. Such diets usually reflect cultural evolution. Presumably, early humans depended on a range of both plant and animal products. While we depend on animal products for vitamin B_{12}, vitamin C is best supplied by fruits and vegetables. The essential fatty acids may be best obtained from plants, probably from leaves. Dietary diseases of modern humans suggest that we are not well suited for the high-fat–high-protein diet of carnivores. Our need for plant fiber indicates that plants contribute to a healthy human diet; our intestines contain bacteria capable of digesting hemicelluloses and hydrated celluloses typical of vegetables and fruits. Our nutritional needs are consistent with a dietary pattern broadly similar to chimpanzees; that is, a diet high in plant products, particularly fruit and leaves, with regular supplements of moderate amounts of animal protein.

Human Physiology and Behavior in Relation to Dietary Toxins. There are two cardinal rules of survival for a hungry person in a strange environment (Garcia and Hankins 1975). The first: eat nothing that tastes unpleasant, particularly if it is bitter or if it smells disgusting. The second: eat only a small amount of any novel food, then wait to see if it is agreeable to your stomach. The interactions of physiology and behavior are of particular importance in our dealings with plant chemicals. Natural selection has provided us with chemical senses that are biologically relevant, and human behavior differs in response to these differences in perception. Neophobic behavior allows us to avoid toxic compounds, to learn from nonlethal experiences, and to make physiological adjustments that allow us to deal with toxins. Studies of human physiology and psychology provide important evidence for understanding our interactions with the chemical world. Most of the relevant information from these fields of investigation is presented elsewhere in this volume, but it is worthwhile to consider it all together in an evolutionary context.

Sensory perception. Taste, odor, and food preferences are expected to relate to the evolutionary diet of humans in ways analogous to the dietary adaptations of other primates. Our distaste for bitter substances as their concentration increases above threshold level is consistent with the necessity, and our capacity, to avoid bitter toxic compounds in plant foods. In studies of the kind of suprathreshold taste preferences described in detail in chapter 5, humans show little preference for sour and bitter substances, ambiguous preference for salt (which seems to relate to our dietary requirement for some sodium), and a strong preference for intensely sweet substances. Our strongest preferences are for energy-rich simple sugars.

Honey, which comprises 40 percent fructose and 33 percent glucose (plus 20 percent water), is essentially pure simple sugar. It thus maximizes our strongest preference and minimizes our lowest. It is the most intensely sweet substance widely available to humankind, and it is actively sought by traditional people around the world (Crane 1980).

Chagnon (1968) recounts in an amusing anecdote how he learned of the esteem for honey of the Yanomamö Indians of Venezuela. Puzzled by the low productivity of some men he had hired to cut logs for a house, he decided to supervise them more closely. To his dismay he discovered that his employees spent most of their time searching through the forest until a tree containing a bee hive was found; then they laboriously chopped it down, carved it up, and consumed their reward. Thereafter, Chagnon decided to pay his employees by the log rather than by the day.

Although honey may be a long-standing human food, the taste for simple sugar comes from the importance of fruit in our evolutionary history. Our appreciation for ripe as opposed to unripe fruit corresponds to our preference for sweet over sour compounds. Vegetables, in contrast to fruit, lack the same taste cues to stimulate feeding and may have some of the unpalatable flavors associated with bitterness and sourness. Bitterness is associated with toxic constituents of plant material encountered in an evolutionary context.

While humans' innate preference for the sweet taste of simple sugars appears adaptive as a predictor of the energy value of potential foods, signals for the high-energy content of foods such as fat and complex carbohydrates are not registered by sensory receptors. Fat appears to be perceived primarily by texture. While fats alone are unpleasant (Schiffman and Dackis 1975), they are vehicles for other tastes. As dietary constituents fats are highly palatable, and combined with sugar they can be highly desirable (Drewnowski and Greenwood 1983). Starch is bland, and its ingestion is probably controlled in relation to the relatively rapid sense of satiety evoked by starchy foods. That fat, starch, and protein intake are not regulated by sensory perception suggests that they were not of the same significance as foods with simple sugars, organic acids, or toxic compounds in the evolution of the human diet.

As discussed above, the general sophistication of human taste may relate primarily to our omnivorous nature and the importance of recognition and learned responses in determining human diet. While stimuli such as sweet, salt, and bitter have great biological relevance, our hedonic responses to even these tastes are affected by experience and learning. For example, initially unpalatable substances such as the alkaloid capsaicin from chili peppers may become preferred with exposure (Rozin 1982). Our behavior toward food items is complicated by the diverse stimuli they offer and by our experiences with them. Our learning abilities are intimately connected with the contribution of culture in determining human diet.

The role of odor in human food preferences may also have a direct evolutionary basis. It is significant that a disgusting odor can be a warning against food poisoning. The smell of rotten objects—whether food or dead bodies—is universally aversive, while fruits and grilled meat or fish are generally preferred (Schleidt et al. 1988). However, preferences and nonpreferences associated with odor may reflect a large learning component.

Detoxication enzymes. As we saw in chapter 2, humans have an array of detoxication enzymes which enable us to respond to a large number of naturally occurring chemicals. The nature of our detoxication systems may have been tuned during periods when our ancestors shared the leaf-eating specialization of most primates. The nonspecificity of these enzymes is consistent with the primate pattern of feeding on a wide range of plant species. In the face of the unpredictability of encountering any of the thousands of plant secondary compounds that exist in the plant kingdom, the system gives us the capability for dealing with small amounts of many compounds rather than large amounts of a single one. Biosynthesis of specific enzymes can be induced for protection as needed. Bacterial enzymes detoxify many compounds, and detoxication systems function in conjunction with gastrointestinal microflora.

Human neophobic behavior is a component of our detoxication system. If only small amounts of unknown foods are sampled, the detoxification system is more likely to be able to cope initially; if a food is nutritionally worthwhile, although mildly toxic, higher levels of detoxifying enzymes can be synthesized (see chapter 2).

Gastrointestinal malaise. If the initial experience with a food is detrimental, vomiting may prevent serious damage from occurring. Like vomiting, geophagy is a response to gastrointestinal malaise caused by the inevitable incidences of poisoning in an unpredictable chemical world. Gastrointestinal malaise is the strongest enforcer of conditioned aversion learning, a phenomenon that is particularly strong in mammals.

Conditioned responses. Several recent studies of human conditioned aversions have confirmed the pervasiveness and strength of this kind of learning (see chapter 2). Midkiff and Bernstein (1985) suggest that the greatest percentage of food aversions among modern North Americans is to protein-rich foods, particularly meat (including fish) products. Most simply this observation could be a reflection of modern diets—plant poisons are relatively rare in most foods in urban society, while poisoning from bacterially contaminated foods is relatively common. Proteins do provide a rich medium for bacterial growth.

Bernstein et al. (1984) reported that rats in experimental studies developed learned aversions to proteins but not to carbohydrates. Thus protein sources may have more salience for aversion learning than other nutrients, and that salience may be relatively independent of palatability. Aversion to protein sources could be based on taste, odor, and/or postingestion properties of these foods. Humans are able to taste many free amino acids and small peptides (Torii 1986). In addition, bacterial decomposition of specific protein substrates could produce characteristic and recognizable flavor profiles to which an aversion could be conditioned.

Human disgust for decomposed protein foods may be innate (Schleidt et al. 1988) and/or learned. Aversions associated with proteins may have an evolutionary significance such as enhancing human avoidance of contaminated meats when they were encountered in scavenging activities. The role of odor in conjunction with taste in conditioning strong aversions (Bermúdez-Rattoni et al. 1986) may have special importance in such situations.

As I have already mentioned, humans' liking for foods that are rich in protein, fats, and complex carbohydrates is not reflected in our sensory perception in the same way as our predilection for foods containing simple sugars. While proteins can have complex flavors and be perceived through different receptors, postingestion consequences and learning mechanisms may be more important in determining protein consumption. This lack of sensory specificity corresponds to the salience of protein-rich foods for conditioned learning. The inclusion of concentrated protein sources in primate diets is coincident with the trend toward behavioral flexibility associated with increasing omnivory. Protein perception uses taste receptors in a less specific way than taste perception for simple sugars and toxic allelochemicals. It is perhaps a secondary development related more to our mental development and capacity for learning.

The value of foods containing starch, which itself is bland, can be learned readily. The more rapid satiation that occurs as starch is metabolized to glucose makes it easier to condition ingestion of this substance compared with protein or fat (Booth 1982). Fat is digested more slowly that either starch or protein. Our poor postingestion mechanisms for regulating fat intake have serious dietary consequences for persons in affluent societies.

Flavor variety in the diet. While humans may achieve satiety in relation to a particular food, their appetites are rekindled by offerings of different foods; and when a variety of foods are offered at a meal, consumption increases (see chapter 1). This behavioral response to variety is suggestive of the foraging behavior of chimpanzees and other primates who eat a number of potentially toxic leaves and other plant foods but by selecting variety appear to minimize the amount of any one kind of poison ingested. Human behavior here appears in coordination with our broad detoxification system. As well, a preference for variety in the diet ensures the consumption of a range of essential amino acids, vitamins, and other nutrients.

Allelochemicals in small amounts may serve to stimulate feeding, and condiments and spices that would be toxic in large amounts can improve the palatability of foods. We may actually have an intrinsic liking for small amounts of certain compounds such as tannins (see chapter 2). I have suggested that the sophisticated palate of humans is likely essential for making the kinds of feeding decisions consistent with our omnivorous behavior. While our taste receptors tell us initially to reject most secondary compounds, our capacity to learn from the consequences of ingestion, our behavioral flexibility, and our sophistication in recognizing specific compounds and foods allow us to obtain nutrients from a variety of potential food sources.

Our predilection for variety in the diet appears in contradiction to our neophobia and the apparent importance of conditioned aversions in the diet. Rozin (1982) discusses the basic dilemma of omnivores: "the opposing tendencies to explore new sources of foods and to fear new foods as possible poisons." Thus humans both prefer familiar foods and at the same time desire variety. Variety is likely to be in terms of familiar tastes, and new substances will still be consumed cautiously and with relative infrequency.

In summary, the comparative evidence from humans and primates indicates that during our long evolution as plant-eating primates we acquired capabilities of dealing effectively with most secondary compounds. If plant chemistry constrained food choice by our ancestors, it only meant that they, like other primates, had to forage selectively. When presented with a variable flora, humans and primates have the physiological, perceptual, behavioral, and learning capacities to obtain a positive nutritional balance and deal with moderate levels of most toxins while avoiding those that are highly deleterious. Like omnivorous relatives such as baboons, we consume a diet with a portion of leaves, fruit, tubers, seeds, and animal products, but, most important, we have the flexibility to change our diet in response to availability.

The variable quality and quantity of protein and nonnutrient substances in plants often make animal foods preferable nutrient sources. However, in spite of any toxins they contain, vegetative parts of plants are necessary. As well, it is possible that at certain points in time bacterial toxins and pathogens associated with meat (see chapter 2) may have provided a far greater restraint to human food procurement than plant toxins. The problems posed by plant allelochemicals, while real, are not insurmountable. The issue is not use or nonuse of plants but rather using those species of plants that are energetically most efficient. Instead of simply avoiding biologically active compounds from plants, we can and do consume them with impunity, and this has important implications in relation to the importance of these compounds in the evolution of medicine.

PLANTS AND PLANT CHEMICALS IN THE HUMAN DIET THROUGH HISTORY

Humans seek sustenance from a large portion of the carbohydrate, protein, and fat sources that the world has to offer. Although modern humans consume more meat than other primates, higher primates are all herbivores first and omnivores second. If our prehominid ancestors were physiologically and behaviorally like chimpanzees, their dietary needs were provided by high-energy sources of sugars such as fruits, with protein coming from leaves and animal products. Both insects and small warm-blooded animals were likely sources of animal protein. Most of their vitamin C was obtained from plants, with their vitamin B_{12} supplied by regular additions of animal products.

Our taste and feeding preferences motivate us to select nutritionally dense foods, and our ancestors probably always preferred items such as meat, fruit, and tubers over leaves and other plant products. However, vegetative parts of plants are more predictable in the time and place of their occurrence, and our ancestors likely relied on them to meet a portion of their regular nutritional needs and to surpass periods of seasonal and unpredicted shortage. Many different leaf species provide similar (although relatively minimal) rewards in nonstructural carbohydrates, and they may have similar quantities but different types of toxins. Thus our desire for variety in the diet corresponds with the dependence of our ancestors on leaves. Tubers, toxic seeds such as acorns, and cycads provide abundant energy sources in the form of nonstructural carbohydrates and are a worthwhile reward when they can be exploited in quantity. Physiological methods of detoxication and behavioral mechanisms for avoiding toxicity have made it possible to neutralize the toxicity of many of these plant products. Crises in food supply undoubtedly forced our ancestors to periodically eat high amounts of marginal foods. Plant toxins would still be a threat to our survival, and the biological defensive mechanisms against toxicity would be challenged under certain circumstances.

The ability of our ancestors to exploit dispersed resources in forest and savanna environments depended on the mental development associated with foraging efficiently over a large territory (Milton 1981). The trend toward greater mental development that started with vegetarian primates may have increased as hominids turned to the higher-quality protein offered by meat sources. A shift in behavior and intellectual capacity coincident with scavenging and hunting may represent an important secondary manifestation of the evolution of primates, with consequences for the evolution of human social structure, communication and language, and material culture.

Cultural developments such as tool use, which preceded the appearance of the genus *Homo* some two million years ago, may themselves have had a profound impact on the biological evolution of our genus in what has been characterized as a feedback loop (Tobias 1981). Increased visual and motor skills, perhaps brought about by reliance on tools, are reflected in changes in the morphology of hominid eyes and hands and perhaps are also linked with the increased cranial capacity that separates our genus from the australopithecines (Tobias 1981). The development of language is intimately associated with our larger brain, and whether this is a cause or an effect relationship, the greater capacity to store, retrieve, and communicate information has been key to the success of hominids. The hominid line appears to have reached its current intellectual capacity about 100,000 years ago with the emergence of *Homo sapiens*, and our subsequent evolution has primarily involved elaboration of our material and intellectual culture. The acquisition of knowledge about plants for foods, medicines, and poisons may have become more sophisticated as part of this process.

While foraging adaptations for procuring high-quality foods may have brought about increased intelligence, it is also true that as brain size increased, the quality of the diet would have to be higher to support it (Milton 1988). Brain tissue is energetically expensive to maintain. Concurrent with the evolution of increased cranial capacity in the genus *Homo*, our species changed morphologically and physiologically in ways consistent with a dependence on high-quality foods (Milton 1987). Our dental morphology became less robust, and our small intestine became larger.

The survey of detoxification discussed in chapter 3 (Johns and Kubo 1988) is consistent with a view that energetically dense foods motivated human technological efforts in relation to the use of plant foods. Our increased intelligence would have made this possible, while our increased brain size may have made it necessary.

While ancient *Homo sapiens* certainly exploited large mammals, evidence supporting the importance of meat earlier than 100,000 years ago is scanty (Isaac and Crader 1981; Gordon 1987). Although early hominids used animal products, we have no evidence that they were dependent on meat as a major part of their diet. While we generally assume that our ancestors would have sought to maximize protein intake through consuming animal flesh, this may be too simplistic a point of view. Situations may occur where animal protein is not a desired food. Speth (1987) presents a metabolic argument supported by ethnographic data that under periods of seasonal stress humans would seek animal fat and plant products in preference to meat.

The earliest hominids were probably not bold hunters. In fact, a sizable body of opinion suggests that they were scavengers (Binford 1984) who derived a large portion of dietary protein and fat from carrion and bone marrow. It is difficult to determine what happened to the overall consumption of plants when meat procurement started to become more sophisticated. It is possible that plants were sufficiently easy to obtain that no improvements were necessary to retain a high proportion of them in the diet. If plants were a positive part of the diet, either regularly or periodically, there is no reason they would ever have been abandoned in a wholesale fashion. Early hominids likely continued to focus on the same plant materials as our primate relatives, with a greater reliance on tubers. The digging stick has been suggested as a fundamental part of the hominid tool kit that would have made underground tubers such as *Dioscorea* yams available (Coursey 1973). Among our primate relatives, baboons utilize shallow tubers that can be obtained with little effort. These, however, are generally more protected by chemical and physical (thorns) defenses than tubers that avoid predation by burying themselves deeper. A digging stick that would enable hominids to obtain these latter tubers would be a major advance in food procurement, although it would not totally overcome the toxicity problem. Because wooden implements, unlike stone tools, are not preserved in archaeological sites, we can only speculate about the early existence of digging sticks.

Most of the plant-processing techniques discussed in chapter 3 are used with tubers and seeds—plant parts that provide nutrient rewards in the form of dense carbohydrate. In contemporary settings among peoples living a traditional life-style such as in Australia

(Kraybill 1977), pounding tools are also employed in conjunction with the roasting and leaching of toxic plants such as yams and cycad nuts, and similar detoxification processes could date to the earliest uses of pounding stones. Leaching may have considerable antiquity as a detoxification technique (Harris 1977). Evidence for the controlled use of fire dates back to between 0.5 and 1 million years ago (Clark and Harris 1985; Isaac 1984), and it is likely that cooking would have served a detoxification function by this date. Leaves were not likely to have been effectively prepared by roasting, although their nutrient contribution could have been enhanced through grinding. It is likely that as hominids' use of meat and starchy plant products increased, the dietary importance of leaves declined. The exploitation of tubers through processing would have been at some nutritional cost, and leaves probably retained importance as sources of vitamins and minerals.

Contrary to current thinking, cooking may have played a minimal role in overcoming the constraints posed by plant chemical defenses. Cooking—i.e., roasting—would have a variable effect on reducing the toxicity of plant foods. For example, parsnip roots (*Pastinaca sativa*) contain toxic psoralens that, like potato glycoalkaloids, are not destroyed by cooking (Ivie et al. 1981), and while some chemicals would be appreciably detoxified by roasting, many plant foods would have been made only marginally less toxic. However, by making them more digestible, cooking could favorably alter the nutrient/ toxin ratio.

The primary contribution of cooking may have been in controlling the microbiological toxins associated with the increased use of meat. Cooking reduced the risk of infection from organisms such as *Salmonella,* and of microbial toxins from *Clostridium botulinum* and other organisms (see chapter 2). Although most food poisons such as salmonella are self-limiting, they do cause severe gastrointestinal and neurological disturbances. Botulism toxins can be found in animal carcasses and would be a threat to scavengers; heating the meat would greatly reduce the risk of poisoning. While it is apparent that animal consumption was important two million years ago, cooking did not become important until a million years later (Gordon 1987). Was cooking unnecessary for meat exploitation, or did the consumption of meat only gradually become more important?

The increased supply of protein from meat sources improved the nutrient/toxin relationship and minimized the need for detoxifying plants before consumption. While reliance on alternative foods could have made plants relatively less important in the diet, improved protein consumption could, in fact, have enabled greater quantities of plants to be consumed because well-nourished individuals with fully functioning detoxication enzymes would be better able to tolerate plant allelochemicals. With improved nutrition from animal foods, plants that might have advantageous prophylactic and therapeutic properties could be consumed with greater impunity. This possibility has important implications in relation to the control of infectious disease and to the origins of pharmacological medicine, two issues that I discuss in the following chapter.

Although *Homo sapiens* increased their reliance on nutritionally dense foods, the ratios of various contributions to the diet are not presently determinable. Trace element analyses show that there was likely a shift to greater consumption of plant foods around ten thousand years ago by people living in the Middle East. The increased dependence on seed foods, either preceding or coincident with the origin of agriculture, is the likely cause of this difference. Earlier reduction in the physical robustness of the human species in the period preceding agriculture, around fifteen to thirty thousand years ago, among populations in the Middle East were not paralleled by a dietary shift. Schoeniger (1982) suggests that if robustness reflects physical activity, then a reduction in stature could have been associated with improvements in the means of procuring and processing foods. Plant foods may have been made more digestible through improvements in grinding and other processing technology. Many detoxification techniques may have developed during this time (cf. Hayden 1981a). While detoxification techniques involving water, such as boiling and leaching, are clearly important in the techniques used around the world today, their origins, like other detoxification techniques, are difficult to place in time.

The domestication of animals and plants may have improved the availability of both protein and carbohydrates to early agriculturalists. Increased food production is generally thought to be one of the important reasons for and results of agriculture. However, evidence is mounting that early agriculturalists often suffered considerable nutritional deficiencies (Cohen and Armelagos 1984). Skeletal remains have shown that nutritional diseases such as osteoporosis increased with the shift from gathering/hunting to agriculture. Some researchers have suggested that protein became more limited at this time, although meat supplied by domesticated animals as well as hunting would likely have been plentiful under most circumstances. Major nutritional problems may have arisen for several reasons. The lack of variety in the diet, particularly the lack of variety in noncereal plant products, could directly cause malnutrition. Social stratification and uneven distribution of food resources, particularly meat, may have put lower-status individuals at risk (Gordon 1987). Increased loads of parasites and diseases related to changes in life-style, increased association with animals, and higher population density (see below) would have a cost in nutritional status. Finally, the inherent ecological instability of agricultural systems should be considered. While agriculture would even out some of the unpredictabilities of the natural availability of food, ecosystems containing low species variability, such as planted fields, are more likely to be affected in catastrophic ways by environmental forces such as drought and disease. Plagues of ergotism became a threat once large-scale grain production became the basis of subsistence (Davis and Diener 1978), and aflatoxin poisoning may have been associated with storage of quantities of seed crops from early times. Crop failures may have pushed people to rely on marginal foods, many of which have potentially toxic levels of secondary compounds.

Many agriculturalists overcame deficiencies in their grain-based diets through various cultural mechanisms for balancing nutrients. These probably developed empirically over considerable periods of time. Probably the best known of these is the exploitation of maize, beans, and squash by people in Mesoamerica. This combination of foods provides a balance of amino acids. Similarly, rice, pulses, and milk products provide a balanced protein source for subsistence farmers in India. Processing techniques may make certain nutrients available. For example, the use of alkali with maize is an effective way of making niacin in the maize available and improving the amino acid content of the digestible protein fraction (Carpenter 1981). In societies that traditionally treat maize with alkali there is little or no pellagra—a dietary deficiency disease that is common in other areas where maize is also a major staple.

As well, agriculturalists continue to rely on wild plant foods. I have recorded more than sixty-five wild food species exploited by the Luo of western Kenya (Johns and Kokwaro 1990). In general agricultural peoples recognize and rely on a greater variety of wild plants than their gatherer/hunter neighbors (Hayden 1981b). Wild foods are essential during famine times, and although agriculturalists are very likely to encounter toxic secondary compounds in their quest for food, under most circumstances they appear to be capable of selecting the most appropriate plants.

Movement into new environments has probably always posed a problem for people depending on wild foods. Peters and O'Brien (1981) argue that the African flora is relatively uniform in the geographical distribution of the genera that are eaten by humans and higher primates today, and suggest, perhaps naïvely, that this uniformity should have facilitated the movement of hominids into new habitats during periods of territorial expansion. However, species within a plant genus can vary greatly in their chemical properties and degree of toxicity (e.g., the variation in potato glycoalkaloids discussed in chapter 4). The Lusitu tragedy illustrates that the risks of toxicity in an unfamiliar environment have not been eliminated. In 1958–60 in what is now Zambia, following the forced movement of people of the Tonga tribe from the site of a new dam to an unfamiliar area and habitat, fifty-six people died in a mysterious outbreak (Gadd et al. 1962). All of those affected were women and children. Although the cause of the deaths was never proven unequivocally, the conclusion of the medical researchers examining the case was that the deaths were due to poisoning. No evidence existed for deliberate poisoning, and the ultimate conclusion was that the deaths were probably caused by the ingestion of minor food plants that are commonly used as sauces, condiments, and snacks. Some significant details are provided by the following description of the events surrounding the deaths:

> It should be noted that the outbreak occurred at the height of the hot dry season and ceased when the rains were established. During this time temperatures of 110–120°F [43–49°C] in the shade are common and last for weeks on end, the wild leafy vegetables and fruits used by the people for relish are dry and withered, and the women and children must forage far and wide in the bush for what they can find for the pot—roots, herbs, nuts and leaves. This foraging would be done in family groups, and during such expeditions it is possible that "snacks" of edible raw vegetable matter might be taken—the only time when the women and children would be likely to eat anything not shared by the men. Deliberate poisoning of the "women's pot" having been excluded, this appeared to be the only theory which fitted the facts of the outbreak, and thereafter investigations were directed towards the finding of a vegetable poison which, in their ignorance of the flora of their new locality, these poor people might be eating during forages in order to assuage hunger or slake thirst. (Gadd et al. 1962)

Knowledge of plants and plant properties would appear to be a valuable commodity for both gatherer/hunters and agriculturalists.

DIET, HUMAN CULTURE, AND THE CHEMICAL-ECOLOGICAL BALANCE IN THE LATE TWENTIETH CENTURY

The industrial revolution had profound effects on the human diet. The new economic systems that coincided with changes in technology produced states of deficiency and increased health stress and disease for a new urban underclass, as well as for subjugated peoples in areas colonized by industrial powers. Malnutrition that persists in developing as well as developed countries has antecedents in the social and economic changes associated with the transition to the modern era. Nevertheless, it is a great paradox that the more unique dietary problems arising from the industrial revolution are problems of affluence rather than scarcity. This situation has become most pronounced in the twentieth century and involves several issues that relate back to our evolutionary relations with plants and plant allelochemicals. Each issue reflects the interconnection of biological and cultural adaptations characteristic of human existence.

PLANT ALLELOCHEMICALS VERSUS PESTICIDES

The decrease in our consumption of secondary chemicals in plant foods has had coincident ecological manifestations. The reduction in levels of allelochemicals in crop plants has been carried to extremes through the skills of geneticists and plant breeders. As well, global agriculture, particularly large-scale mechanized production, has concentrated increasingly on a few crops with the widest acceptability and the greatest yield. These tend to be plants with relatively low levels of allelochemicals. However, a problem of major ecological significance is created when plants are deprived of their natural defenses, particularly when these plants are grown in the large monocultures favored by industrialized agriculture.

The replacement of natural chemical defenses with chemical pesticides is increasingly seen as inadequate and hazardous. Pesticides are costly, and numerous insect pests have developed resistance to them. From a chemical-ecological perspective, the greater irony is that our efforts to reduce toxic chemicals in plants have brought about a situation where we are exposing ourselves through consumption of treated plant products (and perhaps more seriously through contamination of the environment) to toxins that have consequences for our health at least as serious as those we have removed.

NUTRITION AND MODERN FOOD SELECTION

Affluence offers humans the possibility of many choices in what they eat and maximized consumption of those things they most desire. Not unpredictably, most people in industrialized societies choose the substances that our evolutionarily determined senses tell us we should want—dense foods, particularly those high in sugar, fat, and protein. Few individuals in industrialized society have trouble meeting their caloric and protein needs. While periods of abundance in traditional settings are offset by periods of seasonal stress (Hayden 1981b), in today's world scarcity can be avoided, and *ad libitum* consumption is the norm. It is not surprising that obesity is one of the most common health problems in industrial societies. Humans share the behavioral preference of other omnivorous primates for animal meat. While humans may still be living with the digestive physiology and behavioral preferences of omnivores, many human populations consume several times more meat than any of our omnivorous primate relatives. Diseases with a partially dietary etiology such as coronary heart disease, adult-onset diabetes mellitus, and some types of cancer may be the result (Hamilton et al. 1978).

Fat is directly relevant to some of our most prevalent contemporary dietary problems. Our sensory apparatus finds fats in food to be favorable. However, because we lack good postingestion mechanisms for limiting feeding, overconsumption of this material, which is in high concentration in animal products, is common.

Food consumption is enhanced not only by nutritional density but also by variety, and the variety of foods available in an affluent society contributes to patterns of overconsumption. In evolutionary terms, variety leads to the consumption of a range of foods, particularly plant foods, and with them a range of essential nutrients. This behavior, however, was defined by a diet higher in fiber and lower in animal fat and protein than that now available. Where choice is maximized, items with lower caloric value that contain considerable amounts of vitamins and minerals can be avoided, and the balance we need is less likely to be achieved.

Rozin (1982) suggests that flavors that typify particular cuisines can have a biological foundation related to the omnivore's dilemma of wanting new foods but fearing at the same time that they may be toxic. Familiar flavors provide a sense of safety and at the same time variety in flavor provides us with a sense of newness. The cultural ways that flavorings are manipulated to meet the goals of safety and newness can be both consistent with and contrary to biological adaptation.

FOOD PROCESSING AND NUTRITION

Food-processing technologies are carried to extremes in industrialized societies, and many foods are unrecognizable visibly and/or nutritionally in comparison to the plants they came from.

Consider, for example, the fate of the potato in the modern world. My interest in the potato is not strictly intellectual, and I have a special fondness for potato chips. A popular advertisement, "I bet you can't eat just one," had special significance for me after I had lunch a few years ago with a researcher from one of the world's largest manufacturers of potato chips. I eagerly anticipated the opportunity to learn the secrets of the perfect chip. The answer was simple. My colleague's research was directed at finding the right combination of salt and fat to maximize the feeding compulsions of myself and others like me.

In such cases manufacturers cater to and exploit human desires in the production of foods whose primary virtue is maximization of sensory cues of sweetness, fats, and salt. While not all manufacturing processes emphasize palatability at the expense of nutrition, those that do are recognized as a problem. When unprocessed foods alone were available to people, variety that included sweet and high-protein foods as only a part of a range of other items facilitated the consumption of vitamins and minerals. However, when taste cues can be varied in foods that remain more or less uniform in nutrient content, these cues are misleading. This is particularly problematic when sensory satiety results in the avoidance of more nutritional foods, particularly fruits and vegetables. Supplements of vitamins, minerals, and now fiber are increasingly put back into processed foods.

In defense of food processing it should be pointed out that qualities such as convenience, storability, and low cost can contribute positively to the well-being of many people. The challenge for industry is to incorporate lessons from nutrition as well as behavior into product formulation.

While the above discussion points out the present-day conflicts between cultural developments and our biological nature, culture historically has enhanced human adaptability rather than diminished it. Through culture humans have developed cuisines that meet criteria of palatability, familiarity, and variability. Thus cultural patterns of diet can be positive and reflect our nutritional needs. Culture offers humans the possibility of surpassing the limitations of our biology. As we gain greater understanding of our biological characteristics, we have the prospect of adapting our dietary patterns to the complexities of life in the modern world in ways comparable to those by which humans have adapted themselves to particular environmental conditions throughout our history.

8 The Dietary Basis for the Origin of Human Medicine

The physician looks with another eye on the medicinal herb
than the grazing ox which swoops it in with the common grass.
—JOSEPH GRANVILLE, *The Vanity of Dogmatizing*, 24, 1661.

Medicine is "the science and art dealing with the maintenance of health and the prevention, alleviation, or cure of disease" (Gove 1968), and as such it is an important component of human culture. Medicine can be more narrowly defined as a substance used in treating disease or affecting well-being. Many medicinal substances are plants or plant products. Both traditional and modern medical systems contain considerable knowledge about the properties of plants and the ways these properties can be used to affect well-being. As the system of knowledge and practices by which humans manipulate their own health, medicine is a very important component of human interactions with plant chemicals.

In a chemical-ecological sense, obtaining adequate nutrition and responding to allelochemicals are two sides of the same coin. As we saw in chapter 7, herbivores must achieve a balance between the positive effects of nutrients and the deleterious cost of overcoming plant chemical defenses. On the other hand, plant allelochemicals may have positive health effects. While nutrients from plants contribute to health, familiarity with allelochemicals with both positive and negative consequences has been an important cultural force enabling humans to select plants adaptively. While medicine is a complex component of culture, the following discussion of analogies for medicine in animal behavior indicates that the pharmacological use of plants is a fundamental characteristic of our species that has a biological basis of chemical-ecological importance.

CHEMICAL-ECOLOGICAL BASIS OF MEDICINE

Asklepios, a son of Apollo, was the Greek god of medicine. In conferring the name *Asclepias* on the genus of plants that includes the common milkweed, Linnaeus followed the nomenclature used by

Dioscorides in his *Materia Medica* (Gunther 1959; Hort 1938). While *Asclepias* spp. are rarely used as human remedies today, an insect, the monarch butterfly (*Danaus plexippus*), might be considered the true advocate of the medicinal value of this plant, and of the god after which it was named. Monarch butterflies, which feed on *Asclepias* spp., incorporate the plant's cardiac glycosides into their own tissues to defend themselves against attack by birds (Brower 1969). Cardiac glycosides provoke a violent emetic response in blue jays and other birds, who then develop a conditioned aversion response to the visually recognizable pattern of this species. Many other insects similarly concentrate plant secondary compounds in their bodies as a defense against predators (Duffey 1980). These creatures are able to overcome the toxic potential of these compounds and have specialized mechanisms and morphological structures for sequestering them.

Chemical ecologists characterize phenomena of this type as three-trophic-level interactions. The flow of energy between different trophic levels is a basic biological principle. Plants, the primary producers, utilize the sun's energy to fix carbon; they are consumed by herbivores, which in turn are preyed upon by carnivorous animals and parasites. Three-trophic-level interactions involving plant secondary compounds are increasingly recognized by chemical ecologists (Price et al. 1980; Barbosa and Saunders 1985) as being important in the relationships of hosts and parasites. It is in these interactions such as sequestration that the evolutionary basis for medicine can be found. While three-trophic-level interactions take various forms, those that involve effects of plant defensive compounds consumed by herbivores on the survival of predators and parasites are most relevant to understanding the origins of medicine.

Like the monarch butterfly, human ancestors may also have been subject to attack by large predators many millennia ago. Although some animals might find us distasteful, however, this would probably not be because of any chemical defense we possess. Like most large mammals, instead of being consistently threatened by predators larger than ourselves, we are attacked by various parasitic microorganisms and invertebrates that consume us from the inside out.

Direct analogies to human use of medicinal plants can be found in systems where plant chemicals present in the diet of herbivores directly control parasites. Again the best examples of three-trophic-level systems of this nature come from entomology, where plant-insect-parasite interactions have been studied because of the possible conflict between chemical and biological methods of controlling insect pests. Various observers have shown that host plant quality affects the survivorship of various insect parasites and thus the success of the insect herbivore (Barbosa and Saunders 1985; Vinson and Iwantsch 1980). Campbell and Duffey (1979) directly demonstrated that α-tomatine in artificial diets fed to *Heliothus zea* negatively affected the survivorship of *Hyposoter exiguae,* a wasp parasitic on *Heliothus.* In such cases the secondary compounds of the plant may affect both herbivore and parasite, and if a herbivorous insect is to use these compounds to advantage, it must ingest a quantity balanced between toxicity to itself and to the parasite. While it has not been demonstrated that insects deliberately select plants for their parasite-controlling properties, eating plants with some toxicity appears to be an adaptive behavior. Although animals other than insects are subject to parasitic attack, the role of plant chemicals in controlling parasites of noninsects is a largely unstudied area. It is difficult to distinguish a pharmacological from a nutritional adaptation in most studies of animal behavior.

Boppré (1984) proposed applying the term *pharmacophagous* to insects "if they search for certain secondary plant substances directly, take them up, and utilize them for specific purpose other than primary metabolism or (merely) foodplant recognition." Pharmacophagy was not defined with the overt intention of explaining the origins of medicine. On the contrary, it takes a human concept—"drugs"—and uses it to explain a phenomenon observed in nature. Nonetheless, this term broadly recognizes the conceptual link between animal use of chemicals for defense and other nonnutritive purposes and the ingestion of medicine.

D. H. Janzen has often led the way in studies of plant-herbivore interactions by offering insights and speculations that stimulate the work of others. In relation to vertebrates' use of plants for "self-medication," he broke the ground for chemical ecologists in a paper (Janzen 1978) that includes a large number of anecdotes suggesting that mammals deliberately ingest plants for their secondary compounds. The animals he discusses include elephants, black-and-white colobus and red colobus monkeys, Indian bison, pigs, civets, jackals, Indian tigers, bears, wild dogs, rhinoceros, Indian mole rats, and Indian desert gerbils. Significantly, Janzen suggests that the purpose for ingestion of all of the unusual plants eaten in these cases is the elimination of intestinal parasites, although he does speculate that animals might eat plants as pain killers, chemoprophylactics for schistosomiasis, or antidotes for poison. Although, as Janzen suggests, these or other uses of plants could have a coevolutionary basis, antiparasiticity appears to be the primary nonnutrient function that has evolved in animal interactions with plant chemicals.

Individuals stressed by high parasite loads are at a competitive disadvantage relative to other members of a group, especially when it comes to obtaining essential nutrients. In turn, malnourished animals are likely to succumb more readily to disease. Recent research in evolutionary biology has pointed out the importance of parasite infection in determining a male animal's reproductive success (Freeland 1981; Hamilton and Zuk 1982); the crucial importance of parasites in evolutionary terms suggests that behaviors that involve plant use for control of infection would be strongly selected for.

Many species of birds that occupy the same nest sites over a long term incorporate fresh vegetation into their nests. The young of these birds are at risk of incurring large parasite loads, and Clark and Mason (1985) have provided data to support their hypothesis that European starlings use chemicals, particularly mono- and sesquiterpenes, in fresh vegetation as fumigants against parasites and pathogens. Starlings are able to discriminate among the different volatiles emitted from plant materials (Clark and Mason 1987).

The ingestion of plants by carnivorous animals is interesting. Dogs and cats often eat grass, apparently to induce vomiting, and cats are well known to induce behavioral changes through the use of catnip, a plant that contains nepetalactone, a cyclopentanoid monoterpene related to the iridoid glycosides (Harney et al. 1978). Other animals such as horses and donkeys may use stimulants in natural settings (Siegel 1986).

Janzen's suggestion that animals might ingest plant allelochemicals to counter the toxic effects of other classes of chemicals has received some support recently. Mice given a choice of foods containing tannins or saponins did not experience the signs of toxicity associated with consuming one of these toxins alone. "Simultaneous consumption of tannins and saponins in the right proportions may promote chemical interaction that inhibits the toxins' absorption from the intestinal tract," and this type of interaction may influence the evolution of the feeding behavior of herbivores (Freeland et al. 1981). Along similar lines, tannins inhibit the glucosidases responsible for cyanogenesis, and herbivores feeding on cyanogenic glyco-side–containing plants may selectively consume diets to minimize the toxicity due to these compounds (Goldstein and Spencer 1985).

PRIMATE STUDIES

Recent reports on the apparent use of plants for medicine by chim-panzees in Gombe and Mahale national parks in Tanzania have sug-gested in a dramatic way that medicine is not a unique human trait. The manner in which chimpanzees use leaves of *Aspilia* spp. (As-teraceae) has many unusual aspects when compared to their normal use of leaves consumed as part of the diet (Wrangham and Nishida 1983; Wrangham and Goodall 1987). Young leaves of various mem-bers of this genus are selected slowly and carefully and are swallowed individually. Sometimes the animals close their lips over a leaf for several seconds before deciding whether or not to detach it from the stem. Leaves are rolled around on the tongue for several seconds before they are swallowed. The leaves are not chewed, and when found in the feces only their surface cells are damaged. This behavior is usually observed early in the morning and is practiced more often by males than females. Chimpanzees show a similar pattern in their utilization of the leaves of *Hibiscus aponeurus* and *Strychnos* spp. (Wrangham and Goodall 1987).

An obvious interpretation of the use of *Aspilia* is that it serves a medicinal function. A crucial question is whether this represents only another animal behavior like those listed above, or whether it is a culturally transmitted practice similar to human medicine. Two questions arise. First, if this is a medicinal practice, what is its function? Second, how do these animals know how to use these particular leaves for this purpose?

Wrangham and his colleagues have focused their work on the first question. They have provided circumstantial evidence that this plant is important for its antiparasitic properties. Infections by gastroin-testinal parasites cause some debility in chimpanzees (Flynn 1973), and the use of *Aspilia* for endoparasite control is sensible. *Aspilia mossambicensis* and *A. pluriseta,* two of the species used by chim-panzees, contain thiorubrine-A and -B, two dithiocyclohexadiene polyines (Rodriquez et al. 1985). These compounds have potent anti-biotic, phototoxic, antifungal (Towers et al. 1985), and antiviral (Hud-son et al. 1986) activities against a number of organisms. The pho-totoxic (enhanced by light) properties of thiorubrines, which are probably important for the topical applications of *Aspilia* spp. (cf. Wat et al. 1980), are clearly unimportant for controlling intestinal parasites. Conversely, the specific use of the plant early in the morning might be explained as the animals' efforts to avoid damage to their lips and other exposed tissues from chemicals whose effects become greater as they are exposed to more light.

Wrangham and Goodall (1987) document the medicinal uses of *Aspilia* by humans: "Leaves are reported to be used primarily for topical or stomach condition: 9 out of 20 reports of leaf use for topical application (wounds, burns, rashes, ringworm, conjunctivitis, etc.), 7 for stomach problems (pain or worms), and 4 for other complaints (cough, fever)." The cases of treatment of stomach troubles stand out in relation to theories about the antiparasitic role of *Aspilia* use in chimpanzees. However, considering that gastrointestinal, dermatological, and respiratory treatments are the most common purposes for which herbal medicines in general are applied (see below) the significance of this ethnobotanical data must be qualified.

Several other studies have considered the positive effects of secondary compounds in primate diets. Leaves that are high in tannins may be eaten by colobine monkeys to counteract bloat and to help detoxify alkaloids by precipitation (Oates 1977). Glander (1980) hypothesized that howler monkeys use chemical cues in food to affect birth spacing so as to produce offspring at the optimal time for maximum survivability. In an attempt to substantiate primate self-medication, Phillips-Conroy and Knopf (1986) studied the possibility that dietary selection of *Balanites aegyptiaea* by baboons in Ethiopia was a behavioral adaptation to control schistosomiasis. This plant, which contains the steroidal saponin diosgenin, is toxic to the snails that are the intermediate host as well as to the cercariae of *Schistosoma mansoni*. However, in feeding studies with mice, animals fed a diet containing diosgenin showed an enhanced rather than decreased level of schistosomiasis compared with controls.

Baboons consume certain plants that have been classed as "euphorics" (Hamilton et al. 1978). While they are consumed in low amounts, they are eaten in a persistent fashion. The plants, which include *Datura innoxia, D. stramonium*, and *Euphorbia avasmontana*, are highly toxic and noted for hallucinogenic properties. Alternatively, they may serve a role in controlling parasites analogous to that hypothesized for *Aspilia* (cf. Rodriquez et al. 1982).

Associations made between plants and their perceivable effects on various body functions are clearly important for understanding how chimpanzees learned to use *Aspilia* or other plants for their medicinal properties. A stimulatory effect of *Aspilia* leaves on the nervous system has been suggested and could be easily understood in terms of the kind of conditioned learning that characterizes animal interactions with plant chemicals (see chapter 2). This kind of learned behavior could then be maintained in a cultural fashion. Euphorics could be among the plants used in this way. If, on the other hand, parasite control is the primary function of this behavior, it is harder to explain how the animals know to do this. Positive effects on well-being would come from reducing parasite loads, although the feedback is less rapid than that associated with direct effects on the nervous system, blood pressure, blood sugar, et cetera. Conscious awareness of parasite loads combined with awareness of the consequences of ingesting a particular plant are not likely to be learned through conditioning-type experiences. It is difficult (although not impossible) to imagine chimpanzees making these kinds of associations.

If parasite control is the purpose of *Aspilia* use, the interactions of chimpanzees with this plant (or similar plants) would seem to be a genetically determined biological adaptation no different from similar three-trophic-level interactions involving insects, birds, or other animals. Chimpanzees clearly have some cultural adaptations that parallel our own. They seem capable of limited symbolic communication and manipulate the environment through the use of rudimentary tools. *Aspilia* use by chimpanzees has not been reported elsewhere in Africa, and this behavior does seem to be part of the cultural knowledge of the chimpanzees of Gombe. This pharmacophagic behavior may not be necessary in other areas of Africa where a greater concentration of allelochemicals may be routinely consumed in the diet. It is likely, then, that the general ingestion of specific leaves for a nondietary purpose is an evolved behavior that is expressed in situations where parasite control is called for; the specific focus on *Aspilia* would be a culturally transmitted response to a biological need.

HUMAN MEDICINE

BIOLOGICAL AND CULTURAL BASIS

Humans are subject to many of the same parasites and infectious diseases that afflict our primate relatives (Cockburn 1971; Flynn

1973), and our hominid ancestors probably were affected by them in much the same way as primates in their natural habitats are today. Details of the evolution and history of human diseases are treated only generally here (see Cockburn 1971; Armelagos and Dewey 1970). While parasites are common in nature, they have relatively little effect on healthy individuals and are rarely fatal; they can be debilitating and can contribute to death in unhealthy, infant, or aged individuals. Control of parasitic and infectious diseases through the use of plants may have played a role in the evolution of human plant selection. The constant input of biologically active secondary compounds in the diet may have served a prophylactic role and been part of the normal digestive ecology of healthy individuals and populations. Plants may have been particularly important in the control of infections such as parasitic helminths, for which little immunity can be acquired (Dunn 1968). Particular plants like *Aspilia* could have been ingested for this purpose.

I argue that such use of plants is the biological prototype for human medicine. Similarly, the ingestion of clay to adsorb plant toxins has a biological basis. However, unless these behaviors have a cultural component, they are, strictly speaking, not medicine. When humans began to rely on learned knowledge and to communicate it from generation to generation, their use of plants evolved from a biological, genetically determined behavior to a cultural practice. Elaboration in the use of plants for medicine is dependent on increased cognitive skills and the resultant development of speech associated with the enlargement and reorganization of the brain seen during the evolution of hominids (Tobias 1981). Language refined the universal capacity for learned behavior into a mechanism for ensuring the survival of future generations. Accumulated knowledge of the toxic and pharmacological properties of plants would gradually increase human adaptation to a particular environment and would be very important in migration and adaptation to new environments. Cultural innovations gave humans a greater capability to control their environment; self-medication using plants can be one aspect of learned behavior transmitted in a cultural fashion.

As humans expanded into new environments they were confronted with new threats to health, particularly new parasites and infectious diseases, new toxic plants, and new physiological stresses. Where parasites and host coevolve, a relationship emerges in which the host's immunity maintains parasites at a sublethal level. Organisms for which there was no biological resistance were encountered in new habitats, and as the density of human populations increased, diseases could spread more readily. Diets high in animal products and concentrated carbohydrate from processed tuber and seed sources may have lacked the natural prophylactic effects of diets high in a range of plant allelochemicals from leaves and nonprocessed foods. Cultural practices in the form of herbal medicine may have played a critical role in rectifying the balance between human hosts and their parasites and the possible role of plant secondary compounds in this balance. In particular, the long evolutionary association of humans with secondary chemicals from leaves would be helpful in time of need.

As we saw in the preceding chapter, an improved nutritional status, particularly in relation to protein, could have increased the ability of humans to withstand toxins, but at the same time it could have fostered the deliberate ingestion of high concentrations of plant compounds to counter the negative effects of a changing life-style. A capability to tolerate high levels of allelochemicals could have led to the consumption of roots and barks, which tend to be more toxic but likewise more pharmacologically potent than leaves.

THE ARCHAEOLOGICAL AND HISTORICAL RECORD OF MEDICINE

Although the uses of plants and clay by animals as apparent forms of self-medication suggest that humans too have always employed these substances in a similar fashion, the rate at which pharmacological practices became more sophisticated is a matter of speculation. It is unlikely that the development of medicine from a biological phenomenon to the elaborate cultural practice we see today occurred as one abrupt change at the end of the Paleolithic period. Like the origin of agriculture, the development of pharmacology was undoubtedly more continuous in nature.

Increased sophistication of medical practices likely arose along the same exponential road as all other aspects of material culture. Evidence of magic can be seen in ritual burials and art that was present as early as Neanderthal times. Medicine and magic are closely interwoven aspects of many cultures, and it is not unlikely that developments in medicine would have accompanied such practices. Empirical medicine with its biological foundation may have preceded such practices.

Surgical techniques are a likely offshoot of hunting and butchering, and Laughlin (1963) suggests that the necessary anatomical knowledge for surgical medicine arose with early hominids' increased dependence on meat. However, as Landy (1977) pointed out in an editorialized reprint of Laughlin's (1961) paper ("Acquisition of Anatomical Knowledge by Ancestral Man"), surgery would have been the treatment of last resort because of pain, surgical shock, and infection. Simple procedures such as realigning broken bones, relocating displaced joints, removing foreign bodies, and delivering babies are less traumatic.

As plants and plant secondary chemicals preceded animal foods as determinants of human evolution, so it is reasonable to assume that herbal medicine—with its direct analogues in animal behavior—was the original form of medicine. However, which of the surgical or pharmacological roads of medicine is of greatest antiquity is really a meaningless issue. Early rudimentary practices would have provided only a predisposition for the elaborate medical practices that arose later in history. The use of plant secondary compounds and hunting technology go hand in hand (for example in the use of arrow poisons) and are complementary components of medical knowledge.

Archaeology offers only a scanty record of medicinal plant use. Plants used for medicine are no more likely to be preserved than those used for food, and because of this lack of material evidence medicine in prehistoric societies has been ignored or presumed not to exist. The first direct record of the nonfood use of plants is from a Neanderthal burial at Shanidar cave in northern Iraq that occurred approximately sixty thousand years ago (Solecki 1975). Clusters of pollen of a number of plant species with possible medicinal value were discovered in a pattern indicating that their flowers were purposefully placed into the grave (Leroi-Gourhan 1975). The Neolithic lake dwellers of central Europe cultivated or gathered over two hundred different species of plants (Sigerist 1951). Among these are several such as *Papaver somniferum* (opium poppy) that have well-known medicinal properties. Seeds of *P. somniferum* could have been used for food rather than medicine. The tombs of ancient Egypt, where plant products were used for mummification itself, offer important insights into the use of plants for their chemical properties. Incenses such as frankincense and myrrh were prepared from gums and resins of particular trees and shrubs (Dimbleby 1967).

The written records resulting from the development of complex civilization provide the first conclusive evidence of cultural transmission of medical knowledge. Although medical knowledge would have gradually accumulated prior to the Neolithic period, it is also likely that this period brought about rapid developments in medicine for two reasons. First, the stresses associated with a shift in diet based on agricultural production and an increasingly urban life-style made the need greater. Second, the development of writing provided a means of codifying medical knowledge and ideas.

MEDICINE AMONG GATHERER/HUNTERS

In the absence of a good archaeological record of medicine, studies of medical patterns among extant groups of humans provide the best basis for understanding this stage in medical evolution. Studies of health and disease in modern gatherer/hunters, particularly those populations removed from contact with outsiders, may reflect disease patterns of humans in the past. Gatherer/hunters have been observed to be well nourished in comparison with neighboring agriculturalists and urban dwellers (Dunn 1968), and when living in a stable environment free from outside disturbances, they are generally in a good state of health (Wadsworth 1984). Although parasitic and infectious diseases contribute to the mortality of gatherer/hunters, the incidence and severity of these maladies are dependent on the diversity and complexity of particular ecosystems. It is therefore impossible to make generalizations about the importance of parasites and infectious diseases in modern gatherer/hunters. Such diseases can be, and possibly were in the past, important selection agents in gatherer/hunter populations (Dunn 1968). While these diseases may have been an evolutionary regulator of population size, in well-adapted gatherer/hunters population size and disease incidence were in balance.

Gatherer/hunters employ considerable sophistication in the rational use of surgical techniques (Ackerkneckt 1953) and herbal treatments. Laughlin (1961) discusses surgical techniques among the Aleuts of the North Pacific. The pragmatic approach of the Aleuts to surgical medicine is paralleled by the practical focus of their herbal pharmacopoeia (Bank 1952).

Similar practical approaches to treating disease are seen in many cultures. In spite of an absence of recognized life-threatening diseases, Australian Aborigines possess an extensive pharmacopoeia that is used to treat ailments such as skin problems, wounds, pain, parasites, diarrhea, headache, fever, colds, and eye problems (Vaughan 1986). Similar pharmacopoeias are found among traditional cultures worldwide. For example, traditional practices of North American Indians are regarded as rational and effective for treating wounds, sores, fractures, dislocations, coughs and colds, fevers, rheumatism, and pain (Arnason et al. 1981).

A large portion of traditional herbal medicines used today are prepared as infusions, although the consumption of the plants themselves may have a much longer history. The use of infusions would have paralleled the use of boiling techniques, which are generally thought to have developed prior to the Neolithic revolution (Hayden 1981a).

MEDICINE IN PREINDUSTRIAL SOCIETIES

While agriculture increased the availability of food for humans, it also had negative repercussions in terms of human health and nutrition. In turn these deficiencies may have made humans more susceptible to disease. Increased population concentration presented the greatest risk to health as it increased densities above the limits imposed by various parasitic and infectious diseases. Although many paleopathological studies show that parasites and infectious diseases were a problem in ancient civilizations, the archaeological record does not show unequivocally a greater incidence of infection among Neolithic farmers than among their gatherer/hunter forebears (Cohen and Armelagos 1984).

The medical practices of agriculturalists living in small villages are not appreciably more complex than those employed by gatherer/hunters. For example, prior to European contact the Maoris of New Zealand used only a few plants as treatments for minor ailments, as purgatives for constipation, as treatment for diarrhea, and as astringents for wounds (Brooker and Cooper 1962; Te Rangi Hiroa 1970). The introduction of the devastating diseases of civilization made it necessary for the Maori to expand their knowledge and use of herbal remedies. Agricultural populations in North America such as the Iroquois did have a fairly sizable pharmacopoeia prior to the arrival of the Europeans, although like most groups with a similar life-style, most of their remedies were for simple ailments. The Iroquois made extensive use of spring tonics and various beverages, many of which may have played an important role in overcoming nutritional deficiencies associated with an agricultural-based diet and severe seasonal stress (Arnason et al. 1981).

Among the Luo, a largely agricultural group in western Kenya, treatments of gastrointestinal, dermatological, and respiratory problems are the most consistently applied (Johns et al. 1990). Similar patterns involving the kinds of diseases described above are still seen in traditional medicine around the world (cf. Younos et al. 1987).

Urbanization that followed the development of agriculture facilitated the spread of parasites and infectious diseases and brought with it health problems associated with sanitation and the contamination of food and water supplies by human fecal matter and the wastes of domesticated animals. Various diseases were associated with animal husbandry and the expansion of agriculturalists to new environments (Armelagos and Dewey 1970). Typhus and malaria were encountered at this stage. The development of remedies was likely stimulated by need, and herbal medicine as we know it today may have arisen parallel to the increase of diseases of civilization. Egyptian herbal medicines placed heavy emphasis on internal diseases, and malaria was added to the list of primary concerns during the Greek era. Historical accounts show that various infectious diseases such as measles, smallpox, cholera, and syphilis have been serious health problems only in more recent times. Syphilis became a major concern in European pharmacopoeias subsequent to its probable introduction from the Americas in the fifteenth century. Isolation, as well as small population densities, protected many human populations from the ravages of infectious diseases until recent times.

While different patterns of disease characterize different environments, gastrointestinal problems are still a major health problem among rural and urban populations in tropical areas. For example, water contamination, poor food preservation, shared utensils, close proximity with livestock, large population density, exposure to human excrement, lack of physical isolation of sick individuals, and high ambient temperature and humidity are factors that contribute to the high incidence and mortality rate of intestinal disease among the Hausa of northern Nigeria (Etkin and Ross 1982). Similarly, gastrointestinal ailments are a major health concern among Maya Indians in Guatemala (Logan 1973). Among the Hausa, the Maya, and the populations of most developing countries herbal remedies are still the primary treatment for gastrointestinal ailments. Conversely, among many groups treatment of gastrointestinal problems is the most important use of plant remedies (Johns et al. 1990; Younos et al. 1987).

It has been estimated that 80 percent of the world's inhabitants today rely on traditional medicine for their primary health care needs (Farnsworth et al. 1985). Most of this knowledge of medicinal practices is still a part of oral traditions (Fig. 8-1). People in traditional cultures around the world have considerable medical knowledge, although in most societies individual experts emerge who focus their attention on medicine. In some cases knowledge of medicines and the practice of medicine may only be communicated during a lengthy apprenticeship. Medical knowledge exists in particular psychosocial and cultural contexts from which it is often difficult to separate. Medicine men and medical societies may be holders of both medical and religious knowledge. Much of the current work of ethnobotanists is to record this part of the human heritage before it is lost. One positive result of the meeting of traditional and modern worlds is the emergence of local institutions in developing countries that are dedicated to continuing traditional medicine within its cultural and social context. As well, scientists in China (cf. Chang et al. 1985), India (Patel 1986), Africa (cf. Sofowora 1982), and Latin America (e.g., Girón et al. 1988) are at the forefront of efforts to evaluate the properties of traditional medicines.

MEDICINE IN LITERATE SOCIETIES

Written records of oral medical traditions did not begin with modern ethnobotanical studies. In fact, writing has had a significant impact on the development of medicine for at least four thousand years (Mann 1984). Written records have helped to codify many medical traditions and probably have facilitated greater elaboration and sophistication in understanding disease and the effects of various remedies. The written medical traditions of the Chinese (Unschild 1977), Indians, and Greco-Persians (Mann 1984) have been the basis for the world's most renowned medical systems.

Greek medicine, which is at the foundation of modern Western medicine and science, was strongly influenced by both Africa and Babylon. The herbal remedies of ancient Egypt are described in exceptional detail by the Ebers Papyrus dated about 1550 B.C. The medicine of Babylon, while dating to before 2000 B.C., does not seem to have placed importance on drugs, although the Assyrians who ruled in Mesopotamia in the first half of the first millennium B.C. greatly consolidated ancient knowledge about herbal remedies. Many of the plants in ancient herbals remain part of recent pharmacopoeias (Mann 1984). Hippocrates, the "Father of Western Medicine," wrote about some 236 plant drugs, while in the fourth century B.C. the *De*

FIG. 8-1. Traditional healers make an important contribution to health care in many parts of the world. This herbalist from Siaya District, Kenya, grinds up a concoction of herbs to be used to treat gastrointestinal infection.

Historia Plantarum of Theophrastus described the medicinal properties of 455 plants. Dioscorides' *De Materia Medica,* written in the first century A.D., described some 600 plants and plant products and was the basis of the Western materia medica until modern times.

The written knowledge of all of these traditions is the basis for sophisticated contemplation about the world. While aspects of a scientific view of the world are found in oral systems of knowledge, scholarship through writing provides a vehicle for the development of scientific thought. Medicine is one of the greatest foci of inquiry, and the origins of scientific medicine are interconnected with the origins of science itself. The transformation from medicine based on supernaturalism and magic to scientific medicine is first evident in the West in the work of Hippocrates. Medicine based on the examination of natural phenomena through critical and systematic application of rational principles was tied to the application of Greek intellectual thought to the biological world by Aristotle. Theophrastus, a student of Aristotle, applied scientific principles to plants and laid the foundation for medical botany.

While the traditions of Greece were fostered by Byzantine and

Islamic scholars, the Renaissance in Europe dramatically revitalized efforts to understand the world based on scientific principles. The invention of the printing press around 1456 greatly facilitated the interchange of ideas and facts of scientific discovery. The appearance of printed herbals late in the fifteenth century triggered the emergence of a new botanical science in Germany in the sixteenth century. Botany and medicine were essentially interconnected into the eighteenth century, and they continue to be linked today. European exploration of the globe and contact with other peoples led to the introduction of many new drugs such as cinchona, curare, and ipecacuanha, and stimulated the development of the pharmacology of plant substances during the eighteenth century.

The connection between botany and medicine which is central to the origins of the science of pharmacology was altered profoundly by the emergence of chemistry. Paracelsus (1493–1541) was a key figure in the renaissance of medicine and also a pioneer chemist. With his emphasis on the medicinal use of substances such as antimony, arsenic, copper sulphate, iron, mercury, and sulfur, as well as tincture of opium, he laid the foundation for chemical pharmacology. While chemical investigations made major breakthroughs after the time of Robert Boyle (1627–91), it was not until the first decade of the nineteenth century that the chemical nature of the active principles of plant drugs was established. The first alkaloid, prepared from opium, was reported by Derosne in 1802, but it was Sertürner who first isolated an alkaloid in its pure state and also described its physiological properties in relation to a pharmacological problem. Sertürner published his paper on this first chemical drug, morphine, in 1806. Before mid-century the alkaloids quinine, aconitine, atropine, berberine, codeine, colchicin, curarine, hyoscyamine, piperine, quinidine, thebaine, and veratrine (actually a mixture of several alkaloids), the phenolic salicin, and an active principle (containing a mixture of cardiac glycosides) from *Digitalis purpurea* had been isolated. The field of organic chemistry, including modern pharmacy and pharmacology, has had a profound impact not only on our scientific understanding of the world but also on the ways in which we manipulate it for our own adaptive purposes.

The impact of science on the lives of large numbers of people began with the industrial revolution of the eighteenth century. The effects of scientific thought and the technologies derived from scientific breakthroughs are something with which the world is still struggling to come to terms. In human affairs, the cultural revolution of science is perhaps comparable to the biological events that resulted in the large-brained organism called *Homo sapiens*. Along similar lines, herbal drugs both before and after Theophrastus have been central to the development of scientific medicine and science itself. Clearly our interactions with plants and plant chemicals are a crucial component of this most recent revolution in human intellectual development.

TRADITIONAL KNOWLEDGE OF PLANT CHEMISTRY AND PHARMACOLOGY IN RELATION TO DISEASE

Considering that the pure active principals of plant medicine were first isolated less than two hundred years ago, and to a large extent remain unidentified today, it is not surprising that knowledge of plant chemistry per se in traditional cultures is limited. Nonetheless, humans (like animals) interact extensively with phytochemicals; this interaction is based on the knowledge from a cumulative experience with plants in general. I will look at this knowledge from two perspectives. First, I will consider various societies' awareness of the plant chemical world and their understanding of the origins of medicine and of drugs. Second, I will take a more functional approach in attempting to understand the process of acquisition of empirical plant knowledge.

Knowledge of Poison. The vast knowledge about plant poisons possessed by humans around the world relates directly to the origins of medicine. Medicines and poisons are two sides of the same coin. They are interchangeable under different circumstances, and both are part of the same store of knowledge about plant secondary principals.

The use of plants for arrow poisons is widespread around the world, and particularly noteworthy in Africa, South America, and Asia. The neuromuscular blocking agent tubocurarine that is widely used in modern anesthesiology was derived from arrow poisons employed in the Amazon region of South America (Gilman et al. 1985). As we saw in chapter 3, placing poisonous plants in streams and ponds to stun or kill fish is a widespread practice. The insecticide rotenone is obtained from species of *Lonchocarpus* and *Derris*, which are still used as fish poisons in many places. The use of fish poisons may have gone hand in hand with the development of hunting (Harris 1977).

Poisons are important in human relations in their role in murder and suicide. Particularly interesting in this regard are the ordeal poisons that were used historically as a form of judgment. While this practice may have existed in biblical times, it is best recorded from sub-Saharan Africa (Robb 1957).

Humans have prescriptions for the detection and avoidance of toxins in plants and other foods. In some cases, animal behaviors were observed in order to obtain knowledge about edible plants. As well, direct feeding to dogs or other domesticated animals was practiced in Tibet (Rechung 1976), Africa (Scudder 1971), and undoubtedly in other places. The king's taster is a more sophisticated bioassay of this traditional practice.

Traditional medicine systems explicitly recognize the relationship between poisonous and medicinal plants. Poisonous plants in appropriate quantities can be part of medical preparations, and traditional herbal medicines—just like modern pharmaceuticals—can have toxic side effects. For example, Wang and Hu (1985) discuss the toxicity and side effects of Chinese medicinal herbs. Because of their potential toxicity, the medicinal application of plants requires sophisticated understanding of the plants' ecology. They must be picked at the proper time or season, and the proper part of the plant must be used. Many cultures have strict taboos surrounding the collection of herbal medicines and plant foods. By way of example Rechung (1976), in his discussion of Tibetan medicine, recounts that plants not properly collected or prepared can produce the opposite of the effect intended: "A medicine taken to cause purging would cause vomiting instead, and one taken to cause vomiting would cause purging." A rational interpretation of this statement might be that if the particular active chemical was more concentrated than expected, vomiting would result as a physiological response to acute toxicity, while the acute principle in smaller quantities than expected would instigate a more gentle purgative response rather than vomiting.

Phytochemists and chemical ecologists have only recently begun to appreciate the natural variability in secondary compounds among plants of the same species or even parts of the same plant (Louda et al. 1987; McKey 1979). Environmental conditions affect a plant's chemical composition (Rhoades 1985) and add to the unpredictability of the composition of a particular plant part that might be sought as a medicine. Traditional healers must deal with this variability, and in some cases it is clear that they do so at considerable risk to themselves and their patients.

William Withering (1741–99) pioneered the use of digitalis in Western medicine. The effective dose of this cardiotonic drug is close to the toxic dose, and this posed a problem before the use of chemical pharmacological preparations. In his classic book published in 1785, *An Account of the Foxglove, and Some of Its Medical Uses: With Practical Remarks on Dropsy, and Other Diseases,* Withering describes his efforts to standardize the condition of the plant to produce a reliable dose.

> [I] . . . was too well aware of the uncertainty which must attend on the exhibition of the *root* of a biennial plant, and therefore continued to use the *leaves.* These I found to vary much as to dose, at different seasons of the year; but I expected, if gathered always in one condition of the plant, viz. when it was in its flowering state, and carefully dried, that the dose might be ascertained as exactly as that of any other medicine; nor have I been disappointed in this expectation.

Antidotes for poison are important components of traditional medicine. The majority of these remedies deal with snakebites, or they are emetics designed to induce the body's natural response of expelling harmful substances through vomiting. Clays can act as toxin adsorbers. Remedies that counter specific symptoms of poisoning are more difficult to understand, but they can be explained using modern pharmacological models (Gilman et al. 1985). For example, cholinergic substances found in many plants can counter the effects of tropane alkaloids such as atropine (Wang and Hu 1985). On the other hand, Wang and Hu recommend atropine as a remedy for cardiac arrhythmia resulting from poisoning by aconitine found in *Aconitum* spp.

Traditional Concepts of Disease. Although prescientific concepts of diseases have innumerable permutations, they generally fall into four categories: diseases may be caused by harmful substances, loss of vital substances, taboo violation, or witchcraft (Adler and Hammett 1973). While supernatural and social forces may be regarded as underlying any or all of these causes of disease in specific cultures, both these concepts and, more important, the treatments associated with them can have a natural basis. Concepts of intrusion and loss of vital substances clearly have a chemical relevance. Vital substances are replaced by ingesting edible and medicinal substances. Most medical systems recognize the importance of a proper diet (although they may differ in what that proper diet consists of) in maintaining health, and food is often part of the therapy.

Intrusion is described in relation to evil spirits, pollution from social misencounters, or natural substances. In the case of supernatural, social, or natural causes for a disease, the same natural remedies are applied. Starving, puking, and purging are remedies widely used for intestinal disorders caused by poisoning or infection, as well as diseases of less rational etiology. The medicine of ancient Egypt centered on the concept of *wḥdw*, a term that attributed disease to a putrefactive essence that spread from the bowels to the bloodstream and hence to the bodily tissue (Mann 1984). The emetics, purges, and enemas for treating these diseases were the most common drugs of the Egyptians and remained part of Western medicine almost until modern times. Contemporary examples of such remedies are numerous. The Hausa, for instance, employ laxatives, diuretics, emetics, sudorifics, and expectorants to ensure that diseases responsible for intestinal disorders leave the body (Etkin and Ross 1982). In the medical practices of the Chontal of Mexico a strong emetic, *córdoba*, is particularly important; it is believed to expel whatever is causing a particular sickness (Turner 1972). (Vomiting as a ritual means of cleansing is discussed in chapter 2.) Whether substances of intrusion be physical or spiritual, cleansing employs the same pharmacological and physiological mechanisms. The emphasis of traditional medical thought and therapy on gastrointestinal traits supports the notion that the treatment of parasitic and digestive diseases is fundamental to the origin of human medicine.

Humoral medical concepts are the basis for traditional medicinal systems in China, India, Latin America, ancient Greece, and the Mediterranean region in later times. Balance in diet and in other aspects of life is sought between opposing influences, which generally fall into categories such as hot and cold, depending on the direction of the imbalance. Both foods and medicines play a role in therapy; hot foods and medicines alleviate cold diseases, and cold substances are likewise employed against the opposite imbalance. Considerable effort has been devoted to providing rational explanations to categories of hot and cold (Foster 1984), although each system must be considered within its cultural context. Nutritional quality may relate to these categories. For example, in China hot foods are "commonly those that are nutritious, concentrated, meaty, fatty, oily, sticky, dry, and cooked for a long time," while cold foods "are usually fruits and vegetables, eaten raw or lightly cooked, and are relatively low in calories" (Gould-Martin 1978). Classification of herbal medicines does not follow chemical lines but is intrinsically interconnected with Chinese theory on disease and therapy (Gould-Martin 1978). While a pharmacological property may exert a certain effect, its classification is meaningful only in juxtaposition to the classification of the diseases it treats (Kaptchuk 1983). In many cultures staple foods that are bland and considered nutritionally "balanced" are classified as neutral (Foster 1984). Herbal medicines are rarely included in this category; herbal therapies are by definition designed to remedy an imbalance and therefore are likely to decidedly represent one pole or the other.

Legends on the Origin of Medicine. Legends on the origin of medicine show indirect recognition of the chemical-physiological bases of disease and medicine. Tibetan medicine corresponds to the ideas discussed above in teaching that the first disease was indigestion (Rechung 1976). In the golden age human beings "lacked nothing and lived as if they were in heaven. One day one man ate the bitumen on the ground because of a habit acquired in a former life." The cure offered by Brahma was to drink boiling water. Warm water is still part of many home remedies to induce vomiting (Lewis and Elvin-Lewis 1977).

Most cultures do not have a clear understanding of the origins of medicine. In the Buddhist tradition which became important in Tibet, the Buddha Sakyamuni taught the first medical text after delivering his first sermon at Sarnath (Rechung 1976). Generally medicine is seen as a benevolent gift of the gods or some legendary historical figure such as Buddha. Such is the case with the Zuni of North America, where individuals are generally unaware of the origins of specific remedies, but their religion and folklore attribute them to the gods (Camazine 1986). The Incas believed that their most important plant, coca (*Erythroxylum coca*), was one of the gifts borne by their divine first king when he emerged from Lake Titicaca (see Weil 1986). Although they also discovered new medicines by observing the plants eaten by sick animals (Riesenberg 1948), among the Ponape of Oceania the best medicines are dreamed, usually by the patient or his/her relatives. In shamanistic traditions, which are widespread in Asia and the Americas, curers enter dream states to diagnose ailments and may arrive at new remedies in this state.

A scholarly work from the secondary century B.C. entitled *Huai-*

nan tzu credits the legendary cultural hero Shen-nung (Divine Peasant) with being the founder of Chinese pharmaceutical studies:

> The people of antiquity consumed herbs and drank water. They gathered the fruits of the tree, and ate the flesh of the clams. They frequently suffered from illnesses and poisonings. Then Shen-nung taught the people for the first time how to sow the five kinds of grain, and to observe whether the land was dry or moist, fertile or stony, located on hills or in the lowland. He tried the tastes of all herbs and (examined) the springs, whether they were sweet or bitter. Thus he informed the people of what they ought to avoid and where they could go. At that time (Shen-nung) encountered on one day 70 (herbs, liquids, etc.) with medicinal effectiveness (*tu*). (Unschild 1977).

Interestingly, Shen-nung is also held to be the founder of agriculture. Chang (1970) suggests that poisonous and medicinal plants could have been among the first domesticates.

Greek legend holds that the knowledge of medicine began with Apollo and was passed on to his son Asklepios. Hippocrates was one of a guild of physicians who worshipped Asklepios (Mann 1984).

Although passed-down knowledge was important in the expansion of medicine, the practice of medicine is not a passive behavior in traditional cultures, and practitioners must make judgments on the proper plants to use. That active testing is essential is apparent in relation to the problems of variable toxicity discussed above. As they did for Shen-nung, the chemical senses must play a major role in human understanding of plant properties. During the Roman period, Galen (A.D. 131–201) wrote in his work *On the Constitution and Powers of Simple Drugs* that taste was one of the best guides to the properties of plants and their medical potential (Sharples 1985). In the Tibetan tradition medicines prepared from herbs are classified into six varieties: sweet, bitter, sour, astringent, acrid, and salty (Rechung 1976). Taste is regarded as a primary attribute and means of evaluating medicinal plants by the Algonkian of North America (Speck 1917). The Salish of British Columbia use odor to verify the identification of edible mushrooms (Turner et al. 1987). Among the Paiute Indians smell was a primary way of recognizing plants, although "informants did not have many descriptive terms for plant odors except to say that a plant smelled like another plant or some other substance, such as horse's urine" (Fowler and Leland 1967). Examples of plant names that recognize the taste or smell properties of plants in the nomenclature of the Okanagan-Colville Indians were discussed in chapter 1. Most languages make similar references to the chemical properties of plants.

ACQUISITION OF EMPIRICAL MEDICAL KNOWLEDGE

Although the control of parasites is the logical outset for medicine, traditional medical practices go far beyond animal analogues and show a sophistication that often perplexes modern observers. A large number of drugs used in modern medicine are, or have been in the past, obtained from plants (Farnsworth et al. 1985). Seventy-four percent of these were discovered as the result of chemical studies to isolate the active substances responsible for the use of the plant in traditional medicine. Thus there is increasing scientific verification for the traditional use of these plants, and the secrets of medicine may slowly be revealed by modern science. It is impossible not to wonder how, without science as we know it, these secrets could ever have been discovered in the first place.

While the ethnobotanical literature records many techniques for the treatment of human ailments, a more ambitious and explicit intent of this volume has been to address the question of how humans came to use plants for medicine. The acquisition of empirical knowledge can be approached in both specific and general ways. Although I consider the use of certain plants, I have provided few satisfying answers to the question of how humans learned to use specific plants for their particular purposes. In the absence of an archaeological or historical record of the testing that must have preceded the use of many plants, it is difficult to provide direct answers to these questions. In-depth examinations of the practices, knowledge, and rationales of individual cultures existing today can provide another kind of approach to the understanding of empirical medical knowledge (Laughlin 1963). The point of view of the people engaged in medical practice, as well as their language, history, and social structure, are crucial to understanding the development of specific systems of medical knowledge. I have drawn specific data from a number of studies of medicine, but in approaching the fundamental question of how humans acquired knowledge about the properties of plants, the answer I offer is a general evolutionary one.

Laughlin (1961, 1963) expressly addressed the origins of empirical knowledge from the perspective of surgical medicine. Anatomical knowledge is a direct offshoot of hunting and the processing of animals. Surgical knowledge among protohistorical groups can be very sophisticated, and archaeological remains from the early Neolithic indicate that bone-setting and trephining have had a long existence (Mann 1984). In terms of understanding the acquisition of new knowledge about medicine, it is the "empirical attitude" that is relevant. The practice of comparative anatomy and the carrying out of autopsies by traditional peoples indicate a strong inclination to expand their understanding of medical problems. The Aleuts of southwestern Alaska, for example, studied comparative anatomy using the sea otter, an animal with morphological and behavioral similarities to humans (Laughlin 1961, 1963).

In the following discussion I relate the acquisition of empirical medical knowledge to the origins of pharmacology in regard to three components. First, it is necessary to reconsider the biological basis for medicine. This foundation is apparent in various human and animal behaviors and in the conditioning learning processes that are determinants of human interactions with the plant world. Second, it is essential to appreciate the common intellectual capabilities of human beings. Interactions between perceptions and rational interpretations of these perceptions are essential for learning, comprehension, and utilization of knowledge. Many clear cases of events with cause-and-effect relationships are the basis for building up a store of medical knowledge. Third, the question of motivation is of utmost significance in understanding the important place of medicinal knowledge in the history of the human species. Along with giving a greater self-awareness, human intellect has put humans face to face with the existential questions of life. The incomprehensibility of death and illness and the tension of social intercourse are highly motivating forces in directing humans to find explanations and solutions for suffering. Magico-religious and empirical practices converge in the practice of medicine.

BIOLOGICAL FOUNDATIONS

The overlap of culture and human biology makes it difficult to say which of our medicinal behaviors have a specific genetic basis. A more important aspect of our biology in relation to the evolution of medicine is our ability to learn. Behavioral aspects of human interactions with plant allelochemicals can be understood in relation to the conditioned responses discussed in chapter 2. Plant chemicals induce many physiological effects that could be learned through taste conditioning or simple associations involving other senses. Active principles that alter the sense of well-being are likely to have conditioning effects, and plants that positively affect well-being may be eaten preferentially. The mechanisms by which these effects take place can be mediated through a number of physiological responses associated with the nervous system. Plant principles could be perceived through providing relief for a number of disorders such as those involved with blood pressure, thermal regulation, or blood sugar. In some cases actions can be distinguished as affecting specific physiological processes, while in other cases they simply affect well-being. In order for learning to occur, physiological effects must impinge in some way on the nervous system.

Direct effects on the nervous system are likely to be learned most efficiently. Plant substances that reduce pain include analgesics such as salicin and morphine and local anesthetics such as cocaine (Gilman et al. 1985). Stimulants such as caffeine and strychnine may also have important effects on well-being.

Psychoactive properties of plants are impossible to ignore and are obviously learned easily. However, the relation of their effects to well-being depends largely on cultural context. Rodriquez et al. (1982) point out that hallucinogens and chemicals that kill gastrointestinal parasites are often one and the same. They suggest that hallucinogenic uses of plants actually originated from efforts to control parasites.

Chemicals with psychopharmacological effects may also occur in the diet. Benzodiazepines are widely used sedative-antianxiety drugs, and the recent report of trace amounts of diazepam (Valium) and related compounds as constituents of wheat and potatoes (Wildman et al. 1988) is intriguing.

Other plants provide stimuli to the nervous system through indirect effects on bodily processes. Anti-inflammatory agents will reduce pain resulting from various ailments. Antirheumatism and antiarthritis remedies are probably learned via this route. Arthritis is one of the oldest afflictions of humans and animals (Ackerkneckt 1953), and theoretically, as with natural selection for behaviors that control parasites, selection may also have favored behaviors that seek out plants that provide relief from this ailment.

Plants that affect the gastrointestinal tract would appear to be likely candidates for conditioned learning because the important conditioned taste aversions are mediated through this system. Relief of nausea and associated internal malaise could be accomplished by emetics, purgatives, and digestive stimulants in particular cases.

Effects on the circulatory system through coronary stimulation or the alteration of blood pressure are perceivable and could have positive benefits. Diuretics such as caffeine may alter blood pressure, while cardiac glycosides found in plants such as *Asclepias* spp. are powerful heart toners and could be easily associated with ameliorative effects in situations of need. The Fox Indians of North America used *Asclepias incarnata* as a diuretic (H. H. Smith 1928), while *Digitalis purpurea* was used in Europe as a cardiotonic agent. *Digitalis lanata*, which is the actual source of important modern drugs such as digoxin used to treat cardiac arrhythmias and congestive heart failure, was not actually itself used in this way (Farnsworth et al. 1985).

Effects on body temperature are readily perceived, and the value of plants that reduce fever (antipyretics) were probably easily learned. Salicin is an effective antipyretic. *Cinchona,* which was used by South American Indians as a fever remedy, contains the alkaloid quinine. As well as being an antipyretic, quinine is highly toxic to malarial parasites. Similarly, the important new antimalarial drug qinghaosu was traditionally used in China to treat fever (Klayman 1985). It is not necessary to know anything about the biological basis for malaria to treat this disease effectively, just as it is unimportant to know whether particular plants cure the disease or simply remove the symptoms.

Conditions of low energy and weakness may be associated with a number of conditions with organic or psychological causes. A hypoglycemic agent might provide significant relief for a metabolic disorder such as diabetes mellitus. Many plants are reported to have this property (Farnsworth and Segelman 1971). Nutrients or stimulants might provide real relief for chronic conditions of poor health, and at least a more positive outlook on life in other cases of illness. Various tonics and panaceas such as ginseng are used by large numbers of people around the world. Subtle effects produced by these plants, while not unfounded, require great effort to substantiate (Lewis 1986; Ng and Yeung 1986; Shibata et al. 1985).

It is particularly difficult to understand how the uses of these plants developed. Anyone who remembers trying a first cigarette as a teenager must find it hard to understand how the use of tobacco could have developed. Nevertheless, millions of people are addicted to the very subtle "beneficial" (in the short term) effects of nicotine.

It might be possible to test ways in which medicinal behaviors are conditioned in humans, although there are many practical and ethical reasons why experiments cannot be done. As we saw in chapter 1, human capacity for conditioned preferences, which would seem more salient in learning relevant to medicinal plants, is considerably less than our capacity for conditioned aversions.

In Western industrialized societies there may not be any association of the positive postingestional effects of drugs with increased liking. In a study directed at this issue Pliner et al. (1985) concluded that while food aversions are mediated by nausea, the relief of nausea does not appear to increase the preference for drugs. However, this study was limited in scope, and as well was outside the cultural and biological contexts in which traditional medicines were and are used. One factor recognized by these authors is that subjects had strong assumptions about the relative appropriate hedonic response of medicines and foods. In cultures where food and medicines are not regarded as so distinct, these assumptions may be different. Again, it is essential to recognize the importance of culture in mediating human interactions with the biological world.

LEARNING THROUGH CONSCIOUS ASSOCIATIONS

All of the medicinal uses of plants discussed above could be, and may be, learned by animals. The treatment of many other diseases, while not learned through direct physiological mechanisms, could be learned through observation and cognitive associations between cause and effect. The evolution of intellect and language is crucial for the development of this kind of process. Each of the previous groups of conditioned effects could also be recognized and reinforced in this manner. Many of the following cases were probably best learned where conscious awareness of a problem existed and where a solution to the problem was sought.

The more sophisticated manner in which humans (compared with animals) use plants as anthelmintics and amebicides is consistent with a recognition of the antibiotic activities of plants. Parasites, particularly those visible in stools such as roundworms, hookworms, and whipworms, would be deliberately controlled if they were recognized as a problem and if an effect on their numbers could be observed. Human treatment of wounds and skin infections with plants is straightforward and usually effective. For example, plants that contain phototoxic polyacetylenic compounds are widespread in the treatment of skin disorders (Wat et al. 1980). Also important in the treatment of wounds are tannins. The astringent properties of these common compounds have rational and observable effects in many medicinal conditions (Gilman et al. 1985).

The use of plants to affect sexual function—for example, as contraceptives, menses prohibitors, abortives, aphrodisiacs, and emmenagogues—could have developed with our ancestors' increasing awareness of the natural actions of plant secondary compounds on sexual function (Glander 1982). Dietary constituents trigger reproduction in rodents, and Glander (1980) hypothesized that such a system is present in howler monkeys. The idea that plants affect sexual function in humans or nonhuman primates is only interesting conjecture, but the interactions of humans with plant allelochemicals are more complex than we generally appreciate, and ideas on medically relevant permutations of this interaction should be approached with an open mind.

Plants that directly treat easily recognized ailments such as diarrhea and coughs are widely employed in folk medicine (Lewis and Elvin-Lewis 1977). Likewise the effects of diuretics and expectorants can be learned through simple cause-and-effect observations.

Most of the remedies described in this and the previous section correspond closely with those used by gatherer/hunters and traditional agriculturalists to treat simple ailments. Until the progress of scientific medicine in the last century and a half, Western medicine offered nothing superior to these remedies. Illnesses that are much more complicated in their causes are impossible to understand without in-depth study of physiology, biochemistry, microbiology and so on; such studies are outside the experience of traditional herbalists, and much of the folk medicine relating to complex diseases is based on superstition and nonrational concepts.

Nonetheless, effective cures such as those discussed above for malaria could be, and sometimes were, discovered without understanding the etiology of a disease. Most important in such situations are the careful observation of symptoms and the recognition of patterns of symptoms and the responses of conditions to a particular remedy. Again, it is not necessary to understand the biological cause of an ailment to make it go away. Chinese traditional medicine, for example, while highly effective in many ways is often unexplainable by Western science. It must be viewed as a whole. Careful observations and the integration of knowledge into an internally consistent system have built an empirically sound system of understanding and practical treatments.

While the information content on which various cultures formulate their explanations of illness and their attempts to rectify its occurrence may vary, the intellectual methods by which humans process the information do not vary significantly. Within their own environment and cultural context humans build up elaborate stores of information relevant to their survival. We see this reflected in folk taxonomies. Within their individual contexts, humans' ability to understand and control important aspects of their world is remarkably parallel.

Many of the physiologically verifiable effects of plants discussed above may be employed for the treatment of diseases for which they offer no cure. Nonetheless they produce effects that may be within cultural expectations of effective treatment (cf. Ortiz de Montellano 1975). The remedies for intestinal disorders discussed above in relation to traditional concepts of disease meet expectations related to the perceived cause of the disease. Similarly, the cultural expectation of the Zuni Indians of the western United States is that treatments for stomachache should induce vomiting: "This expectation may have its basis in earlier times when certain gastrointestinal illnesses were probably caused by the ingestion of spoiled or toxic foods" (Camazine 1986). Therefore any emetic, whether it provided direct relief of stomachache or not, would meet expectations. By Zuni definitions this is an effective treatment; even if it does not make a physical contribution to health, such a plant could play a positive role in a psychosomatic fashion.

MOTIVATION FOR MEDICINE

Needless to say, not all diseases can be dealt with successfully in a rational manner. Cancer, for example, is an intractable human affliction of long standing, and many herbal remedies have been used in its treatment. Few, if any, have any effect. Podophyllotoxin and other lignans from *Podophyllum peltatum* are among the few anticancer drugs from plant sources that can be traced back to folk medicine (Farnsworth et al. 1985). *P. peltatum* was used by the Penobscot Indians of Maine to treat cancer (Hartwell 1967). Many folk remedies cannot be accepted as effective when viewed by scientific criteria; but many modern cancer remedies when viewed by scientific criteria can also not be accepted as effective. However, in cases where diseases defy cures, humans continue to press for solutions and explanations.

Most diseases affect older individuals—those beyond the reproductive life span. Old age, in which the homeostasis of the body begins to break down, may be considered a disease—a disease for which there is no cure. Nonetheless, much of the focus of human medicine through time has been on both the symptoms and the cause of aging. In the face of our inevitable mortality, human self-awareness and intellectual capacities have resulted in a characteristic anxiety about death and dying, and this anxiety can relate to social and economic concerns (Turner 1972). In most subsistence societies there are insufficient surpluses, and sick and aged individuals are at risk of losing their role in society and becoming a burden. While the same methods of chemical manipulation that are applied to other perturbations of health are applied to the disease of aging, the metaphysical nature and biological inevitability of this condition have introduced concerns of existence and brought about the intricate interrelationship that we see between religion and medicine in most human cultures.

Many traditional concepts of disease, particularly those concerned with taboo violation and witchcraft, search for explanations in human social relations. Humans are social animals, and the health and happiness of individuals are intimately related to their relationships with society. Religion seeks to find explanations and relief for suffering in its codes for social and moral conduct, and in its teachings about the purpose of life seeks to give meaning as well as to control the course of life. Magic solutions and superstitions may offer explanations and are often employed in traditional medicine to control disease. Medicine is concerned with both controlling illness and understanding and removing its causes. Medicine, religion, and magic are inseparable in many societies and are totally separated in few, if any. Together they provide a driving force for seeking medical solutions as well as a framework for understanding the world.

Cultural concepts may lead the search for solutions in particular directions. The doctrine of signatures, whereby some morphological feature of a plant indicates its utility, is widely applied in folk medicine, while plants with strong tastes or smells may be regarded as useful remedies. As Camazine (1986) points out, the latter types of plants were used to drive out evil influences that were often held to be responsible for disease. Since many diseases are self-limiting and not life threatening, many such remedies may have been regarded as effective when an illness subsided on its own (Camazine 1986). Many of the effective remedies of folk medicine were probably discovered in spite of such misconceptions. If humans are motivated to search the natural world for cures, "serendipity and millennia of trial and error would result in the use of folk remedies with a valid scientific basis" (Camazine 1986; cf. Johns et al. 1990).

Where rational medicine fails to provide solutions, religion can offer alternatives in the form of acceptance and consolation. Where effective treatments brought about by rational medicine are unexplainable, religion and culture play an important social function in examining the causes of illness and helping individuals come to terms with the place of illness in their lives. People today accept modern medicine for its perceived effectiveness but still look to traditional sociocultural explanations for the causes of disease (Lieban 1976).

Even modern Western medicine, which is characterized by a greater degree of scientific principles than traditional medicine as it was practiced in European history or elsewhere, retains an implicit nonrational component. Psychosomatic effects are widely acknowledged as being important in the timing of episodes of disease and playing a crucial role in healing. Most diseases are not life threatening, and removing symptoms can make one at least feel better; if a plant product succeeds in improving the body's ability to function and take care of itself, it can affect survival positively. Social interactions are essential for our well-being. So also does the reconciliation with the meaning of life which is sought in philosophical contemplations and religious practices play a part in the maintenance of a healthy individual. Traditional medicinal systems often recognize better than modern technological medicine that health involves satisfying all of the physical, emotional, social, and ontological needs of the person.

Although the interaction of mind and body can be mediated at a chemical level, it is extremely difficult conceptually or practically to ascertain their meeting point. Psychosomatic factors appear to affect the etiology of disease through an "extensive network of psychologic, central nervous system, neurotransmitter, endocrine, immune, and other biologic systems" (Stein 1986). Psychological stresses can lead to disease states with serious organic manifestations. Bronchial asthma, for example, has a complex etiology that can involve psychological (Creer and Kotses 1983) as well as biological factors. It is routinely treated with theophylline, and even when the condition is brought on by anxiety, this natural alkaloid found in tea reduces the severity of attack through its action as a bronchodilating agent. Smoke from the leaves of jimsonweed (*Datura stramonium*) is also an effective bronchial dilator (Hertz 1979).

Stress-related secretions of hormones appear to play a role in altering immunity. An immunoregulatory role has been suggested for β-endorphins (Shavit et al. 1984) based on experiments with rats in which the suppression of immunity through stress-induced release of natural opioid peptides in the brain was mimicked by the administration of morphine. Morphine, of course, is a naturally occurring constituent of *Papaver somniferum*, and it is intriguing to speculate about possible roles that other constituents of herbal medicine may have relative to the etiology or treatment of psychosomatic ailments. A large number of individuals in modern society self-medicate with plants and plant products (e.g., cocaine, marijuana). Is this just herbal medicine for the fatigue and stresses resulting from modern living?

The role of nutrition in mediating immune responses is discussed in chapter 7. Sound nutrition is an apparent contributor to defense against stress-related ailments.

In attempting to determine the chemical basis for health and disease and the origins of medicine it is necessary to separate the rational and scientifically verifiable from that which is not. However, the social, ritual, and magical components of medicine cannot truly be separated from biological aspects of the origin of medicine. In fact, in most studies of folk medicine it is the task of distinguishing the two which is primary. By doing so we can consider each in juxtaposition to the other.

INTERRELATIONSHIP OF MEDICINE AND FOOD

If the biological basis of medicine lies in our consumption of secondary chemicals in food, the consumption of pharmaceutical agents is an intrinsic human behavior. Many traditional cultures do not make a strong distinction between food and medicine, and nutritional and pharmacological properties of plants are not readily distinguishable. Proper diet as a requirement for health along with prophylactic uses of herbal medicines are emphasized in many systems of traditional medicine. Many plants that are generally considered food have important therapeutic roles (Etkin and Ross 1982), and diet is part of the healing process in both traditional and modern medicine. Foods used during therapy may make important nutritional contributions. Food composition—for example, in the nature of the amino acids present

(Spring 1986) or the presence of pharmacologically active constituents (Wildmann et al. 1988)—may affect a person's mood and behavior. From a chemical-ecological perspective, poison, stimulant,
medicine, condiment, beverage, and food form a continuum that is
the basis of the following summary.

Beverages play a special role that overlaps both food and medicine.
Infusions or teas are often the preferred way of consuming medicine.
Herbal teas are often consumed routinely, and the distinction between teas consumed as a refreshing drink and teas used as medicine
is unclear. Beverages containing caffeine and other stimulants may
have been more important in the past as medicines. Teas may make
important nutritional contributions in vitamins and minerals. They
may be more palatable than intact plant material, and nutrients in
teas may be more easily ingested when they are extracted from fiber
and digestibility-reducing substances.

Spices and condiments play an important role in the food preparation of many cultures. Their characteristic flavors result from the
particular secondary chemicals they contain. These chemicals may
serve a number of functions. Those with antibiotic properties may
help preserve food against spoilage. Their flavors improve the palatability of food, and for this reason may have increased in importance with humans' greater dependence on grain-based diets. Specific
combinations of spices characterize various cuisines and contribute
to the cultural determination of food preferences. As we have seen, in
the evolutionary past the flavor of phytochemicals likely stimulated
feeding and contributed to the consumption of varied diet. Condiments may help stimulate appetite in this way or through direct
physiological effects. For example, capsaicin, the active ingredient of
chili peppers, activates the gastrointestinal system by stimulating
salivation, gastric secretion, and gut motility (Tyler 1987).

Many plants used as condiments also have ancillary medicinal
uses, and their deliberate uses as food additives may have originated
in this way. An excellent example of a medicinal condiment is *Chenopodium ambrosoides,* the wormwood. Under names such as *epazote* (Mexico) or *paico* it is one of the most important anthelmintics
in Latin America and is widely used for this purpose around the
world. While the extract of this plant is a powerful anthelmintic
agent, as the plant is generally used by adults its efficacy appears
minimal (Kliks 1985). However, it is important to recognize that in
situations where reinfection is high, no anthelmintic agent can have

a long-term effect. Wormwood is also an important condiment in many Latin American dishes (Kliks 1985), and it is believed to contribute to prophylaxis against worms. Although it may make a contribution at these doses, its regular ingestion may also help to make its somewhat repugnant taste acceptable in preparation for times when larger doses are called for.

Wormwood contains the terpene ascaridol, which in high doses may be harmful. Although in therapeutic doses it may cause some side effects, the repugnant taste of high amounts make it difficult for adults to poison themselves. In a study of anthelmintics in Oaxaca, Mexico, Girardi (1987) reported that "children given this tea may develop an aversion to its taste and smell and will no longer be able to eat *epazote* as a condiment." Children may be more at risk of poisoning from this plant. Epazote's role in food, as a medicine, and as a poison is a good example of the chemical continuum from poison to food.

Many plants consumed as food are recognized as having pharmacological properties in addition to nutritional value, and under certain circumstances food plants are purposely consumed for this purpose. Etkin and Ross (1982) emphasize the important relationship between food and medicine in their paper on Hausa plant use. Of 107 plants used for gastrointestinal disorders, 53 were used as dietary constituents in other contexts. The authors suggest that when used in the diet these plants may contribute something physiologically beneficial above and beyond their nutritive properties. Similar patterns of plant use are seen across Africa and around the world. The añu and maca that were discussed in chapter 4 are good examples of foods with important medicinal associations. Likewise *Euphorbia lancifolia,* which is used as a salad green in Central America, is believed to be a powerful galactagogue (Rosengarten 1978).

From a chemical-ecological perspective, the two most important cultural stages of the evolution of human diet—the advent of technology and the origins of agriculture—are also crucially interconnected with the origins of medicine. An increased utilization of animal protein, believed to be associated with the appearance of the genus *Homo,* would have made it possible to eat toxic plants with greater impunity. Plants, then, theoretically could have been more readily ingested by early hominids for medicinal purposes separate from nutritional need. The availability of nutritionally dense carbohydrates associated with the origins of agriculture was accompanied by reduction in toxic chemicals in domesticated plants. However, these events are also associated with increases in disease associated with nutritional deficiencies and urbanization. Humans' reliance on wild plants for both broadening the sources of essential nutrients and for medicinal agents probably increased appreciably during this time. While allelochemicals consumed by animals may play a pharmaco-phagic role, humans' technological adaptations allow us to distinguish the nutritional and pharmacological properties of plants both conceptually and/or in practice.

THE ROLE OF LEAVES IN HUMAN EVOLUTION

I believe leaves are the key to understanding the evolution of both the human diet and medicine. As we saw in chapter 7, all of our closest primate relatives consume considerable amounts of leaves and are able to use leaves as an important dietary constituent in spite of the presence of secondary chemicals. We evolved from plant-eating primates, and human dietary requirements for vitamins C and A, folic acid, and linoleic and α-linolenic acids can best be filled from fruit and leaves. As well, leaves can readily provide a large portion of our protein needs.

Nevertheless, leaves have been generally discounted as an important component of the human diet. Toxicity arguments that have been used to minimize the evolutionary importance of plants in the human diet often focus on leaves. If we are as adapted for avoiding toxins as our behavioral and physiological characteristics suggest, then perhaps leaves are not so formidable after all. The real reason we do not consume more leaves is probably that the amount of energy they provide relative to their nondigestible bulk is not great enough by itself to meet the needs of a large-brained animal (Milton 1988).

Interestingly, many of the leafy vegetables consumed in Africa today are plants consumed by other primates (Bohrer 1977). Studies of primate feeding and speculation on early hominid food procurement (Peters and O'Brien 1981) indicate that young leaves are likely to make a bigger contribution to the diet than mature leaves. Digestibility rather than lesser toxicity is most likely to make young leaves preferred foods. Grinding and cooking offer ways to make leaves more digestible, and there are places in the world where these techniques continue to allow leaves to compete with richer foods as significant parts of the diet.

Of particular interest among African food plants are various leafy vegetables containing high levels of secondary chemicals (Table 8-1). These are usually prepared so as to partially, although not necessarily totally, eliminate toxicity. Although these plants are domesticated, high levels of allelochemicals may be deliberately retained in them for pharmacological purposes. Alternatively, I have suggested that secondary compounds are important for resistance to insects and so contribute to adequate yields of these crops. These two ecological explanations are not mutually exclusive. I have been told by Luo and Luhyia informants in western Kenya that leafy vegetables such as *Solanum americanum* and *Crotalaria brevidens* are important as protectors against "malaria" (broadly meaning fevers, including malaria caused by *Plasmodium* spp.) and intestinal parasites. Among the Kamba of Machakos District, Kenya, *S. americanum (S. nigrum)* is said to be effective for serious diarrhea and vomiting (Kloos et al. 1986). Further investigations of the chemical properties and biological activities of minor food plants are necessary in order to understand the interconnection of diet and medicine.

The widespread importance of leaves as medicines and condiments suggests that rather than being avoided, leaves are often sought out for their allelochemicals. The possible role of human salivary proline-rich proteins in tannin detoxification and our hypothetical "taste for tannins" are consistent with an evolutionary history as consumers of tannin-containing plants. Rather than being relevant components of the human diet only insofar as it was desirable to avoid them, allelochemicals in leaves may have played a major role in our dietary ecology. As the hominid diet changed to nutritionally dense meat, concentrated carbohydrates, and refined foods, leaves likely did become proportionally less important. In addition, processing led to the reduction of allelochemicals in the diet. In this sense herbal medicine can be seen as an attempt to replace the allelochemicals that in an evolutionary context were always part of our diet and of our ecological relationship with plants and parasites.

In bioassays of the efficacy of gastrointestinal remedies applied in the traditional medicine of the Luo of western Kenya (Johns et al. 1990) to kill the parasite responsible for giardiasis, it was striking that most of the effective remedies were derived from the bark of the roots or stems of trees and shrubs, not the leaves, seeds, or starchy tubers (unpublished results). Typically bark has the highest levels of toxic chemicals, and, generally speaking, most herbivores avoid bark. However, culturally determined improvements in the nutrient qual-

Table 8-1. Leafy Vegetables from Sub-Saharan Africa with Chemical Variability

Species	Allelochemicals
Crotalaria brevidens Benth.	pyrrolizidine alkaloids
Gynandropsis gynandra (L.) Briq.	glucosinolates
Solanum americanum Mill.	glycoalkaloids
Vernonia amygdalina Del.	elemanolide lactones

ity of the diet allow humans to consume plant parts that would be too toxic for other animals.

As traditional patterns of dietary and medicinal behavior break down, the consumption of biologically active allelochemicals also diminishes (Iwu 1986). Simplistically speaking, one might speculate that with modern pharmaceuticals humans continue to rectify the basic biological balance between nutrients, infectious disease, and nonnutrient organic chemicals by deliberately ingesting greater levels of the latter, albeit in highly sophisticated ways.

A LOOK TO THE FUTURE

The chemical-based medicine of the twentieth century has provided effective and highly specific remedies for many human afflictions. However, many modern pharmaceuticals are also potentially more toxic than botanical drugs of previous centuries. Drug interactions are recognized as a serious problem (May et al. 1987). The evolutionary principles that govern the adaptations of the human species to our environment apply today as they have through history, and we ignore the nature of such biological interactions at our own peril. In the relationship between chemicals and disease resistance, the ecological balance that is represented in Figure 7-1 must exist in modern systems just as it does in traditional ones. Nutrition, drugs, and disease interact in both modern and traditional settings in ways we are just beginning to appreciate.

The interaction of drugs and diet is recognized to affect the success of drugs, but the dynamics of the interaction of herbal remedies in complex mixtures are harder to understand. Because they contain small amounts of many different chemicals, people point to herbal drugs as posing fewer risks than modern drugs. Certainly herbal drugs are more "natural" medicines in the sense that their use is closer to the way biologically active chemicals were consumed in an evolutionary context. However, this does not preclude their potential for causing serious acute and chronic toxicity (Saxe 1987). For both traditional and modern drugs that are used routinely over an extended period there are concerns of long-term side effects that are often not well understood. We continue to fight an age-old chemical-ecological battle in our uses of drugs and pesticides; that is, we are constantly confronted with the double bind of toxicity to ourselves inherent in our attempts to control other organisms.

Ecological approaches to medicine can give us a greater theoretical understanding of health and disease. Concepts of balance and homeostasis form the philosophical basis from which many traditional medical systems understand health and look for the cause of disease. Chemical and biochemical processes are primarily understood by scientists in terms of concepts of equilibrium and homeostasis that, in fact, have roots in the humoral medicine of Persia and Greece. Likewise, balance is an essential concept in a chemical-ecological model of human environmental interactions. Humoral and homeopathic forms of traditional medicine, while differing in details and level of understanding, have parallels to chemical models of physiological processes. A chemical ecology model is a powerful tool for bridging the gap between traditional and modern science and for understanding the nutritional and pharmacological bases of human health.

Understanding the chemical-ecological aspects of human interactions with plants gives a powerful perspective from which to tie together our understanding of evolutionary aspects of health and disease. Human physiological and intellectual capacities are sufficiently sophisticated that we are not simply obliged to avoid or accept xenobiotic compounds. Detoxication enzymes, taste perception, geophagy, and learning all interact to allow us not only to avoid nonnutrient compounds but also to consume them for our overall well-being. Plants are used not simply in spite of toxins but also because of them. Thus the situation is not simply one of either/or, and we need a better understanding of the complex effects of biologically active chemicals on human physiology and ecology. The continuum of poison, psychoactive agent, medicine, condiment, beverage, food is as important today as it was in the past.

Hypotheses about the positive role of allelochemicals in human diet and the synergistic balance between nutritional and pharmacological properties of foods are being tested by chemical ecologists, either from their particular perspectives on animal-plant interactions or in relation to understanding the evolution of human medicine and diet. Studies in human biology and animal biology have usually proceeded along parallel, although somewhat separate, courses. Although human chemical ecology is a field based on animal models, introducing the concept of medicine into chemical ecology has profound implications for that field as a whole.

Each of the nine hypotheses presented in the opening chapter can be the basis of further study in this area. Of these hypotheses, the ninth—that high levels of secondary compounds are retained in the diet for ecological reasons—offers the greatest potential for expanding our understanding of human interactions with the chemical environment. The idea that medicine has its basis in the positive contribution to survival made by allelochemicals in the diet is highly conjectural. The hypothesis that these chemicals were important dietary constituents before humans came to depend on more nutritionally dense and processed foods is difficult to test directly. Nevertheless, field and laboratory studies with humans and with animals can be undertaken to test many of these ideas.

One of the biggest obstacles to progress in this area is our minimal knowledge of the chemical complexity of the botanical world. Studies in natural products chemistry have only begun to scratch the surface on the diversity of organic chemicals in plants. Particularly lacking are extensive studies of the chemistry and nutritional properties of medicinal plants and traditional foods (but see Duke 1985). Rapid advances in chemical analysis stimulated by the revolution in computer technology offer increased hope of producing significant advances in the data available in these areas. New drugs and new products of economic potential arising from this research could complement studies of an evolutionary nature.

We also lack basic biochemical knowledge of the effects of naturally occurring compounds on the human body. Natural products that have pharmacological potential are frequently studied, but the physiological effects of the vast majority of allelochemicals in our food are poorly known. Synergistic interactions between different pharmacological agents (Gilman et al. 1985) and man-made toxins are recognized but not necessarily well understood. Synergistic interactions are also recognized among biologically active constituents of plants (Berenbaum 1985; Izaddoost and Robinson 1987), but the complexities of this area are difficult to approach. Extensive testing of the effects of dietary plants on parasites using in vivo and in vitro assays are the first step toward answering some of the fundamental questions raised here.

Psychophysiological studies have made important contributions to the understanding of behavioral aspects of human interactions with dietary chemistry. If allelochemicals have specific roles in determining feeding preferences in humans and folivorous primates, the roles of naturally occurring compounds could be studied in relation to questions of appetite and satiety. Further investigations into primate taste and smell perception could provide valuable insights into the role of food chemistry in determining feeding behavior and diet. Most studies of detoxification enzymes are concerned with synthetic xenobiotics. Greater attention needs to be directed at understanding the actions of our detoxification mechanisms on phytochemicals and considering the natural substrates of these enzymes.

The abilities of humans to deal with the synthetic chemicals introduced into our environment by modern industry are dependent on the physiological characteristics of our species. We need a better understanding of our interactions with plant allelochemicals in the evolutionary process. The role of environmental chemicals in the etiology of cancer is of great concern in industrialized nations. Ames (1983; Ames et al. 1987) has fueled the controversy in this area by suggesting that naturally occurring toxins in plant foods are just as carcinogenic as the pesticides, food additives, and wastes that are produced by modern industry. A crucial question in dealing with environmental toxins is why, among the large number of persons exposed, only certain individuals show a toxic reaction (Goedde 1986). The importance of environmental factors such as diet and drug therapy and the role of natural selection in bringing about different frequencies of genetic polymorphisms are important for understanding this issue.

Reports of potentially positive interactions of mammals with allelochemicals in food plants suggest that these chemicals must be viewed more broadly than just as potential toxins. Potential positive effects of tannins (Mole and Waterman 1985) and phytic acid (Thompson et al. 1987) on digestive processes, the possible involvement of tannins in detoxification of other plant chemicals (Freeland et al. 1985; Goldstein and Spencer 1985) and in the control of dental caries (Kakiuchi et al. 1986), the antioxidant properties of flavonoids (Robak and Gryglewski 1988), and the detection of psychoactive compounds in foodstuffs (Wildmann et al. 1988) all suggest that we should look for greater subtlety in our interactions with allelochemicals.

Health and disease are complex phenomena that cannot be adequately explained by single-factor hypotheses. Science is often no better at teasing apart the interrelated variables involved in the ecology of illness than is traditional medicine. We are only beginning to understand our interactions with nutrient and nonnutritive chemicals, their effects on us, and our responses to them. A chemical-ecological perspective offers a context for increasing this understanding.

Because plant allelochemicals make up a minimal (although not unimportant) component of the highly refined diets of modern industrialized societies, biomedical studies of the diet and medicine of traditional cultures provide the best opportunities for considering human-plant interactions in an evolutionarily relevant context. In the rapidly changing world, these opportunities for examining the diverse ways in which people use plants, particularly wild plants, are rapidly disappearing. Knowledge of the diversity of human adaptations is essential for understanding the past and directing the future. Diversity is the raw material for evolution, and the evolutionary path we follow has little chance to succeed without it.

Traditional systems of subsistence in stable situations, such as that of the Aymara on the Andean altiplano, are often more ecologically sound and culturally acceptable than systems based on imported technologies. In spite of the hardships inherent in survival in marginal environments, it is my opinion that traditional settings offer better alternatives for most people to maintain health than conditions offered by urban slums. For the majority of the world's people the best alternatives for the future involve merging the traditional and modern technologies and cultures. It is essential to recognize the oneness of all humans while encouraging our diversity in cultural and ecological adaptations. The role of ethnobiology in the modern world must be more than to just record this diversity for posterity; it must recognize the evolutionary foundation in the traditional patterns of using plants and animals, seek to evaluate these systems using the resources of modern science, and strive to facilitate the merging of empirical wisdom of traditional ways of interacting with the world with the positive contributions of modern science. This path runs two ways, and the rewards to be gained from the study of the world's biological and cultural diversity can benefit all of humankind.

Appendix 1 Classification of Traditional Plant Processing Techniques

1. No special detoxification techniques applied (subdivide as in Coursey, 1973)
2. Special detoxification techniques applied
   2.1 Detoxification by heat
      2.11 Unspecified cooking
         2.111 Cooking of whole pieces
            2.1111 Cooking without the addition of salt, lye, or acid
            2.1112 Cooking with the addition of salt
            2.1113 Cooking with the addition of lye
            2.1114 Cooking with the addition of acid
            2.1115 Cooking after drying
            2.1116 Cooking after soaking
         2.112 Cooking after comminution (subdivide as for 2.111)
         2.113 Cooking after peeling (subdivide as for 2.111)
      2.12 Boiling, stewing, etc. (subdivide as for 2.11)
      2.13 Roasting, baking (subdivide as for 2.11)
      2.14 Frying (subdivide as for 2.11)
      2.15 Steaming (subdivide as for 2.11)
   2.2 Detoxification by solution
      2.21 Soaking in static water
         2.211 Soaking or leaching of whole pieces
            2.2111 Followed by unspecified cooking
            2.2112 Followed by boiling
               2.21121 Simple boiling (subdivide as for 2.111)
               2.21122 Repeated boiling in changes of water (subdivide as for 2.111)
            2.2113 Followed by roasting or baking
            2.2114 Followed by frying
            2.2115 Followed by steaming
            2.2116 Followed by drying
            2.2117 Followed by fermentation
            2.2118 Followed by pickling
         2.212 Soaking or leaching after comminution (subdivide as for 2.211)

2.213 Soaking or leaching after cooking and comminution (subdivide as for 2.211)
2.214 Soaking or leaching after cooking (subdivide as for 2.211)
2.215 Soaking or leaching after boiling with lye (subdivide as for 2.211)
2.216 Soaking or leaching after freezing (subdivide as for 2.211)
2.217 Soaking or leaching after drying (subdivide as for 2.211)
2.218 Soaking or leaching after peeling or cutting (subdivide as for 2.211)

2.22 Soaking with change(s) in water (subdivide as for 2.21)
2.23 Soaking in running water (subdivide as for 2.21)
2.24 Leaching (subdivide as for 2.21)
2.25 Soaking in salt water (subdivide as for 2.21)
2.26 Soaking with the addition of ashes or lye (subdivide as for 2.21)
2.27 Soaking with the addition of acidic substances (subdivide as for 2.21)
2.28 Boiling
  2.281 Boiling of whole pieces
    2.2811 Simple boiling
      2.28111 Without salt, lye, or acid
      2.28112 With salt
      2.28113 With lye
      2.28114 With acid
      2.28115 After drying
    2.2812 Repeated boiling in changes of water (subdivide as for 2.2811)
  2.282 Boiling after comminution (subdivide as for 2.281)
  2.283 Boiling after peeling (subdivide as for 2.281)
2.3 Detoxification by fermentation
  2.31 Spontaneous fermentation
    2.311 Fermentation of whole pieces
      2.3111 Without previous treatment
        2.31111 Followed only by washing
        2.31112 Followed by washing and heat treatment
        2.31113 Followed by heat treatment
        2.31114 Followed by comminution
        2.31115 Followed by drying
      2.3112 After cooking (subdivide as for 2.3111)
      2.3113 After boiling with lye (subdivide as for 2.3111)
      2.3114 After soaking (subdivide as for 2.3111)
      2.3115 With addition of salt (subdivide as for 2.3111)
    2.312 Fermentation after comminution (subdivide as for 2.311)
  2.32 Fermentation with use of inoculum from earlier preparations (subdivide as for 2.31)
2.4 Detoxification by adsorption
  2.41 Addition of clay
    2.411 Addition to whole pieces

2.4111 Addition during soaking
    2.41111 Addition without previous treatment
    2.41112 Addition after cooking
2.4112 Addition during boiling (subdivide as for 2.4111)
2.4113 Addition during cooking (subdivide as for 2.4111)
2.4114 Addition during comminution (subdivide as for 2.4111)
2.4115 Addition to consumed product (subdivide as for 2.4111)
2.412 Addition after comminution (subdivide as for 2.411)
2.42 Addition of charcoal (subdivide as for 2.41)
2.43 Soaking in wet mud
    2.431 Soaking of whole pieces
    2.432 Soaking after comminution
2.5 Detoxification by drying
    2.51 Sundrying
        2.511 Drying of whole pieces
            2.5111 Sundrying followed by cooking
            2.5112 Sundrying followed by soaking
            2.5113 Sundrying followed by fermentation
            2.5114 Sundrying followed by comminution
        2.512 Drying after comminution (subdivide as for 2.511)
    2.52 Kiln or hot-air drying (subdivide as for 2.51)
2.6 Detoxification by physical processing
    2.61 Peeling
    2.62 Grating or rasping
    2.63 Squeezing
    2.64 Pounding
    2.65 Grinding
    2.66 Cutting
2.7 Detoxification by pH change
    2.71 Lye or lime added
    2.72 Acidic substance added

Appendix 2

Traditional Methods of Plant Detoxification. Each report is categorized for the seven basic techniques outlined in chapter 3 and coded according to the schema in Appendix 1 (from Johns and Kubo 1988).

Araceae

Plant name	Plant part[1]	Location	1	2	3	4	5	6	7	Code
Alocasia macrorhiza Schott. (Colocasia macrorhiza)			cyanogenic glycosides							
			proteinase inhibitors							
	(St)	Australia	X	X				X		2.132, 2.212
	(T)	Philippines	X							2.121, 2.131
	(T)	New Caledonia	X							2.13
Amorphophallus abyssinicus N.E.Br.	(T)	S. Africa	X	X						2.2711
A. aphyllus (Hook.) Hutch.	(T)	W. Africa	X	X			X		X	2.235, 2.1125, 2.235
A. campanulatus (Roxb.) Blume.	(T)	India	X	X						2.2211
A. dracontioides N.E.Br.	(T)	Asia, Philippines	X							2.121, 2.13
			saponins							
A. glabra F.M. Bailey	(T)	W. Africa	X	X				X		2.222222
A. lyratus Kunth	(TFSt)	Australia	X							2.13
	(T)	India	X							2.121
Anchomanes difformis (Bl.) Engl.	(Rh)	W. Africa	X	X	X				X	2.31134
A. welwitschii Rendle	(T)	W. Africa	X	X						2.2211
Arisaema amurense Maxim	(R)	Asia	X	X				X		2.2111
A. curvatum Kunth.	(T)	India	X	X	X					2.31112
A. triphyllum (L.) Torr. (Arum triphyllum)	(T)	E. North America	X	X			X	X		2.11, 2.5122, 2.132
			cyanogenic glycosides							
Arisarum vulgare Targ.	(R)	N. Africa		X						2.22
Arum maculatum L.	(R)	Europe	X							2.11
	(L)	Europe	X				X			2.28115

Plant name	Plant part[1]	Location	Chemistry Processing classification[2]							Code
			1	2	3	4	5	6	7	
Calla palustris L.	(R)	Europe, N. Asia, North America	X					X	X	2.112
Colocasia antiquorum Schott.	(L)	Australia	X	X						2.2812
C. esculenta (L.) Schott.	(T)	W. Africa	X	X						2.224
	(T,St)	Pacific						X		2.61
	(T)	Australia	X							2.12, 2.13
C. indica Hassk.	(T)	S. Asia	X							2.11
Lysichiton americanum Hulten & St. John	(R)	W. North America	X							2.121
Peltandra virginica Rafin. (*Arum virginica*)	(T)	E. North America	X				X	X		2.514, 2.131, 2.12
Plesmonium margaritiferum Schott.	(T)	India	X	X					X	2.28114
Stilochiton lancifolia Kotschy & Peyr.	(L)	W. Africa	X	X						2.28112
	(Rh)	W. Africa		X					X	2.2611
Symplocarpus foetidus Nutt.	(R)	E. North America					X	X		2.131
Typhonium angustilobium F.V. Muell.	(T)	Australia	X					X		2.132
T. brownii Schott.	(T)	Australia	X					X		2.132
T. trilobatum (L.) Schott.	(Rh)	West Africa	X				X			2.11, 2.5

Cycadaceae

Cycas sp. MAM (methylazoxymethanol) glycosides

Plant name	Plant part[1]	Location	1	2	3	4	5	6	7	Code
Cycas sp.	(S)	Australia	X	X				X		2.211, 2.214
C. circinalis L.	(S)	Guam	X	X				X		2.2221
C. media R. Br.	(S)	Australia	X	X	X		X	X		2.2127, 2.233, 2.132
	(F)	Australia	X	X	X		X	X		2.2127
C. revoluta Thunb.	(St)	Trop. Asia	X					X		2.132, 2.64
C. rumphii Miq.	(S)	Bay of Bengal	X							2.121
	(St,S)	Oceania	X	X					X	2.2112

Plant name	Plant part[1]	Location	Chemistry Processing classification[2]							Code
			1	2	3	4	5	6	7	
Dioscoreaceae										
Dioscorea spp.				saponins, alkaloids						
D. alata L.	(T)	E. Africa	X	saponins			X	X		2.28222, 2.31211
D. bulbifera L.	(T)	Africa	X	saponins, alkaloid: dioscorine				X		
D. cochleari-apiculata De Wild	(T)	Pacific	X	X				X		2.2812, 2.231, 2.243, 2.2
D. dumetorum Pax.	(T)	Africa		alkaloids, phenanthrenes		X		X	X	2.211, 2.213, 2.22, 2.43, 2.25
D. preussii Pax.	(T)	W. Africa		X				X		2.232, 2.221
D. sansibarensis Pax.	(T)	Africa	X	X	alkaloids			X		2.213, 2.253, 2.233
Tamus communis L.	(St)	Europe, Persia, N. Africa	X	X	saponins, phenanthrenes					2.2712
Euphorbiaceae										
Elateriospermum tapos Blume.	(S)	Malaysia, Indonesia	X	igenane-type diterpene esters						2.121, 2.131
Euphorbia lathyrus L.	(S)	S. Europe	X						X	2.251

Plant name	Plant part[1]	Location	Chemistry Processing classification[2]							Code
			1	2	3	4	5	6	7	
Hevea brasiliensis Muell.-Arg.	(S)	S.E. Asia		cyanogenic glycosides	X					2.311
Jatropha curcas L.	(S)	Mexico	X							2.131
	(S)	Trop. America, Asia						X		2.61
J. multifida L.	(S)	Trop. America, Asia						X		2.61
Manihot esculenta Crantz. many methods of detoxification practiced worldwide				cyanogenic glycosides						
	(T)	Malay Peninsula	X	X				X		2.233
	(T)	Congo				X				2.43
Fabaceae										
Abrus precatorius L.	(F,S)	Bay of Bengal	X	lectins, alkaloids						2.131
Acacia albida Delile	(S)	S. Africa	X	X				X	X	2.28123
Canavalia obtusifolia DC	(S)	Australia	X	X				X		2.2113
Cassia occidentalis L.	(S)	W. Africa	X							2.131
Castanospermum australe A. Cunningh. & Fraser	(S)	Australia	X	X	octahydroindolizine alkaloid			X		2.234
Crotalaria spp.				pyrrolizidine alkaloids						
C. mucronata Desv.	(S)	E. Trop. Asia	X	X	X					2.2147

			Chemistry	Processing classification[2]							
Plant name	Plant part[1]	Location		1	2	3	4	5	6	7	Code
Entada phaseoloides Mern.	(S)	Australia	saponins, entegenic acid	X	X				X		2.233
Erythrina veriegata L.	(S)	Trop. Asia	erythrina alkaloids	X							2.131, 2.111
Intsia retusa Kurz.	(F)	E. Malaysia		X	X				X		2.214
Lupinus albus L.	(S)	Italy	quinolizidine alkaloids, saponins	X							2.121
L. hirsutus L.	(S)	Europe	quinolizidine alkaloids	X							2.121
L. littoralis Dougl.	(R)	British Columbia	quinolizidine alkaloids	X							2.121, 2.151
L. luteus L.	(S)	Italy	quinolizidine alkaloid	X							2.121
L. mutabilis Sweet	(S)	Andes, South America	quinolizidine alkaloids, saponins	X	X						2.234, 2.211
L. termis Forsk.	(S)	Mediterranean	quinolizidine alkaloids	X							2.211
Mucuna spp.											
M. cochinchinensis A. Chev.	(S)	Trop. Asia	proteinase inhibitors, canavanine	X							2.211
Neorautanenia spp.			rotenoids, pterocarpans								
N. mitis (A. Rich.) Verd.	(R)	S. Africa		X	X				X	X	2.2182

Plant name	Plant part[1]	Location	Chemistry Processing classification[2]							Code
			1	2	3	4	5	6	7	
Olneya tesota A. Gray	(S)	Mexico	X	X						2.2812
Pachyrrhizus erosus Urb. (*P. angulatus* Rich.)	(T)	China	X							2.11
Prosopis spp.			nonprotein amino acids							
P. pubescens Benth.	(S)	W. North America	X		X	X				2.13, 2.3, 2.41
P. juliflora DC	(S)	W. North America				X		X		2.41141
Tamarindus indica L.	(S)	S. Africa	hemagglutinins							
			X	X				X	X	2.28123
Fagaceae										
Quercus spp.			tannic acid							
	(S)	E. North America							X	2.261
	(S)	W. North America		X				X	X	2.28123, 2.242
	(S)	W. North America		X				X		2.4123
	(S)	Persia		X		X		X		2.232
	(S)	Japan	X							2.281
	(S)	Sardinia	X			X				2.4123
Q. chrysolepis Liebm.	(S)	W. North America				X				2.431

Plant name	Plant part[1]	Location	Chemistry Processing classification[2]							Code
			1	2	3	4	5	6	7	
Q. garryana Dougl.	(S)	W. North America				X				2.431
Zamiaceae										
Macrozamia spp.			MAM glycosides							
M. fraseri Miq. (M. macdonnelli)	(S)	Australia	X	X						2.213, 2.2116
M. miquelii A. Dc.	(StS)	Australia	X							2.214
M. spiralis Miq.	(S)	Australia	X	X				X		2.112, 2.212
	(S)	E. Australia	X					X		2.112
Zamia sp.			MAM glycosides							
Zamia chigua Cuatr.	(S)	Australia		X			X			2.2116
	(S)	Panama	X	X						2.2112
Z. furfuracea Ait.	(R)	Mexico, Central America								2.
Zamia pumila L. (Z. integrifolia)	(T)	Caribbean		X						2.2

NOTES:

[1] Plant part: F = fruit; L = leaf; R = root; Rh = rhizome; S = seed; St = stem; T = tuber.

[2] 1 = heating; 2 = solution; 3 = fermentation; 4 = adsorption; 5 = drying; 6 = physical processing; 7 = pH change.

Bibliography

Ackerkneckt, E. H. 1953. Paleopathology. In *Anthropology today*, ed. A. L. Kroeber, 120–26. Chicago: University of Chicago Press.

Adler, H. M., and V. B. O. Hammett. 1973. The doctor-patient relationship revisited. *Ann. Intern. Med.* 78:595–98.

Alcorn, J. B. 1982. Dynamics of Huastec ethnobotany: Resources, resource perception, and resource management in Teenek, Tsabaal, Mexico. Ph.D. diss., University of Texas, Austin.

———. 1984. *Huastec Mayan ethnobotany*. Austin: University of Texas Press.

Ambrose, S. H., and M. J. DeNiro. 1986. Reconstruction of African human diet using bone collagen carbon and nitrogen isotope ratios. *Nature* 319:321–24.

Ames, B. N. 1983. Dietary carcinogens and anticarcinogens. *Science* 221:1256–64.

Ames, B. N.; R. Magaw; and L. S. Gold. 1987. Ranking possible carcinogenic hazards. *Science* 236:271–80.

Andersen, J. R., et al. 1988. Decomposition of wheat bran and ispaghula husk in the stomach and the small intestine of healthy men. *J. Nutr.* 118:326–31.

Anderson, G. H., and J. L. Johnston. 1983. Nutrient control of brain neurotransmitter synthesis and function. *Can. J. Physiol.* 61:271–81.

Anderson, K. E.; A. H. Conney; and A. Kappas. 1986. Nutrition as an environmental influence on chemical metabolism in man. In *Ethnic differences in reactions to drugs and xenobiotics*, ed. W. Kalow, H. W. Goedde, and D. P. Agarwal, 39–54. New York: Alan R. Liss.

Andrewartha, K. A., and I. W. Caple. 1980. Effects of changes in nutritional copper on erythrocyte superoxide dismutase activity in sheep. *Res. Vet. Sci.* 28:101–4.

Anell, B., and S. Lagercrantz. 1958. Geophagical customs. *Studia Ethnographica Upsaliensia* 17:1–84.

Applebaum, S. W., and Y. Birk. 1979. "Saponins." In *Herbivores: Their interactions with secondary plant metabolites*, ed. G. A. Rosenthal and D. H. Janzen, 539–66. New York: Academic Press.

Ardrey, R. 1976. *The hunting hypothesis*. New York: Atheneum.

Armelagos, G. J., and J. R. Dewey. 1970. Evolutionary response to human infectious diseases. *BioScience* 157:638–44.

Arnason, T.; R. J. Hebda; and T. Johns. 1981. Use of plants for food and medicine by native people of eastern Canada. *Can. J. Bot.* 59:2189–2325.

Asprey, G. F., and P. Thornton. 1955. Medicinal plants of Jamaica, part III. *West Indian Med. J.* 4:69–144.

Bailey, F. L. 1940. Navaho foods and cooking methods. *Amer. Anthropol.* 42:270–90.

Bajaj, K. L.; G. Kaur; and M. L. Chadha. 1979. Glycoalkaloid content and other chemical constituents of the fruits of some eggplant (*Solanum melongena* L.) varieties. *J. Plant Foods* 3:163–68.

Baker, D.; R. Keeler; and W. Gaffield. 1987. Lesions of potato sprout and extracted potato sprout alkaloid toxicity in Syrian hamsters. *Clin. Toxicol.* 25:199–208.

Baker, P. T., and M. A. Little. 1976. *Man in the Andes.* Stroudsburg, Pa.: Dowden, Hutchinson, and Ross.

Bank, T. P. 1952. Botanical and ethnobotanical studies of the Aleutian Islands. II: The Aleuts. *Papers Mich. Acad. Sci. Arts Lett.* 38:415–31.

Barbosa, P., and J. A. Saunders. 1985. Plant allelochemicals: Linkages between herbivores and their natural enemies. In *Recent advances in phytochemistry.* Vol. 19, *Chemically mediated interactions between plants and other organisms,* ed. G. A. Cooper-Driver, T. Swain, and E. E. Conn, 107–37. New York: Plenum Press.

Barnwell, G. M.; J. Dollahite; and D. S. Mitchell. 1986. Salt taste preference in baboons. *Physiol. Behav.* 37:279–84.

Barr, M., and E. S. Arnista. 1957. Adsorption studies on clays. 1: The adsorption of two alkaloids by attapulgite, halloysite, and kaolin. *J. Amer. Pharm. Ass.* 46:486–89.

Barrett, S. A. 1952. Material aspects of Pomo culture, part one. *Bull. Pub. Mus. City Milwaukee* 20:1–260.

Bartoshuk, L. M. 1978. History of taste research. In *Handbook of perception.* Vol. 7A, *Tasting and smelling,* ed. E. C. Carterette and M. P. Friedman, 3–18. New York: Academic Press.

Bastien, J. W., and J. M. Donahue. 1981. *Health in the Andes.* American Anthropological Association, Special Publication no. 12.

Bate-Smith, E. C. 1972. Attractants and repellents in higher animals. In *Phytochemical ecology,* ed. J. B. Harborne, 45–56. London: Academic Press.

Beauchamp, G. K. 1987. The human preference for excess salt. *Amer. Sci.* 75:27–33.

Beauchamp, G. K., and M. Bertino. 1985. Rats (*Rattus norvegicus*) do not prefer salted solid food. *J. Comp. Physiol.* 99:240–47.

Beckman, G. 1978. Enzyme polymorphism. In *Biochemical genetics of man,* ed. D. J. H. Brock and O. Mayo, 186–253. London: Academic Press.

Beidler, L. M. 1987. Vertebrate taste receptors. In *Perspectives in chemoreception and behavior,* ed. R. F. Chapman, E. A. Bernays, and J. G. Stoffolano, 47–58. New York: Springer-Verlag.

Beisel, W. R. 1982. Synergism and antagonism of parasitic disease and malnutrition. *Rev. Inf. Dis.* 4:746–50.

Bender, B. 1978. Gatherer-hunter to farmer: A social perspective. *World Archaeol.* 10:261–89.

Benn, M. 1977. Glucosinolates. *Pure Appl. Chem.* 49:197–210.

Berenbaum, M. 1985. Brementown revisited: Interactions among allelochemicals in plants. In *Recent advances in phytochemistry.* Vol. 19, *Chemically mediated interactions between plants and other organisms,* ed. G. A. Cooper-Driver, T. Swain, and E. E. Conn, 139–69. New York: Plenum Press.

Berlin, B.; D. Breedlove; and P. Raven. 1973. General principles of classification and nomenclature in folk biology. *Amer. Anthropol.* 75:214–42.

Berlin, B.; D. E. Breedlove; and P. H. Raven. 1974. *Principles of Tzeltal plant classification.* New York: Academic Press.

Berlin, B., and P. Kay. 1969. *Basic color terms: Their universality and evolution.* Berkeley: University of California Press.

Bermúdez-Rattoni, F., et al. 1986. Flavor-illness aversions: The role of the amygdala in the acquisition of taste-potentiated odor aversions. *Physiol. Behav.* 38:503–8.

Bernstein, I. L. 1978. Learned taste aversions in children receiving chemotherapy. *Science* 200:1302–3.

Bernstein, I. L.; L. E. Goehler; and D. P. Fenner. 1984. Learned aversions to proteins in rats on a dietary self-selection. *Behav. Neurosci.* 98:1065–72.

Bernstein, I. L., and R. A. Sigmundi. 1980. Tumor anorexia: A learned food aversion? *Science* 209:416–18.

Bernstein, I. L.; M. V. Vitiello; and R. A. Sigmundi. 1980. Effects of tumor growth on taste-aversion learning produced by antitumor drugs in the rat. *Physiol. Psych.* 8:51–55.

Bertino, M., and M. M. Chan. 1986. Taste perception and diet in individuals with Chinese and European ethnic backgrounds. *Chem. Sens.* 11:229–41.

Bertonio, B. 1612. Vocabulário de la lengua Aymara. Francisco del Canto for the Compania de Jesus, Juli.

Bickel, M. H. 1969. Pharmacology and biochemistry of *N*-oxides. *Pharm. Rev.* 21:325.

Binford, L. R. 1984. *Faunal remains from Klasies River mouth.* New York: Academic Press.

Birch, L. L., and D. W. Marlin, 1982. I don't like it; I never tried it: Effects of exposure on two-year-old children's food preferences. *Appetite* 3:353–67.

Bird, R. McK. 1984. South American maize in Central America? In *Pre-Columbian plant migration,* ed. D. Stone, 39–65. Papers of the Peabody Museum of Archaeology and Ethnology, vol. 76. Cambridge: Harvard University Press.

Birk, Y., and I. Peri. 1980. Saponins. In *Toxic constituents of plant foodstuffs.* 2d ed., ed. I. E. Liener, 162–82. New York: Academic Press.

Bjerve, K. S.; S. Fischer; and K. Alme. 1987. Alpha-linolenic deficiency in man: Effect of ethyl linolenate on plasma and erythrocyte fatty acid composition and biosynthesis of prostanoids. *Amer. J. Clin. Nutr.* 46:570–76.

Black, D. A. K. 1956. Reevaluation of tierra sigillata. *Lancet* 2:883–84.

Boeckh, J. 1980. Ways of nervous coding of chemosensory quality at the input level. In *Proceedings of the Seventh International Symposium on Olfaction and Taste and of the Fourth Congress of the European Chemoreception Research Organization,* ed. H. van de Starre, 113–22. London: IRL Press.

Bohrer, V. L. 1977. West African dietary elements as relicts of hominid evolution. *J. Anthropol. Res.* 33:121–32.

Bolt, G. H. 1979. *Soil chemistry.* B: *Physico-chemical models.* Amsterdam: Elsevier.

Bolton, R. 1984. The hypoglycemia-aggression hypothesis: Debate versus research. *Curr. Anthropol.* 25:1–53.

Booth, D. A. 1982. How nutritional effects of foods can influence people's dietary choices. In *Psychobiology of human food selection,* ed. L. M. Barker, 63–84. Westport, Conn.: AVI.

Boppré, M. Redefining "pharmacophagy." 1984. *J. Chem. Ecol.* 10:1151–54.

Borchers, E. A., and R. T. Ṭaylor. 1988. Inheritance of fruit bitterness in a cross of *Cucurbita mixta* × *C. pepo. HortScience* 23:603–4.

Bousfield, J. 1979. The world seen as a colour chart. In *Classifications in their social context,* ed. R. F. Ellen and D. Reason, 195–220. New York: Academic Press.

Braidwood, R. J. 1960. The agricultural revolution. *Sci. Amer.* 203:130–48.

Brattsten, L. B. 1979. Biochemical defense mechanisms in herbivores against plant allelochemicals. In *Herbivores: Their interactions with secondary plant metabolites,* ed. G. A. Rosenthal and D. H. Janzen, 199–269. New York: Academic Press.

Bray, T. M.; S. Kubow; and W. J. Bettger. 1986. Effect of dietary zinc on endogenous free radical production in rat lung microsomes. *J. Nutr.* 116: 1054–60.

Bresnick, E. 1980. Induction of the enzymes of detoxication. In *Enzymatic basis of detoxication.* 2 vols., ed. W. B. Jakoby, 1:69–83. New York: Academic Press.

Briggs, L. T. 1976. Dialectal variation in the Aymara language of Bolivia and Peru. Ph.D. diss., University of Florida, Gainesville.

Brindley, G. W., and G. Brown, eds. 1980. *Crystal structures of clay minerals and their X-ray identification.* Mineralogy Society, Monograph no. 5, London.

Brink, A. J.; G. J. H. Rall; and J. C. Breytenbach. 1977. Pterocarpans of *Neorautanenia edulis* and *N. amboensis. Phytochemistry* 16:273–76.

Bristol, M. L. 1969. Tree *Datura* drugs of the Colombian Sibundoy. *Bot. Mus. Leaf. Harvard University* 22:165–227.

Bristol, M. L.; W. C. Evans; and J. F. Lampard. 1969. The alkaloids of the genus *Datura,* section *Brugmansia.* Part VI: Tree datura drugs (*Datura candida* cvs.) of the Colombian Sibundoy. *Lloydia* 32:123–30.

Brock, R. G. 1984. El Niño and world climate: Piecing together the puzzle. *Environment* 26:14–20, 37–39.

Brooker, S. G., and R. C. Cooper. 1962. *New Zealand medicinal plants.* Auckland War Memorial Museum Handbook. Auckland: Unity Press.

Brower, L. P. 1969. Ecological chemistry. *Sci. Amer.* 220 (2):22–29.

Brown, C. H. 1984. *Language and living things.* New Brunswick, N.J.: Rutgers University Press.

Browne, J. E., et al. 1980. Acid-base equilibria of tetracycline in sodium montmorillonite. *J. Phar. Sci.* 69:811–15.

Brücher, H. 1977. *Tropische Nutzpflanzen.* Berlin: Springer-Verlag.

Brücher, O., et al., 1969. Comparison of phytohaemagglutinins in wild beans and in common beans. *Phytochemistry* 8:1739–43.

Brush, S. 1980. Potato taxonomies in Andean agriculture. In *Indigenous knowledge systems and development,* ed. D. W. Brokencha, D. M. Warren, and O. Werner, 37–47. Lanham: University Press of America.

Brush, S. B. 1986. Farming systems research. *Human Organiz.* 45:220–28.

Brush, S. B.; H. J. Carney; and Z. Huamán. 1981. Dynamics of Andean potato agriculture. *Econ. Bot.* 35:70–88.

Buchmann, S. L.; C. E. Jones; and L. J. Colin. 1977. Vibratile pollination of *Solanum douglasii* and *S. xanti* (Solanaceae) in southern California. *Wasmann J. Biol.* 35:1–25.

Buck, A. A.; T. T. Sasaki; and R. I. Anderson. 1968. *Health and disease in four Peruvian villages.* Baltimore: Johns Hopkins University Press.

Bumstead, M. P. 1985. Past human behavior from bone chemical analysis—Respects and prospects. *J. Human Evol.* 14:539–51.

Bunting, B. T. 1967. *The geography of soil.* London: Hutchinson University Press.

Burk, M. C., and E. M. Pao. 1976. Methodology for large-scale surveys of household and industrial diets. In *Home economics research report,* no. 40, 55–64. Washington, D.C.: U.S. Department of Agriculture.

Burkill, H. M. 1985. *The useful plants of west tropical Africa.* Vol. 1. Kew: Royal Botanic Gardens.

Burkitt, D. P.; A. R. P. Walker; and N. S. Painter. 1972. Effect of dietary fibre on stools and transit-times, and its role in the causation of disease. *Lancet* 30:1408–11.

Butler, L. G., et al. 1986. Dietary effects of tannins. In *Plant flavonoids in biology and medicine,* ed. V. Cody, E. Middleton, and J. B. Harborne, 141–58. New York: Alan R. Liss.

Bye, R. A. 1976. Ethnoecology of the Tarahumara of Chihuahua, Mexico. Ph.D. diss., Harvard University, Cambridge.

Cabanac, M. 1979. Sensory pleasure. *Quart. Rev. Biol.* 54:1–29.

Cagan, R. H. 1973. Chemostimulatory protein: A new type of taste stimulus. *Science* 181:32–35.

Cain, W. S. 1980. Chemosensation and cognition. In *Proceedings of the Seventh International Symposium on Olfaction and Taste and of the Fourth Congress of the European Chemoreception Research Organization,* ed. H. van der Starre, 347–57. London: IRL Press.

Caldwell, J. 1980. Comparative aspects of detoxication in mammals. In *Enzymatic basis of detoxication,* 2 vols., ed. W. B. Jakoby, 1:85–114. New York: Academic Press.

Camadro, E. L., and S. J. Peloquin. 1982. Selfing rates in two wild polyploid *Solanums. Amer. Potato J.* 59:197–204.

Camazine, S. 1986. Zuni Indian medicine: Folklore or pharmacology, science or sorcery? In *Folk medicine: The art and the science,* ed. R. P. Steiner, 23–39. Washington, D.C.: American Chemical Science.

Campbell, B. C., and S. S. Duffey. 1979. Tomatine and parasitic wasps: Potential incompatibility of plant antibiosis with biological control. *Science* 205:700–702.

Carpenter, K. J. 1981. Effects of different methods of processing maize on its pellagragenic activity. *Fed. Proc.* 40:1531–35.

———. 1986. *The history of scurvy and vitamin C.* Cambridge: Cambridge University Press.

Carson, M. S., and T. K. Smith. 1983. Role of bentonite in prevention of T-2 toxicosis in rats. *J. Anim. Sci.* 57:1498–1506.

Carvalho, A.; J. S. Tango; and L. C. Monaco. 1965. Genetic control of the caffeine content of coffee. *Nature* 205:314.

Castetter, E. F. 1935. Ethnobiological studies in the American Southwest. 1: Uncultivated native plants used as sources of food. *Univ. New Mexico Bull.,* Biological Series 4:1–62.

Chagnon, N. A. 1968. *Yanomamo: The fierce people.* New York: Holt, Rinehart, and Winston.

Chagnon, N. A.; P. Le Quesne; and J. M. Cook. 1970. Algunos aspectos de uso de drogas, comercio y domesticacion de plantas entre los indigenas yanomamo de Venezuela y Brazil. *Acta Científica Venezolana* 21:186–93.

———. 1971. Yanomamo hallucinogens: Anthropological, botanical, and chemical findings. *Curr. Anthropol.* 12:72–74.

Chamberlain, A. F. 1903. Primitive taste-words. *Amer. J. Psych.* 14:146–53.

Chan, L., and N. A. Croll. 1979. Medical parasitology in China: An historical perspective. *Amer. J. Chin. Med.* 7:39–51.

Chandel, R. S., and R. P. Rastogi. 1980. Triterpenoid saponins and sapogenins: 1973–1978. *Phytochemistry* 19:1889–1908.

Chandra, R. K. 1988. Nutrition, immunity, and outcome; past, present and future. *Nutr. Res.* 8:225–37.

Chang, H. M.; H. W. Yeung; W.-W. Tso; and A. Koo. 1985. *Advances in Chinese medicinal materials research.* Singapore: World Scientific.

Chang, K.-C. 1970. The beginnings of agriculture in the Far East. *Antiquity* 44:175–85.

Chapman, R. F., and W. M. Blaney. 1979. How animals perceive secondary compounds. In *Herbivores: Their interactions with secondary plant metabolites,* ed. G. A. Rosenthal and D. H. Janzen, 161–98. New York: Academic Press.

Chen, X., et al. 1986. Effect of selenium deficiency on the chronic toxicity of adriamycin in rats. *J. Nutr.* 116:2453–65.

Choudhury, B. 1976. Eggplant. In *Evolution of crop plants,* ed. N. W. Simmonds, 278–79. London: Longman.

Christiansen, J. A. 1977. The utilization of bitter potatoes to improve food production in the high altitude of the tropics. Ph.D. diss., Cornell University, Ithaca.

Christiansen, J. A., and N. R. Thompson. 1977. The utilization of "bitter" potatoes in the cold tropics of Latin America. In *Proceedings of the Fourth*

Symposium of the International Society for Tropical Root Crops, ed. J. Cock, R. MacIntyre, and M. Graham. Ottawa: International Development Research Centre.

Clark, J. D., and J. W. K. Harris. 1985. Fire and its roles in early hominid lifeways. *Afr. Archaeol. Rev.* 3:3–27.

Clark, L., and J. R. Mason. 1985. Use of nest material as insecticidal and anti-pathogenic agents by European starling. *Oecologia* (Berlin) 67:169–76.

Clark, L., and J. R. Mason, 1987. Olfactory discrimination of plant volatiles by the European starling. *Anim. Behav.* 35:227–35.

Clegg, E. J. 1978. Fertility and early growth. In *The biology of high altitude peoples*, ed. P. T. Baker, 65–115. Cambridge: Cambridge University Press.

Cock, J. H. 1982. Cassava: A basic energy source in the tropics. *Science* 218:755–62.

Cockburn, T. A. 1971. Infectious diseases in ancient populations. *Curr. Anthropol.* 12:45–62.

Cody, V.; E. Middleton; and J. B. Harborne, eds. 1986. *Plant flavonoids in biology and medicine*. New York: Alan R. Liss.

Cohen, M. N., and G. J. Armelagos. 1984. Paleopathology at the origins of agriculture. In *Paleopathology at the origins of agriculture*, ed. M. N. Cohen and G. J. Armelagos, 585–601. Orlando: Academic Press.

Conklin, H. C. 1980. Folk classification: A topically arranged bibliography of contemporary and background references through 1971. Department of Anthropology, Yale University, New Haven.

Conn, E. E. 1979. Cyanide and cyanogenic glycosides. In *Herbivores: Their interactions with secondary plant metabolites*, ed. G. A. Rosenthal and D. H. Janzen, 387–412. New York: Academic Press.

Correll, D. S. 1962. *The potato and its wild relatives*. Renner: Texas Research Foundation.

Corry, J. E. L.; D. Roberts; and F. A. Skinner, eds. 1982. *Isolation and identification methods for food-poisoning organisms*. London: Academic Press.

Coursey, D. G. 1973. Cassava as a food: Toxicity and technology. In *Chronic cassava toxicity*, ed. B. Nestel and R. MacIntyre, 27–36. Ottawa: International Development Research Centre.

Cox, P. A. 1980. Two Samoan technologies for breadfruit and banana preservation. *Econ. Bot.* 34:181–85.

Crane, E. 1980. *A book of honey*. Oxford: Oxford University Press.

Creasey, W. A. 1985. *Diet and cancer*. Philadelphia: Lea and Febiger.

Creer, T. L., and H. Kotses. 1983. Asthma: Psychologic aspects and management. In *Allergy: Principles and practice*. 2 vols., ed. E. Middleton, C. E. Reed, and E. F. Ellis, 1015–36. St. Louis: C. V. Mosby.

Cummings, J. H. 1984. Microbial digestion of complex carbohydrates in man. *Proc. Nutr. Soc.* 43:35–44.

Cushing, F. H. 1920. Zuni breadstuff. *Indian Notes and Monographs* 8:1–673.

Danford, D. E. 1982. Pica and nutrition. *Ann. Rev. Nutr.* 2:303–22.

Dart, R. A. 1953. The predatory transition from ape to man. *Int. Anthropol. Ling. Rev.* 1:201–18.

Darwin, C. 1890. *The variation of animals and plants under domestication.* 2d ed. London: John Murray.

Davies, A. G., and I. C. Baillie. 1988. Soil-eating in red leaf monkeys (*Presbytis rubicunda*) in Sabah, northern Borneo. *Biotropica* 20:252–58.

Davis, N. D., and U. L. Diener. 1978. Mycotoxins. In *Food and beverage mycology,* ed. L. R. Beuchat, 397–444. Westport, Conn.: AVI.

Davis, R. G. 1978. Increased bitter taste detection thresholds in Yucatan inhabitants related to coffee as a dietary source of niacin. *Chem. Senses Flav.* 3:423–29.

Davis, T., and R. A. Bye. 1982. Ethnobotany and progressive domestication of *Jaltomata* (Solanaceae) in Mexico and Central America. *Econ. Bot.* 36: 225–41.

De Lucca D., M. 1983. *Diccionario Aymara-Castellano Castellano-Aymara.* La Paz, Bolivia: Comision de Alfabetizacion y Literatura en Aymara.

Denton, D. 1982. *The hunger for salt.* Berlin: Springer-Verlag.

de Rosa, G., et al. 1980. Regulation of superoxide dismutase activity by dietary manganese. *J. Nutr.* 110:795–804.

DeSimone, J. A.; G. L. Heck; and L. M. Bartoshuk. 1980. Surface active taste modifiers: A comparison of the physical and psychophysical properties of gymnemic acid and sodium lauryl sulfate. *Chem. Senses* 5:317–30.

de Wet, J. M. J., and J. R. Harlan. 1975. Weeds and domesticates: Evolution in the man-made habitat. *Econ. Bot.* 36:225–41.

Diether, V. G. 1977. The taste of salt. *Amer. Sci.* 65:744–51.

Dimbleby, G. W. 1967. *Plants and archaeology.* New York: Humanities Press.

Dixon, J. B., et al., eds. 1977. *Minerals in soil environment.* Madison, Wisc.: Soil Science Society of America.

Dodds, K. S. 1965. The history and relationships of cultivated potatoes. In *Essays on crop plant evolution,* ed. J. Hutchinson, 123–41. Cambridge: Cambridge University Press.

Domjan, M. 1975. Ingestional aversion learning: Unique and general processes. *Adv. Study Behav.* 11:205–11.

Drewnowski, A., and M. R. C. Greenwood. 1983. Cream and sugar: Human preferences for high-fat foods. *Physiol. Behav.* 30:629–33.

Duffey, S. S. 1980. Sequestration of plant natural-products by insects. *Ann. Rev. Entom.* 25:447–77.

Duke, J. A. 1985. *CRC handbook of medicinal herbs.* Boca Raton, Fla.: CRC Press.

Dunn, F. L. 1968. Epidemiological factors: Health and disease in hunter-gatherers. In *Man the hunter,* ed. R. B. Lee and I. DeVore, 221–28. Chicago: Aldine.

Dupire, M. 1987. Des goût et des odeurs: classifications et universaux. *L'Homme* 27:5–25.

Eastwood, M. 1984. Dietary fiber. In *Nutrition Reviews' present knowledge in nutrition.* 5th ed., 156–75. Washington, D.C.: The Nutrition Foundation, Inc.

Eaton, S. B., and M. Konner. 1985. Paleolithic nutrition: A consideration of its nature and current implications. *New Eng. J. Med.* 312:283–89.

Edmonds, J. M. 1977. Taxonomic studies of *Solanum* section *Solanum* (Maurella). *Bot. J. Linn. Soc.* 75:141–78.

Ehrlich, P. R., and P. H. Raven. 1964. Butterflies and plants: A study in coevolution. *Evolution* 18:586–608.

Eidlitz, K. 1969. Food and emergency food in the circumpolar area. *Studia Ethnographica Upsaliensia* 32:1–175.

El-Olemy, M. M., and J. Reisch. 1979. Isolation of batatsin I from *Dioscorea dumetorum* rhizomes. *Z. Naturforsch.* 34:288–89.

Erickson, J. M., and P. Feeny. 1974. Sinigrin: Chemical barrier to black swallowtail butterfly: *Papilio polyxenes. Ecology* 55:103–11.

Erickson, R. P. 1982. Studies on the perception of taste: Do primaries exist? *Physiol. Behav.* 28:57–62.

Etkin, N. L., and P. J. Ross. 1982. Food and medicine and medicine as food. *Soc. Sci. Med.* 16:1559–73.

Ettenberg, A., and N. White. 1978. Conditioned taste preferences in rat induced self-stimulation. *Physiol. Behav.* 21:363–68.

Evans, D. A. P. 1986. Acetylation. In *Ethnic differences in reactions to drugs and xenobiotics*, ed. W. Kalow, H. W. Goedde, and D. P. Agarwal, 209–44. New York: Alan R. Liss.

Faboya, O. O. P. 1983. The mineral content of some green leafy vegetables commonly found in the western part of Nigeria. *Food Chem.* 12:213–16.

Fallon, A. E., and P. Rozin. 1983. The psychological bases of food rejections by humans. *Ecol. Food Nutr.* 13:15–26.

Farnsworth, N. R., and A. B. Segelman. 1971. Hypoglycemic plants. *Tile Till* 57:52–55.

Farnsworth, N. R., et al. 1985. Medicinal plants in therapy. *Bull. World Health Org.* 43:965–81.

Fazzalari, F. A., ed. 1978. *Compilation of odor and taste threshold values data.* Philadelphia: American Society for Testing and Materials.

Feigin, M. B.; A. Sclafani; and S. R. Sunday. 1987. Species differences in polysaccharide and sugar taste preferences. *Neurosci. Biobehav. Rev.* 11:231–40.

Feinberg, A., and M. S. Zedeck. 1980. Production of a highly reactive alkylating agent from the organospecific carcinogen methylazoxymethanol by alcohol dehydrogenase. *Cancer Res.* 40:4446–50.

Ferrell, R. E.; D. E. Vermeer; and W. S. Le Blanc. 1985. Chemical and mineralogical composition of geophagical materials. In *Trace substances in environmental health.* Vol. 19, 47–55. Columbia: University of Missouri.

Festing, M. F. W. 1987. Genetic factors in toxicology: Implications for toxicological screening. *CRC Critical Reviews in Toxicology* 18:1–26.

Fewkes, J. W. 1896. A contribution to ethnobotany. *Amer. Anthropol.* 9:14–21.

Flatz, G., and H. Rotthauwe. 1973. Lactose, nutrition, and natural selection. *Lancet* 1973:16–17.

Fletcher, P. J. 1986. Conditioned taste aversion induced by tryptamine: A temporal analysis. *Pharmacol. Biochem. Behav.* 25:995–99.

Flynn, R. J. 1973. *Parasites of laboratory animals.* Ames: Iowa State University Press.

Food and Agriculture Organization. 1983. *FAO production yearbook.* Vol. 37. Rome: United Nations Food and Agriculture Organization.

Ford, R. I. 1968. An ecological analysis involving the population of San Juan Pueblo, New Mexico. Ph.D. diss., University of Michigan, Ann Arbor.

————. 1978. Ethnobotany: Historical diversity and synthesis. In *The nature and status of ethnobotany,* ed. R. I. Ford, 3–49. Ann Arbor: Museum of Anthropology, University of Michigan.

————. 1981. Ethnobotany in North America: A historical phytogeographical perspective. *Can. J. Bot.* 59:2178–88.

Foster, G. M. 1984. The concept of "neutral" in humoral medical systems. *Med. Anthropol.* 8:180–94.

Fowler, C. S., and J. Leland. 1967. Some northern Paiute native categories. *Ethnology* 6:381–404.

Freeland, W. J. 1981. Parasitism and behavioral dominance among male mice. *Science* 213:461–62.

Freeland, W. J.; P. H. Calcott; and L. R. Anderson. 1981. Tannins and saponin: Interaction in herbivore diets. *Biochem. Syst. Ecol.* 13:189–93.

Freeland, W. J., and D. H. Janzen. 1974. Strategies in herbivory by mammals: The role of plant secondary compounds. *Amer. Nat.* 108:269–89.

Freis, E. D. 1986. Antihypertensive agents. In *Ethnic differences in reactions to drugs and xenobiotics,* ed. W. Kalow, H. W. Goedde, and D. P. Agarwal, 313–22. New York: Alan R. Liss.

Frisancho, A. R. 1981. *Human adaptation.* Ann Arbor: University of Michigan Press.

Gadd, K. G., et al. 1962. The Lusitu tragedy. *Cent. Afr. J. Med.* 8:491–508.

Gade, D. W. 1975. *Plants, man, and the land of the Villcanota Valley of Peru.* The Hague: Junk.

Galdikas, B. M. F. 1988. Orangutan diet, range, and activity at Tanjung Puting, central Borneo. *Int. J. Primatol.* 9:1–37.

Galindo Alonso, J. 1982. La papita güera. *Naturaleza* 3:175–80.

Gallaher, D., and B. O. Schneeman. 1986. Nutritional and metabolic response to plant inhibitors of digestive enzymes. In *Nutritional and toxicological significance of enzyme inhibitors in foods,* ed. M. Friedman, 167–84. New York: Plenum Press.

Garb, J. L., and A. J. Stunkard. 1974. Taste aversions in man. *Amer. J. Psychiat.* 131:1204–7.

Garcia, J., et al. 1967. Conditioning with delayed vitamin injection. *Science* 155:716–18.

Garcia, J., and W. G. Hankins. 1974. The evolution of bitter and the acquisition of toxiphobia. In *Olfaction and taste,* ed. D. A. Denton and J. P. Coghlan, Olfaction and taste V: 39–45. New York: Academic Press.

Garcia, J.; W. G. Hankins; and K. W. Rusiniak. 1974. Behavioral regulation of the milieu interne in man and rat. *Science* 185:824–31.

Gaud, F. 1911. *Les manja.* Collection de Monographies Ethnographique. Brussels: Albert de Wet.

Geissman, T. A., and D. H. G. Crout. 1969. *Organic chemistry of secondary plant metabolism.* San Francisco: Freeman, Cooper.

Gifford, E. W. 1936. California balanophagy. In *Essays in anthropology.* 87–98. Berkeley: University of California Press.

Gilman, A. G., et al., eds. 1985. *The pharmacological basis of therapeutics.* 7th ed. New York: Macmillan.

Girardi, N. M. 1987. Ethnomedical and biochemical aspects of antiparasitic plants from Oaxaca, Mexico. M.Sc. thesis, University of California, Berkeley.

Girón, L. M., et al. 1988. Anticandidal activity of plants used for the treatment of vaginitis in Guatemala and clinical trial of a *Solanum nigrescens* preparation. *J. Ethnopharmacol.* 22:307–13.

Glander, K. E. 1978. Howling monkey feeding behavior and plant secondary compounds: A study of strategies. In *The ecology of arboreal folivores,* ed. G. G. Montgomery, 561–574. Washington, D.C.: Smithsonian Institution Press.

———. 1980. Reproduction and population growth in free-ranging mantled howling monkeys. *Amer. J. Phys. Anthropol.* 53:25–36.

———. 1982. The impact of plant secondary compounds on primate feeding behavior. *Yearbook Phys. Anthropol.* 25:1–18.

Glaser, D. 1972. Vergleichende untersuchungen uber den Geschmackssin de Primaten. *Folia Primatol.* 17:267–74.

———. 1980. Die Leistungen des Geschmacksorgans beim Menschen im Vergleich zu nichtmenschlichen Primaten. *Lebensmittel-Wissenschaft Technologie* 13:276–82.

Glaser, D., et al. 1978. The taste responses in primates to the proteins thaumatin and monellin and their phylogenetic implications. *Folia Primatol.* 29:56–63.

Glaser, D., et al. 1984. Effects of gymnemic acid on sweet taste perception in primates. *Chem. Senses* 8:367–74.

Glaser, D., and G. Hobi. 1986. Taste responses in primates to citric and acetic acid. *Int. J. Prim.* 6:395–98.

Glaumont, M. 1889. Usages, moeurs et coutomes des Neo-Caledoniens. *Revue d'Ethnographie* 7:73–141.

Glendinning, D. R. 1976. Neo-tuberosum: New potato breeding material. 4: The breeding system of neo-tuberosum, and the structure and composition of the neo-tuberosum gene-pool. *Potato Res.* 19:27–36.

Goedde, H. W. 1986. Ethnic differences in reactions to drugs and other xenobiotics: Outlook of a geneticist. In *Ethnic differences in reactions to drugs and xenobiotics,* ed. W. Kalow, H. W. Goedde, and D. P. Agarwal, 9–20. New York: Alan R. Liss.

Goedde, H. W., and D. P. Agarwa. 1986. Aldehyde oxidation: Ethnic variations in metabolism and response. In *Ethnic differences in reactions to drugs and xenobiotics,* ed. W. Kalow, H. W. Goedde, and D. P. Agarwal, 113–38. New York: Alan R. Liss.

Goldstein, W. S., and K. C. Spencer. 1985. Inhibition of cyanogenesis by tannins. *J. Chem. Ecology* 11:847–58.

Gomez, M. I. 1981. Carotene content of some green leafy vegetables of Kenya and effects of dehydration and storage on carotene retention. *J. Plant Foods* 3:231–44.

Gordon, K. D. 1987. Evolutionary perspectives on human diet. In *Nutritional anthropology,* ed. F. E. Johnston, 3–39. New York: Alan R. Liss.

Gould-Martin, K. 1978. Hot cold clean poison and dirt: Chinese folk medicine categories. *Soc. Sci. Med.* 12:39–46.

Gove, P. B., ed. 1968. *Webster's Third New International Dictionary.* Springfield, Mass.: Merriam.

Grant, V. 1981. *Plant speciation.* 2d. ed. New York: Columbia University Press.

Green, K. F., and J. Garcia. 1971. Recuperation from illness: Flavor enhancement for rats. *Science* 173:749–51.

Greene, L. S. 1974. Physical growth and development, neurological maturation, and behavioral functioning in two Ecuadorian Andean communities in which goiter is endemic. *Amer. J. Phys. Anthropol.* 41:139–52.

———. 1980. Social and biological predictors of physical growth and neurological development in an area where iodine and protein-energy malnutrition are endemic. In *Social and biological predictors of nutritional status, physical growth, and neurological development,* ed. L. S. Greene and F. E. Johnston, 223–56. New York: Academic Press.

Gregory, P. 1984. Glycoalkaloid composition of potatoes: Diversity and biological implications. *Amer. Potato J.* 61:115–22.

Gregory, P., et al. 1981. Glycoalkaloids of wild, tuber-bearing *Solanum* species. *J. Agri. Food Chem.* 29:1212–15.

Grey, G. 1841. Journals of two expeditions of discovery in Northwest and Western Australia during the years 1837, 38 and 39. 2 vols. London: T. and W. Boone.

Griffiths, D. W. 1986. The inhibition of digestive enzymes by polyphenolic compounds. In *Nutritional and toxicological significance of enzyme inhibitors in foods,* ed. M. Friedman, 509–16. New York: Plenum Press.

Grine, F. E. 1981. Trophic differences between "gracile" and "robust" australopithecines: A scanning electron microscope analysis of occlusal events. *S. A. J. Sci.* 77:203–30.

Gunther, R. T. 1959. *The Greek herbal of Dioscorides.* New York: Hafner.

Hackett, A. M. 1986. The metabolism of flavonoid compounds in mammals. In *Plant flavonoids in biology and medicine,* ed. V. Cody, E. Middleton, and J. B. Harborne, 177–94. New York: Alan R. Liss.

Hall, M. J., et al. 1975. PTC blindness and the taste of caffeine. *Nature* 253:442–43.

Halstead, J. A. 1968. Geophagia in man: Its nature and nutritional effects. *Amer. J. Clin. Nutr.* 21:1384–93.

Hambraeus, L. 1982. Naturally occurring toxicants in food. In *Adverse effects of foods,* ed. E. F. P. Jelliffe and D. B. Jelliffe, 13–36. New York: Plenum Press.

Hamilton, W. D., and M. Zuk. 1982. Heritable true fitness and bright birds: A role for parasites. *Science* 218:384–87.

Hamilton, W. J.; R. E. Buskirk; and W. H. Buskirk. 1978. Omnivory and utilization of food resources by chacma baboons, *Papio ursinus. Amer. Nat.* 112:911–24.

Hamilton, W. J., and C. D. Busse. 1978. Primate carnivory and its significance to human diets. *BioScience* 28:761–66.

Hanamura, Y. 1970. The substances that control the feeding behavior and the growth of the silkworm *Bombyx mori* L. In *Control of insect behavior by natural products,* ed. D. L. Wood, R. M. Silverstein, and M. Nakajima, 55–80. New York: Academic Press.

Harborne, J. B. 1988. *Introduction to ecological biochemistry.* 3d ed. London: Academic Press.

Harborne, J. B., and B. L. Turner. 1984. *Plant chemosystematics.* London: Academic Press.

Harding, R. S. O. 1981. An order of omnivores: Nonhuman primate diets in the wild. In *Omnivorous primates gathering and hunting in human evolution,* ed. R. S. O. Harding and G. Teleki, 191–214. New York: Columbia University Press.

Harlan, J. R. 1975. Our vanishing genetic resources. *Science* 188:618–21.

Harlan, J. R.; J. M. J. de Wet; and E. G. Price. 1973. Comparative evolution of cereals. *Evolution* 27:311–25.

Harney, J. W.; I. M. Barofsky; and J. D. Leary. 1978. Behavioral and toxicological studies of cyclopentanoid monoterpenes from *Nepeta cataria. Lloydia* 41:367–74.

Harris, D. R. 1977. Alternative pathways toward agriculture. In *Origins of agriculture,* ed. C. A. Reed, 179–243. The Hague: Mouton.

———. 1989. An evolutionary continuum of people-plant interaction. In *Foraging and farming,* ed. D. R. Harris and G. C. Hillman. London: Unwin Hyman.

Hartwell, J. L. 1967. Plants used against cancer. A survey. *Lloydia* 30:379–85.

Harwood, J. P., et al. 1986. Effect of long-term feeding of soy-based diets on the pancreas of cebus monkeys. In *Nutritional and toxicological significance of enzyme inhibitors in foods,* ed. M. Friedman, 223–37. New York: Plenum Press.

Hawkes, J. G. 1947. On the origin and meaning of South American Indian potato names. *Bot. J. Linn. Soc.* 53:205–50.

———. 1962a. Introgression in certain wild potato species. *Euphytica* 11:26–35.

———. 1962b. The origin of *Solanum juzepczukii* Buk. and *S. curtilobum* Juz. et Buk. *Z. Pflanzenzuch* 47:1–14.

———. 1972a. Evolutionary relationships in wild tuber-bearing *Solanum* species. *Symp. Biol. Hung.* 12:65–69.

———. 1972b. Evolution of the cultivated potato *Solanum tuberosum. Symp. Biol. Hung.* 12:183–88.

———. 1975. Practical problems in exploration: Vegetatively propagated crops. In *Crop resources for today and tomorrow,* ed. O. H. Frankel and J. G. Hawkes, 117–21. Cambridge: Cambridge University Press.

———. 1978. Biosystematics of the potato. In *The potato crop,* ed. P. M. Harris, 15–69. London: Chapman and Hall.

———. 1983. *The diversity of crop plants*. Cambridge, Mass.: Harvard University Press.

———. 1989. The domestication of roots and tubers in the American tropics. In *Foraging and farming*, ed. D. R. Harris and G. C. Hillman. London: Unwin Hyman.

Hawkes, J. G., and J. P. Hjerting. 1969. *The potatoes of Argentina, Brazil, Paraguay, and Uruguay*. London: Academic Press.

Hayden, B. 1981a. Research and development in the Stone Age: Technological transitions among hunter-gatherers. *Curr. Anthropol.* 22:519.

———. 1981b. Subsistence and ecological adaptations of modern hunter/gatherers. In *Omnivorous primates gathering and hunting in human evolution*, ed. R. S. O. Harding and G. Teleki, 344–421. New York: Columbia University Press.

Hays, T. E. 1982. Utilitarian/adaptionist explanations of folk biological classification: Some cautionary notes. *J. Ethnobiol.* 2:89–94.

Hedrick, U. P., ed. 1919. *Sturtevant's notes on edible plants*. Albany: New York Agricultural Experiment Station.

Hegnauer, R. 1962–86. *Chemotaxonomie der Pflanzen*. Band 1–8. Basel: Birkhauser Verlag.

———. 1975. Secondary metabolites and crop plants. In *Crop resources for today and tomorrow*, ed. O. H. Frankel and J. G. Hawkes, 249–65. Cambridge: Cambridge University Press.

Heinrich, B. 1983. Insect foraging energetics. In *Handbook of experimental pollination biology*, ed. C. E. Jones and R. J. Little, 187–214. New York: Scientific and Academic Editions.

Hellekant, G., et al. 1974. On the gustatory effects of miraculin and gymnemic acid in the monkey. *Chem. Senses Flav.* 1:137–45.

Hellekant, G., et al. 1981. Gustatory responses in three prosimian and two simian primate species to six sweeteners and miraculin and their phylogenetic implications. *Chem. Senses* 6:165–73.

Helmer, O. M., and W. E. Judson. 1968. Metabolic studies on hypertensive patients with suppressed plasma renin activity not due to hyperaldosteronism. *Circulation* 38:965.

Heppner, M. J. 1926. Further evidence on the factor for bitterness in the sweet almond. *Genetics* 11:605–6.

Hertz, C. W. 1979. Historial aspects of anticholinergic treatment of obstructive airways disease. *Scand. J. Resp. Dis.* 1979 (S103):105–9.

Hill, E., and J. Evans. 1989. Crops of the Pacific: New evidence from chemical analysis of organic residues in pottery. In *Foraging and farming*, ed. D. R. Harris and G. C. Hillman. London: Unwin Hyman.

Hladik, C. M. 1977a. Chimpanzees of Gabon and Gombe: Some comparative data on the diet. In *Primate ecology*, ed. T. H. Clutton-Brock, 481–501. London: Academic Press.

———. 1977b. A comparative study of the feeding strategies of two sympatric species of leaf monkeys: *Presbytis senex* and *Presbytis entellus*. In *Primate ecology*, ed. T. H. Clutton-Brock, 324–53. London: Academic Press.

————. 1978. Adaptive strategies of primates in relation to leaf-eating. In *The ecology of arboreal folivores,* ed. G. G. Montgomery, 373–95. Washington, D.C.: Smithsonian Institution Press.

————. 1981. Diet and the evolution of feeding strategies among forest primates. In *Omnivorous primates gathering and hunting in human evolution,* ed. R. S. O. Harding and G. Teleki, 10–36. New York: Columbia University Press.

Hladik, C. M., and L. Gueguen. 1974. Geophagie et nutrition minerale chez les primates sauvages. *CR Acad. Sci.* (Paris) 279:1393–96.

Hodge, F. W., ed. 1907. The narrative of Alvar Nuñez Cabeza de Vaca. In *Spanish explorers in the southern United States, 1528–1543,* 12–126. New York: Charles Scribner's Sons.

Hodge, W. H. 1951. Three native tuber foods of the high Andes. *Econ. Bot.* 5:185–201.

Holmstedt, B., et al. 1977. Determination of cocaine in some South American species of *Erythroxylum* using mass fragmentography. *Phytochemistry* 16:1753–55.

Horowitz, R. M. 1986. Taste effects of flavonoids. In *Plant flavonoids in biology and medicine,* ed. R. V. Cody, E. Middleton, and J. B. Harborne, 163–76. New York: Alan R. Liss.

Hort, A., trans. 1938. *The "Critica Botanica" of Linnaeus.* London: The Ray Society.

Hough, W. 1910. Roots. In *Handbook of American Indians north of Mexico.* Bulletin 30, pt. 2, ed. F. W. Hodge, 395–96. Washington, D.C.: Bureau of American Ethnology.

Huamán, Z. 1975. The origin and nature of *Solanum ajanhuiri* Juz. et Buk., a South American cultivated diploid potato. Ph.D. diss., University of Birmingham, England.

————. 1979. Review of planning conference on the utilization of genetic resources: Status of germ plasm maintenance and computerization. In *Exploration, taxonomy, and maintenance of potato germ plasm.* Lima: International Potato Center (CIP), 3:36–39.

Huamán, Z.; J. G. Hawkes; and P. R. Rowe. 1980. *Solanum ajanhuiri:* An important diploid potato cultivated in the Andean altiplano. *Econ. Bot.* 34:335–43.

————. 1982. A biosystematic study of the origin of the cultivated diploid potato, *Solanum × ajanhuiri* Juz. et Buk. *Euphytica* 31:665–76.

————. 1983. Chromatographic studies on the origin of the cultivated diploid potato *Solanum × ajanhuiri* Juz. et Buk. *Amer. Potato J.* 60:36–367.

Hudson, C. M., ed. 1979. *Black drink: A native American tea.* Athens: University of Georgia Press.

Hudson, J. B., et al. 1986. Antiviral properties of thiarubrine-A, a naturally occurring polyine. *Planta Medica* 1:51–55.

Hughes, M. A.; M. A. Dunn; and J. R. Pearson. 1985. A regulatory element controlling the synthesis of the cyanogenic β-glucosidase (Linamarase) of white clover. *Heredity* 55:387–91.

Hunn, E. 1977. *Tzeltal folk zoology: The classification of discontinuities in nature.* New York: Academic Press.

——. 1982. The utilitarian factor in folk biological classification. *Amer. Anthropol.* 84:830–47.

Hunter, J. M. 1973. Geophagy in Africa and in the United States: A culture-nutrition hypothesis. *Geogr. Rev.* 63:170–95.

Hunter, J. M., and R. de Kleine. 1984. Geophagy in Central America. *Geogr. Rev.* 74:157–69.

Hussain, S. F.; H. Guinaudeau; and M. Shamma. 1988. Two spirobenzyliso-quinoline alkaloids from *Corydalis stewartii. J. Nat. Prod.* 51:1136–39.

Hvidberg, E. F., 1986. Ethnic differences in phenytoin kinetics. In *Ethnic differences in reactions to drugs and xenobiotics,* ed. W. Kalow, H. W. Goedde, and D. P. Agarwal, 279–88. New York: Alan R. Liss.

Irvine, F. R. 1952. Supplementary and emergency food plants of West Africa. *Econ. Bot.* 6:23–40.

Isaac, G. L. 1984. The archaeology of human origins: Studies of the lower Pleistocene in East Africa 1971–1981. *Adv. World Archaeol.* 3:1–87.

Isaac, G. L., and D. C. Crader. 1981. To what extent were early hominids carnivorous? An archaeological perspective. In *Omnivorous primates gathering and hunting in human evolution,* ed. R. S. O. Harding and G. Teleki, 37–103. New York: Columbia University Press.

Ivie, G. W.; D. L. Holt; and M. C. Ivey. 1981. Natural toxicants in human foods: Psoralens in raw and cooked parsnip root. *Science* 213:909–10.

Iwu, M. M. 1986. Empirical investigations of dietary plants used in Igbo ethnomedicine. In *Plants used in indigenous medicine and diet: Bio-behavioral approaches,* ed. N. L. Etkin, 131–50. South Salem, N.Y.: Red-grave.

Izaddoost, M., and T. Robinson. 1987. Synergism and antagonism in the pharmacology of alkaloidal plants. In *Herbs, spices, and medicinal plants,* 3 vols., ed. L. E. Craker and J. E. Simon, 2:137–58. Phoenix: Oryx Press.

Jackson, M. T.; J. G. Hawkes; and P. R. Rowe. 1980. An ethnobotanical field study of primitive potato varieties in Peru. *Euphytica* 29:107–13.

Jadhav, S. J.; R. P. Sharma; and D. M. Salunkhe. 1981. Naturally occurring toxic alkaloids in foods. *CRC Critical Reviews in Toxicology* 9:21–104.

Jaffé, W. G. 1980. Hemagglutinins (lectins). In *Toxic constituents of plant foodstuffs.* 2d ed., ed. I. E. Liener, 73–102. New York: Academic Press.

Jakoby, William B., ed. 1980. *Enzymatic basis of detoxication.* 2 vols. New York: Academic Press.

Janzen, D. H. 1978. Complications in interpreting the chemical defenses of trees against tropical arboreal plant-eating vertebrates. In *The ecology of arboreal folivores,* ed. G. G. Montgomery, 73–84. Washington, D.C.: Smithsonian Institution Press.

Janzen, D. H.; H. B. Juster; and I. E. Liener. 1976. Insecticidal action of the phytohemagglutinin in black beans on a bruchid beetle. *Science* 192:795–96.

Jelliffe, E. F. P., and D. B. Jelliffe, eds. 1982. *Adverse effects of foods.* New York: Plenum Press.

Jeraci, J. L., and P. J. Van Soest. 1986. Interaction between human gut bacteria and fibrous substrates. In *CRC handbook of dietary fiber in human nutrition*, ed. G. A. Spiller, 299–303. Boca Raton, Fla.: CRC Press.

Johanson, D. C., et al. 1987. New partial skeleton of *Homo habilis* from Olduvai Gorge, Tanzania. *Nature* 327:205–9.

Johns, T. 1981. The añu and the maca. *J. Ethnobiol.* 1:208–12.

———. 1985. Chemical ecology of the Aymara of western Bolivia: Selection for glycoalkaloids in the *Solanum* × *ajanhuiri* domestication complex. Ph.D. diss., University of Michigan, Ann Arbor.

———. 1986a. Detoxification function of geophagy and the domestication of the potato. *J. Chem. Ecol.* 12:635–46.

———. 1986b. Chemical selection in Andean domesticated tubers as a model for the acquisition of empirical plant knowledge. In *Plants used in indigenous medicine and diet: Biobehavioral approaches*, ed. N. L. Etkin, 266–88. South Salem, N.Y.: Redgrave Publishers.

———. 1989. A chemical-ecological model of root and tuber domestication in the Andes. In *Foraging and farming*, ed. D. R. Harris and G. C. Hillman, 504–19. London: Unwin Hyman.

———. Forthcoming. Clay ingestion and its relationship to plant use and plant domestication. In *Geophagy: Perspectives on a worldwide practice*, ed. D. E. Vermeer and S. H. Katz. Gordon and Breach, forthcoming.

Johns, T., et al. 1982. Anti-reproductive and other medicinal effects of *Tropaeolum tuberosum*. *J. Ethnopharmacol.* 5:149–61.

Johns, T., et al. 1987. Relationships among wild, weed, and cultivated potatoes in the *Solanum* × *ajanhuiri* complex. *Syst. Bot.* 12:541–52.

Johns, T., and S. L. Keen. 1985. Determinants of taste perception and classification among the Aymara of Bolivia. *Ecol. Food Nutr.* 16:253–71.

Johns, T., and S. L. Keen. 1986a. Ongoing evolution of the potato in the altiplano of western Bolivia. *Econ. Bot.* 40:409–24.

Johns, T., and S. L. Keen. 1986b. Taste evaluation of potato glycoalkaloids by the Aymara: A case study in human chemical ecology. *Hum. Ecol.* 14:437–52.

Johns, T., and J. O. Kokwaro. 1990. Plants used for food by the Luo of Siaya District, Kenya. *Econ. Bot.* In press.

Johns, T.; J. O. Kokwaro; and E. K. Kimanani. 1990. Herbal remedies of the Luo of Siaya District, Kenya: Establishing quantitative criteria for consensus. *Econ. Bot.* 44(3).

Johns, T., and I. Kubo. 1988. A survey of traditional methods employed for the detoxification of plant foods. *J. Ethnobiol.* 8:81–129.

Johns, T., and S. F. Osman. 1986. Glycoalkaloids of *Solanum* series *Megistacrolobum* and related potato cultigens. *Biochem. Syst. Ecol.* 14:651–55.

Johns, T., and G. H. N. Towers. 1981. Isothiocyanates and thioureas in hydrolysates of *Tropaeolum tuberosum*. *Phytochemistry* 20:2687–89.

Johnson, C.; R. Beaton; and K. Hall. 1975. Poison-based avoidance learning in nonhuman primates: Use of visual cues. *Physiol. Behav.* 14:403–7.

Jones, V. H. 1941. The nature and scope of ethnobotany. *Chronica Botanica* 6:219–21.

Kakiuchi, N., et al. 1986. Studies on dental caries prevention by traditional medicines. VIII: Inhibitory effects of various tannins on glucan synthesis by glucosyltransferase from *Streptococcus mutans. Chem. Pharm. Bull.* 34:720–25.

Kalow, W. 1986. Caffeine and other drugs. In *Ethnic differences in reactions to drugs and xenobiotics*, ed. W. Kalow, H. W. Goedde, and D. P. Agarwal, 331–41. New York: Alan R. Liss.

Kaneko, K., et al. 1977. Structure of barogenin from *Solanum tuberosum. Phytochemistry* 16:791–93.

Kappas, A.; K. E. Anderson; A. H. Conney; and A. P. Alvares. 1976. Influence of dietary protein and carbohydrate on antipyrine and heophyliline metabolism. *Clin. Pharmacol. Ther.* 20:643–53.

Kaptchuk, T. J. 1983. *The web that has no weaver: Understanding Chinese medicine.* New York: Congdon and Weed.

Kawamura, Y., and M. R. Kare, eds. 1987. *Umami: A basic taste.* New York: Marcel Dekker.

Keeley, L. H., and N. Toth. 1981. Microwear polishes on early stone tools from Koobi-Fora, Kenya. *Nature* 293:464–65.

Kennedy, L. M., and B. P. Halpern. 1980. A biphasic model for action of the gymnemic acids and ziziphins on taste receptor cell membranes. *Chem. Senses* 5:149–58.

Keshinro, O. O. 1983. The free and total folate activity of some commonly available tropical foodstuffs. *Food Chem.* 11:87–93.

Keshinro, O. O., and A. O. Ketiku. 1979. Effect of traditional cooking methods on the ascorbic acid content of some Nigerian leafy and fruit vegetables. *Food Chem.* 4:303–10.

Kingsbury, J. M. 1964. *Poisonous plants of the United States and Canada.* Englewood Cliffs, N.J.: Prentice-Hall.

Klaassen, C. D. 1985. Toxicology. In *The pharmacological basis of therapeutics.* 7th ed., ed. A. G. Gilman et al., 1592–1604. New York: Macmillan.

Klaassen, C. D.; M. O. Amdur; and J. Doull, eds. 1986. *Casarett and Doull's toxicology: The basic science of poisons.* New York: Macmillan.

Klayman, D. L. 1985. *Qinghaosu* (Artemisinin): An antimalarial drug from China. *Science* 228:1049–55.

Kliks, M. 1978. Paleodietetics: A review of the role of dietary fiber in preagricultural human diets. In *Topics in dietary fiber research*, ed. G. A. Spiller and R. J. Amen, 181–202. New York: Plenum Press.

Kliks, M. M. 1985. Studies on the traditional herbal anthelmintic *Chenopodium ambrosiodes* L.: Ethnopharmacological evaluation and clinical field trials. *Soc. Sci. Med.* 21:879–86.

Kloos, H., et al. 1986. Knowledge, perceptions, and health behavior pertaining to *Schistosomia mansoni* related illness in Machakos district, Kenya. *Trop. Med. Parasit.* 37:171–75.

Kraybill, D. R. 1977. Pre-agricultural tools for the preparations of foods in the Old World. In *Origins of agriculture*, ed. C. A. Reed, 485–521. The Hague: Mouton.

Kresel, J. J., and I. Barofsky. 1979. Conditioned saccarin aversion development following administration of hypotensive agents. *Physiol. Behav.* 23:733–36.

Kreulen, D. A. 1985. Lick use by large herbivores: A review of benefits and banes of soil consumption. *Mammal Rev.* 15:107–23.

Kreulen, D. A., and T. Jager. 1984. The significance of soil ingestion in the utilization of arid rangelands by large herbivores, with special reference to natural licks on the Kalahari pans. In *Herbivore nutrition in the subtropics and tropics,* ed. F. M. C. Gilchrist and R. I. Mackie, 204–21. Johannesburg: The Science Press.

Krishnaswamy, K. 1987. Effects of malnutrition on drug metabolism and toxicity in humans. In *Nutritional toxicology,* 2 vols., ed. J. N. Hathcock, 2:105–28. Orlando: Academic Press.

Kruuk, H. 1972. *Hyena: A study of predation and social behavior.* Chicago: University of Chicago Press.

Kubo, I., and F. J. Hanke. 1985. Multifaceted chemically based resistance in plants. In *Recent advances in phytochemistry.* Vol. 19, *Chemically mediated interactions between plants and other organisms,* ed. G. A. Cooper-Driver, T. Swain, and E. E. Conn, 171–94. New York: Plenum Press.

Kubo, I.; J. A. Klocke; and S. Asano. 1983. Effects of ingested phytoecdysteroids on the growth and development of two lepidopterous larvae. *J. Insect Physiol.* 29:307–16.

La Barre, W. 1947. Potato taxonomy among the Aymara Indians of Bolivia. *Acta Americana* 5:83–103.

———. 1948. The Aymara Indians of the Lake Titicaca plateau, Bolivia. *Amer. Anthropol.* 50:1–250.

———. 1951. Aymara biologicals and other medicines. *J. Amer. Folk.* 64: 171–78.

Lackey, C. J. 1978. Pica: A nutritional anthropology concern. In *The anthropology of health,* ed. E. E. Bauwens, 121–29. St. Louis: C. V. Mosby.

Laekeman, G. M., et al. 1983. Isolation and pharmacological characterization of vernolepin. *J. Nat. Prod.* 46:161–69.

Lancaster, P. A., et al. 1982. Traditional cassava-based foods: Survey of processing techniques. *Econ. Bot.* 36:12–45.

Land, D. G., and R. Shepherd. 1988. Scaling and ranking methods. In *Sensory analysis of food.* 2nd ed., ed. J. R. Piggott, 155–85. London: Elsevier Applied Science.

Landy, D., ed. 1977. *Culture, disease, and healing: Studies in medical anthropology.* New York: Macmillan.

Laufer, B. 1930. *Geophagy.* Field Museum of Natural History Anthropological Series, no. 18, 99–198. Chicago: Field Museum of Natural History.

Laughlin, W. S. 1961. Acquisition of anatomical knowledge by ancient man. In *Social life of early man,* ed. S. Washburn, 150–75. Viking Fund Publication in Anthropology no. 31.

———. 1963. Primitive theory of medicine: Empirical knowledge. In *Man's image in medicine and anthropology,* ed. I. Gladston, 116–40. New York: International Universities Press.

Lawless, H. T. 1987. Gustatory psychophysics. In *Neurobiology of taste and smell*, ed. H. E. Finger and W. L. Silver, 401–20. New York: John Wiley and Sons.

Lentner, C., ed. 1981. *Geigy scientific tables*. Vol. 1. West Caldwell, N.J.: Ciba-Geigy, Medical Education Division.

Leopold, A. C., and Ardrey, R. 1972. Toxic substances in plants and the food habits of early man. *Science* 176:512–14.

Leroi-Gourhan, A. 1975. The flowers found in Shanidar IV, a Neanderthal burial in Iraq. *Science* 190:562–64.

Levin, D. A. 1978. Pollinator behavior and the breeding structure of plant pollinators. In *Pollination of flowers by insects*, ed. A. J. Richards, 133–50. Linnean Society Symposium Series, no. 6. London: Academic Press.

Lewellen, T. C. 1981. Aggression and hypoglycemia in the Andes: Another look at the evidence. *Amer. Anthropol.* 22:347–61.

Lewis, W. H. 1986. Ginseng: A medical enigma. In *Plants used in indigenous medicine and diet: Biobehavioral approaches*, ed. N. L. Etkin, 291–305. South Salem, N.Y.: Redgrave.

Lewis, W. H., and M. P. Elvin-Lewis. 1977. *Medical botany*. New York: John Wiley and Sons.

Lieban, R. W. 1976. Traditional medical beliefs and the choice of practitioners in a Philippine city. *Soc. Sci. Med.* 10:289–96.

Liener, I. E. 1979. Phytohemagglutinins. In *Herbivores: Their interaction with secondary plant metabolites*, ed. G. A. Rosenthal and D. H. Janzen, 567–98. New York: Academic Press.

Liener, I. E., ed. 1980. *Toxic constituents of plant foodstuffs*. 2d ed. New York: Academic Press.

Liener, I. E., and M. L. Kakade. 1980. Protease inhibitors. In *Toxic constituents of plant foodstuffs*. 2d ed., ed. I. E. Liener, 7–71. New York: Academic Press.

Lison, M.; S. H. Blondheim; and R. N. Melmed. 1980. A polymorphism of the ability to taste urinary metabolites of asparagus. *Brit. Med. J.* 281:20–27.

Logan, M. H. 1973. Digestive disorders and plant medicinals in highland Guatemala. *Anthropos* 68:537–47.

Logue, A. W.; I. Ophir; and K. E. Strauss. 1981. The acquisition of taste aversions in humans. *Behav. Res. Therapy* 19:319–33.

Louda, S. M.; M. A. Farris; and M. J. Blua. 1987. Variation in methylglucosinolate and insect damage to *Cleome serrulata* (Capparaceae) along a natural soil moisture gradient. *J. Chem. Ecol.* 13:569–81.

Lucas, P. W.; R. T. Corlett; and D. A. Luke. 1985. Plio-Pleistocene hominid diets: An approach combining masticatory and ecological analysis. *J. Human Evol.* 14:187–202.

Luke, B. 1979. *Maternal nutrition*. Boston: Little, Brown.

McCollum, G. D., and S. L. Sinden. 1979. Inheritance study of tuber glycoalkaloids in a wild potato, *Solanum chacoense* Bitter. *Amer. Potato J.* 56:95–113.

McHenry, H. M. 1984. Relative cheek-tooth size in *Australopithecus*. *Amer. J. Phys. Anthropol.* 64:297–306.

McKenna, D. J.; G. H. N. Towers; and F. A. Abbott. 1984. Monoamine oxidase inhibitors in South American hallucinogenic plants: Tryptamine and β-carboline constituents of *ayahuasca. J. Ethnopharmacol.* 10:195–223.

McKey, D. 1978. Soil, vegetation, and seed-eating by black colobus monkeys. In *The ecology of arboreal folivores*, ed. G. G. Montgomery, 423–37. Washington, D.C.: Smithsonian Institution Press.

———. 1979. The distribution of secondary compounds within plants. In *Herbivores: Their interactions with secondary plant metabolites*, ed. G. A. Rosenthal and D. H. Janzen, 56–133. New York: Academic Press.

McKey, D., et al. 1978. Phenolic content of vegetation in two African rain forests: ecological implications. *Science* 202:61–64.

McKey, D. B., et al. 1981. Food selection by black colobus monkeys (*Colobus satanas*) in relation to plant chemistry. *Biol. J. Linn. Soc.* 16:115–46.

MacNeish, R. S., and T. C. Patterson. 1975. *The central Peruvian prehistoric interaction sphere.* Papers of the Robert S. Peabody Foundation for Archaeology. Vol. 7. Andover, Mass.

Maes, F. W., and R. P. Erickson. 1984. Gustatory intensity discrimination in rat NTS: A tool for the evaluation of neural coding. *J. Comp. Physiol. A* 155:271–82.

Mahato, S. B.; A. W. Gunguly; and N. P. Sahu. 1982. Steroid saponins. *Phytochemistry* 21:959–78.

Mamani, M. 1981. El chuño: Preparation, uso almacenamiento. In *La technologia en el mundo andino*, 2 vols., ed. H. Lechtman and A. M. Soldi, 1:235–46. Mexico: Universidad Nacional Autonoma de Mexico.

Mann, A. E. 1981. Diet and human evolution. In *Omnivorous primates gathering and hunting in human evolution*, ed. R. S. O. Harding and G. Teleki, 10–36. New York: Columbia University Press.

Mann, R. D. 1984. *Modern drug use.* Lancaster: MTP Press.

Mattocks, A. R. 1968. Toxicity of pyrrolizidine alkaloids. *Nature* 217:723–28.

May, J. R.; J. T. DiPiro; and J. F. Sisley. 1987. Drug interactions in surgical patients. *Amer. J. Surg.* 153:327–35.

Mazess, R. B., and P. T. Baker. 1964. Diet of Quechua Indians living at high altitude: Nuñoa, Peru. *Amer. J. Clin. Nutr.* 15:341–51.

Mears, J. A., and T. J. Mabry. 1971. Alkaloids in the Leguminosae. In *Chemotaxonomy of the Leguminosae*, ed. J. B. Harborne, D. Boulter, and B. L. Turner, 73–178. London: Academic Press.

Mennicke, W. H.; K. Görler; and G. Krumbiegel. 1983. Metabolism of some naturally occurring isothiocyanates in the rat. *Xenobiotica* 13:203–7.

Meshali, M. M. 1982. Adsorption of phenazopyridine hydrochloride on pharmaceutical adjuvants. *Pharmazie* 37:718–20.

Meydani, M. 1987. Dietary effects on detoxification processes. In *Nutritional toxicology*, 2 vols., ed. J. N. Hathcock, 2:1–39. Orlando: Academic Press.

Midkiff, E. E., and I. L. Bernstein. 1985. Targets of learned food aversions in humans. *Physiol. Behav.* 34:839–41.

Milton, K. 1979. Factors influencing leaf choice by howler monkeys: A test of some hypotheses of food selection by generalist herbivores. *Amer. Nat.* 114:362–78.

———. 1981. Distribution patterns of tropical plant foods as an evolutionary stimulus to primate mental development. *Amer. Anthropol.* 83:534–48.

————. 1986. Digestive physiology in primates. *News in Physiol. Sci.* 1:76–79.

————. 1987. Primate diets and gut morphology: Implications for hominid evolution. In *Food and evolution: Toward a theory of human food habits,* ed. M. Harris and E. B. Ross, 93–115. New York: Temple University Press.

————. 1988. Foraging behavior and the evolution of primate intelligence. In *Machiavellian intelligence: Social expertise and the evolution of intellect in monkeys, apes, and humans,* ed. R. Bryne and A. Whiten, 285–305. Oxford: Clarendon Press.

Milton, K., and M. W. Demment. 1988. Digestion and passage kinetics of chimpanzees fed high and low fiber diets and comparison with human data. *J. Nutrit.* 118:1082–88.

Miracle, M. 1967. *Agriculture in the Congo basin.* Madison: University of Wisconsin Press.

Mitchell, B. K., and J. F. Sutcliffe. 1984. Sensory inhibition as a mechanism of feeding deterrence: Effects of three alkaloids on leaf beetle feeding. *Physiol. Entomol.* 9:57–64.

Mitchell, D.; W. Winter; and C. M. Morisaki. 1977. Conditioned taste aversions accompanied by geophagia: Evidence for the occurrence of "psychological" factors in the etiology of pica. *Psychosom. Med.* 39:402–12.

Mole, S., and P. G. Waterman. 1985. Stimulatory effects of tannins and cholic acid on tryptic hydrolysis of proteins: Ecological implications. *J. Chem. Ecol.* 11:1323–32.

Montgomery, C. G., ed. 1978. *The ecology of arboreal folivores.* Washington, D.C.: Smithsonian Institution Press.

Moore, J. G.; B. K. Krotoszynski; and H. J. O'Neill. 1984. Fecal odorgrams: A method of partial reconstruction of ancient and modern diets. *Digest. Dis. Sci.* 29:907–11.

Moore, R. F. 1983. Effect of dietary gossypol on the boll weevil (Coleoptera: Curculionidae). *J. Econ. Entomol.* 76:696–99.

Moskowitz, H. R. 1974. The psychology of sweetness. In *Sugars in nutrition,* ed. H. L. Sipple and K. W. McNutt, 37–64. New York: Academic Press.

Moskowitz, H. R., et al. 1974. Sugar sweetness and pleasantness: Evidence for different psychological laws. *Science* 184:583–85.

Moskowitz, H. R., et al. 1975. Cross-cultural differences in simple taste preferences. *Science* 190:1217–18.

Moskowitz, H. R., et al. 1976. Effects of hunger, satiety, and glucose load upon taste intensity and taste hedonics. *Physiol. Behav.* 16:471–75.

Murra, J. V. 1975. El control verticál de un máximo de pisos ecológicos en la economia de las sociedades andinas. In *Formaciones económicas y política del mundo andino,* ed. J. V. Murra, 59–116. Lima: Instituto de Estúdios Peruanos.

Myers, C. S. 1904. The taste-names of primitive peoples. *Brit. J. Psych.* 1:117–26.

Mykytowycz, R. 1985. Olfaction—A link with the past. *J. Human Evol.* 14:75–90.

Nachman, M. 1963. Learned aversion to the taste of lithium chloride and generalization to other salts. *J. Comp. Physiol. Psych.* 56:343–49.

Naim, M.; J. G. Brand; and M. R. Kare. 1986. Role of variety of food flavor in fat deposition produced by a "cafeteria" feeding of nutritionally controlled diets. In *Interaction of the chemical senses with nutrition*, ed. M. R. Kare and J. G. Brand, 269–92. Orlando: Academic Press.

Nartey, F. 1978. *Manihot esculenta (cassava)*. Copenhagen: Munksgaard.

National Academy of Sciences. 1973. *Toxicants occurring naturally in foods*. 2d ed. Washington, D.C.: National Academy of Sciences Press.

Ndiokwere, Ch. L. 1984. Analysis of various Nigerian foodstuffs for crude protein and mineral contents by neutron activation. *Food Chem.* 14:93–102.

Needham, J. 1956. *Science and civilization in China*. Vol. 2, *History of scientific thought*. Cambridge: Cambridge University Press.

Nelson, C. E., et al. 1983. Regulation of synthesis and accumulation of proteinase-inhibitors in leaves of wounded tomato plants. *ACS Symp. Ser.* 208:103–23.

Neter, J., and W. Wasserman. 1974. *Applied linear statistical models*. Homewood, IL: R. D. Irwin.

Nishie, K.; W. P. Norred; and A. P. Swain. 1975. Pharmacology and toxicology of chaconine and tomatine. *Res. Commun. Chem. Pathol. Pharmacol.* 12:657–67.

Nishie, K., et al. 1976. Positive inotropic action of Solanaceae glycoalkaloids. *Res. Commun. Chem. Pathol. Pharmacol.* 15:601–7.

Ng, T. B., and H. W. Yeung. 1986. Scientific basis of the therapeutic effects of ginseng. In *Folk medicine: The art and the science*, ed. R. P. Steiner, 139–51. Washington, D.C.: American Chemical Science.

Oates, J. F. 1977. The guereza and its food. In *Primate ecology: Studies of feeding and ranging behaviour in lemurs, monkeys, and apes*, ed. T. H. Clutton-Brock, 275–321. London: Academic Press.

———. 1978. Water-plant and soil consumption by guereza monkeys (*Colobus guereza*): A relationship with minerals and toxins in the diet? *Biotropica* 10:241–53.

Oblitas Poblete, E. 1969. *Plantas medicinales de Bolivia*. Cochabamba: Editorial "Los Amigos del Libro."

Ochoa, C. 1958. *Expedicion colectora de papas cultivadas a la cuenca del Lago Titicaca*. Lima: Programa Cooperativo de Experimentacion Agropecuaria.

———. 1984. Karytaxonomic studies on wild Bolivian tuber-bearing *Solanum*, sect. *Petota*. *Phytologia* 55:17–40.

Ofcarcik, R. P., and E. E. Burns. 1971. Chemical and physical properties of selected acorns. *J. Food Sci.* 36:575–78.

Okada, K. A., and A. M. Clausen. 1982. Natural hybridization between *Solanum acaule* Bitt. and *S. megistacrolobum* Bitt. in the Province of Jujuy, Argentina. *Euphytica* 31:817–35.

Oke, O. L. 1969. Oxalic acid in plants and nutrition. *World Rev. Nutr. Dietetics* 10:262–303.

O'Mahony, M., et al. 1979. Confusion in the use of the taste adjectives "sour" and "bitter." *Chem. Senses Flav.* 4:301–18.

O'Mahony, M., and R. Ishii. 1987. The umami taste concept: Implications for the dogma of four basic tastes. In *Umami: A basic taste,* ed. Y. Kawamura and M. R. Kare, 75–93. New York: Marcel Dekker.

O'Mahony, M., and M. D. Manzanoa. 1980. Taste descriptions in Spanish and English. *Chem. Senses* 5:47–62.

O'Mahony, M., and H. Muhiudeen. 1977. A preliminary study of alternative taste languages using qualitative description of sodium chloride solutions: Malay versus English. *Brit. J. Psych.* 68:275–78.

O'Mahony, M.; T. Tsang; and T. Tsang. 1980. A preliminary comparison of Cantonese and American-English as taste languages. *Brit. J. Psych.* 71: 221–26.

Ommen, G. S. 1986. Susceptibility to occupational and environmental exposures to chemicals. In *Ethnic differences in reactions to drugs and xenobiotics,* ed. W. Kalow, H. W. Goedde, and D. P. Agarwal, 527–46. New York: Alan R. Liss.

Ortiz de Montellano, B. 1975. Empirical Aztec medicine. *Science* 188:215–20.

Osman, S. F., et al. 1978. Glycoalkaloid composition of wild and cultivated tuber-bearing *Solanum* species of potential value in potato breeding programs. *J. Agri. Food Chem.* 26: 1246–48.

Osman, S. F., et al. 1982. 1-β-hydroxyneotigogenin, a sapogenin from *Solanum polyadenium* leaves. *Phytochemistry* 21:472–73.

Osman, S. F.; T. Johns; and K. Price. 1986. Sisunine, a glycoalkaloid found in hybrids between *Solanum acaule* and *Solanum* × *ajanhuiri. Phytochemistry* 25:967–68.

Otero, G. A. 1951. *La piedra magica.* Mexico: Instituto Indigenista Interamericano, Ediciones Especiales.

Pangborn, R. M. 1970. Individual variation in affective responses to taste stimuli. *Psychonomic Sci.* 21:125–26.

Paque, C. 1980. Saharan Bedouins and the salt water of the Sahara: A model for salt intake. In *Biological and behavioral aspects of salt intake,* ed. M. R. Kare, M. J. Fregley, and R. A. Bernard, 31–47. New York: Academic Press.

Parker, L.; A. Failor; and K. Weidman. 1973. Conditioned preferences in the rat with an unnatural need state: Morphine withdrawal. *J. Comp. Physiol. Psych.* 82:294–300.

Patel, N. G. 1986. Ayurveda: The traditional medicine of India. In *Folk medicine: The art and the science,* ed. R. P. Steiner, 41–65. Washington, D.C.: American Chemical Science.

Pauling, L. 1970. Evolution and the need for ascorbic acid. *Proc. Natl. Acad. Sci. U.S.A.* 67:1643–48.

Pederson, C. S., and M. N. Albury. 1969. *The sauerkraut fermentation.* New York State Agriculture Exp. Station Bulletin 824.

Pederson, J. C., and B. L. Welch. 1985. Comparison of ponderosa pines as feed and nonfeed trees for abert squirrels. *J. Chem. Ecol.* 11:149–57.

Pelchat, M. L., and P. Rozin. 1982. The special role of nausea in the acquisition of food dislikes by humans. *Appetite* 3:341–51.

Pennington, C. W. 1963. *The Tarahumar of Mexico: Their environment and material culture.* Salt Lake City: University of Utah Press.

Peters, C. R. 1987. Nut-like oil seeds: Food for monkeys, chimpanzees, humans, and probably ape-men. *Amer. J. Phys. Anthropol.* 73:333–63.

Peters, C. R., and E. M. O'Brien. 1981. The early hominid plantfood niche: Insights from an analysis of plant exploitation by *Homo, Pan,* and *Papio* in eastern and southern Africa. *Curr. Anthropol.* 22:127–40.

Phillips-Conroy, J. E., and P. M. Knopf. 1986. The effects of ingesting plant hormones on schistosomiasis in mice: An experimental study. *Biochem. Syst. Ecol.* 14:637–45.

Pickersgill, B. 1977. Taxonomy and the origin and evolution of cultivated plants in the New World. *Nature* 268:591–95.

———. 1982. Biosystematics of crop-weed complexes. *Kulturpflanze* 29: 377–88.

Picón-Reátegui, E. 1978. The food and nutrition of high-altitude populations. In *The biology of high-altitude populations,* ed. P. T. Baker, 219–49. Cambridge: Cambridge University Press.

Picón-Reátegui, E.; E. R. Buskirk; and P. T. Baker. 1970. Blood glucose in high-altitude natives and during acclimatization to altitude. *J. Appl. Physiol.* 29:560–63.

Pliner, P., et al. 1985. Role of specific postingestional effects and medicinal context in the acquisition of liking for tastes. *Appetite* 6:243–52.

Plowman, T. 1979. Botanical perspectives on coca. *J. Psychedelic Drugs* 11:103–17.

Price, P. W., et al. 1980. Interactions among three trophic levels: Influence of plants on interactions between insect herbivores and natural enemies. *Ann. Rev. Ecol. Syst.* 11:41–65.

Price, T. D.; M. J. Schoeninger; and G. J. Armelagos. 1985. Bone chemistry and past behavior: An overview. *J. Hum. Evol.* 14:419–47.

Purseglove, J. W. 1968. *Tropical crops: Dicotyledons.* London: Longmans.

Ramirez, I. 1986. Intragastric feeding of rats. In *Interaction of the chemical senses with nutrition,* ed. M. R. Kare, 151–65. Orlando: Academic Press.

Rechung Rinpoche Jampal Kunzang. 1976. *Tibetan medicine.* Berkeley: University of California Press.

Reddy, B. S., and G. A. Spiller. 1986. Modification by dietary fiber of toxic or carcinogenic effects. In *CRC handbook of dietary fiber in human nutrition,* ed. G. A. Spiller, 387–90. Boca Raton: CRC Press.

Reed, D. J. 1985. Cellular defense mechanisms against reactive metabolites. In *Bioactivation of foreign compounds,* ed. M. W. Anders, 71–110. Orlando: Academic Press.

Reed, D. J., and M. J. Meredith. 1984. Glutathione conjugation systems and drug disposition. In *Drug nutrients: The interactive effects,* ed. D. A. Roe and T. C. Campbell, 179–224. New York: Marcel Dekker.

Reid, D. P. 1987. *Chinese herbal medicine.* Boston: Shambhala.

Rhoades, D. F. 1985. Pheromonal communication between plants. In *Chemically mediated interactions between plants and other organisms,* ed. G. A. Cooper-Driver, T. Swain, and E. E. Conn, 195–218. New York: Plenum Press.

Rhoades, D. F., and R. G. Cates. 1976. Toward a general theory of plant antiherbivore chemistry. In *Biochemical interactions between plants and insects*, ed. J. W. Wallace and R. L. Mansell, 168–213. New York: Plenum Press.

Richter, C. P. 1936. Increased salt appetite in adrenalectomized rats. *Amer. J. Physiol.* 115:155–61.

Riesenberg, S. H. 1948. Magic and medicine in Ponape. *Southwest. J. Anthropol.* 4:406–29.

Rindos, D. 1984. *The origins of agriculture.* Orlando: Academic Press.

Robak, J., and R. J. Gryglewski. 1988. Flavonoids are scavengers of superoxide anions. *Biochem. Pharmacol.* 17:837–41.

Robb, G. L. 1957. The ordeal poisons of Madagascar and Africa. *Bot. Mus. Leaf. Harv. Univ.* 17:265–316.

Robbins, W. W.; J. P. Harrington; and B. Freire-Marreco. 1916. *Ethnobotany of the Tewa Indians.* Smithsonian Institution, Bureau of American Ethnology, Bulletin 55. Washington, D.C.: Smithsonian Institution Press.

Robinson T. 1979. The evolutionary ecology of alkaloids. In *Herbivores: Their interactions with secondary plant metabolites,* ed. G. A. Rosenthal and D. H. Janzen, 413–48. New York: Academic Press.

Rodman, J. E., and F. S. Chew. 1980. Phytochemical correlates of herbivory in a community of native and naturalized Cruciferae. *Biochem. Syst. Ecol.* 8:43–50.

Rodriquez, E.; J. C. Cavin; and J. E. West. 1982. The possible role of Amazonian psychoactive plants in the chemotherapy of parasitic worms: A hypothesis. *J. Ethnopharmacol.* 6:303–9.

Rodriquez, E., et al. 1985. Thiarubrine-A, a bioactive constituent of *Aspilia* (Asteraceae) consumed by wild chimpanzees. *Experientia* 41:419–20.

Rolls, B. J. 1986. Sensory-specific satiety. *Nutr. Rev.* 44:93–101.

Rolls, B. J., et al. 1981. Variety in a meal enhances food intake in man. *Physiol. Behav.* 26:215–21.

Rosengarten, F. 1978. A neglected Mayan galactagogue Ixbut. (*Euphorbia lancifolia*). *Bot. Mus. Leaf. Harv. Univ.* 26:277–95.

Rosenthal, G. A., and D. H. Janzen, eds. 1979. *Herbivores: Their interaction with secondary plant metabolites.* New York: Academic Press.

Rothman, S. S. 1986. The biological functions and physiological effects of ingested inhibitors of digestive reactions. In *Nutritional and toxicological significance of enzyme inhibitors in foods,* ed. M. Friedman, 19–31. New York: Plenum Press.

Rowland, I. R.; A. K. Mallett; and A. Wise. 1985. The effect of diet on the mammalian gut flora and its metabolic activities. *CRC Crit. Rev. Toxicol.* 16:31–103.

Rozin, P. 1976. The selection of food by rats, humans, and other animals. *Adv. Stud. Behav.* 6:21–76.

———. 1980. Acquisition of food preferences and attitudes to food. *Int. J. Obesity* 4:356–63.

———. 1982. Human food selection: The interaction of biology, culture, and individual experience. In *Psychobiology of human food selection,* ed. L. M. Barker, 225–54. Westport, Conn: AVI.

Rozin, P., and K. Kennel. 1983. Acquired preferences for piquant foods by chimpanzees. *Appetite* 4:69–77.

Rozin, P., and D. Schiller. 1980. The nature and acquisition of a preference for chili pepper by humans. *Motiv. Emot.* 4:77–101.

Ryan, C. A. 1979. Proteinase inhibitors. In *Herbivores: Their interactions with secondary plant metabolites,* ed. G. A. Rosenthal and D. H. Janzen, 599–681. New York: Academic Press.

Said, S. A.; A. M. Shibel; and M. E. Abdullah. 1980. Influence of various agents on adsorption capacity of kaolin for *Pseudomonas aeruginosa* toxin. *J. Pharmaceut. Sci.* 69:1238–39.

Salaman, R. N. 1985. *The history and social influence of the potato.* Cambridge: Cambridge University Press.

Sato, T. 1980. Recent advances in the physiology of taste cells. *Prog. Neurobiol.* 14:25–67.

Saxe, T. G. 1987. Toxicity of medicinal herbal preparations. *Amer. Fam. Phys.* 35:135–42.

Schaefer, O. 1986. Adverse reactions to drugs and metabolic problems perceived in northern Canadian Indians and Eskimos. In *Ethnic differences in reactions to drugs and xenobiotics,* ed. W. Kalow, H. W. Goedde, and D. P. Agarwal, 77–86. New York: Alan R. Liss.

Scheuer, P. J., et al. 1963. The constituents of *Tacca leontopetaloides. Lloydia* 26:133–40.

Schiffman, S. S., and C. Dackis. 1975. Taste of nutrients: Amino acids, vitamins, and fatty acids. *Perc. Psych.* 17:140–46.

Schiffman, S. S.; T. B. Clark; and J. Gagnon. 1982. Influence of chirality of amino acids on the growth of perceived taste intensity with concentration. *Physiol. Behav.* 28:457–65.

Schleidt, M.; P. Neumann; and H. Morishita. 1988. Pleasure and disgust: Memories and associations of pleasant and unpleasant odours in Germany and Japan. *Chem. Senses* 13:279–93.

Schmiediche, P. E.; J. G. Hawkes; and C. M. Ochoa. 1980. Breeding of the cultivated potato species *Solanum* × *juzepczukii* Buk. and *Solanum* × *curtilobum* Juz. et Buk. I. *Euphytica* 29:685–704.

Schneeman, B. O., and D. Gallaher. 1986. Among species response to dietary trypsin inhibitor: Variation in species. In *Nutritional and toxicological significance of enzyme inhibitors in foods,* ed. M. Friedman, 185–87. New York: Plenum Press.

Schoeninger, M. J. 1982. Diet and the evolution of modern human form in the Middle East. *Amer. J. Phys. Anthropol.* 58:37–52.

Schoonhoven, L. M. 1967. Chemoreception of mustard oil glucosides in larvae of *Pieris brassicae. Proc. Kon. Ned. Akad. Wtsch. C* 70:556–68.

Schultes, R. E. 1984. Amazonian cultigens and their northward and westward migrations in Pre-Columbian times. In *Pre-Columbian Plant Migration,* ed. D. Stone, 19–37. Cambridge: Papers of the Peabody Museum of Archaeology and Ethnology, Harvard University, vol. 76.

Schultes, R. E. 1986. Recognition of variability in wild plants by Indians of the northwest Amazon: An enigma. *J. Ethnobiol.* 6:229–38.

Schwanitz, F. 1966. *The origin of cultivated plants.* Cambridge, Mass.: Harvard University Press.

Scott, T. R. 1980. Taste hedonics: A neural assessment. In *Proceedings of the Seventh International Symposium on Olfaction and Taste and of the Fourth Congress of the European Chemoreception Research Organization*, ed. H. van der Starre, 247–50. London: IRL Press.

Scudder, T. 1971. Gathering among African woodland savannah cultivators. A case study: The Gwembe Tonga. *Zambian Papers*, no. 5.

Seetle, R. G., et al. 1986. Chemosensory properties of sour tastants. *Physiol. Behav.* 36:619–23.

Shachak, E. A.; Y. Chapman; and Y. Steinberger. 1976. Feeding, energy flow, and soil turnover in the desert isopod, *Hemilepistus reaumuri. Oecologia* 24:57–69.

Shallenberger, R. S., and T. E. Acree. 1971. Chemical structure of compounds and their sweet and bitter taste. In *Handbook of sensory physiology.* Vol. 4, *Taste*, ed. L. M. Beidler, 221–77. Berlin: Springer-Verlag.

Sharples, R. W. 1985. Theophrastus on tastes and smells. In *Theophrastus of Eresus: On his life and work*, ed. W. W. Fortenbaugh, 2:183–204. New Brunswick, N.J.: Transaction Books.

Shavit, Y., et al. 1984. Opioid peptides mediate the suppressive effect of stress on natural killer cell cytotoxicity. *Science* 223:188–90.

Shibata, S., et al. 1985. Chemistry and pharmacology of *Panax.* In *Economic and medicinal plant research*, ed. H. Wagner, H. Hikino, and N. R. Farnsworth, 1:218–84. London: Academic Press.

Shinnick, F. L., et al. 1988. Oat fiber: Composition versus physiological function in rats. *J. Nutr.* 118:144–51.

Siegel, R. K. 1986. Jungle revelers. *Omni* 8(6):70–74, 100.

Sigerist, H. E. 1951. *A history of medicine.* New York: Oxford University Press.

Silver, W. L. 1987. The common chemical sense. In *Neurobiology of taste and smell*, ed. H. E. Finger and W. L. Silver, 65–87. New York: John Wiley and Sons.

Simmonds, N. W. 1976. *Evolution of crop plants.* London: Longmans.

Simoons, F. J. 1981. Celiac disease as a geographic problem. In *Food, nutrition, and evolution*, ed. D. Walcher and N. Kretchmer, 179–99. New York: Masson.

Sinden, S. L.; K. L. Deahl; and B. B. Aulenbach. 1976. Effects of glycoalkaloids and phenolics on potato flavor. *J. Food Sci.* 41:520–23.

Sinden, S. L.; L. L. Sanford; and S. F. Osman, 1980. Glycoalkaloids and resistance to the Colorado potato beetle in *Solanum chacoense* Bitter. *Amer. Potato J.* 57:331–43.

Sinden, S. L.; L. L. Sanford; and R. E. Webb. 1984. Genetic and environmental control of potato glycoalkaloids. *Amer. Potato J.* 61:141–56.

Small, E. 1984. Hybridization in the domesticated-weed-wild complex. In *Plant biosystematics*, ed. W. F. Grant, 193–210. Toronto: Academic Press.

Smith, G. R. 1982. Botulism in waterfowl. *Symp. Zool. Soc. Lond.* 50:97–119.

Smith, H. H. 1928. Ethnobotany of the Meskwaki Indians. *Bull. Pub. Mus. Milwaukee* 4:175–326.

Smith, S. E., and P. D. O. Davies. 1973. Quinine taste thresholds: A family study and a twin study. *Ann. Hum. Genet.* 37:227–32.

Sofowora, A. 1982. *Medicinal plants and traditional medicine in Africa.* New York: John Wiley and Sons.

Sokolov, J. J.; R. T. Harris; and M. R. Hecker. 1976. Isolation of substances from human vaginal secretions previously shown to be sex attractant pheromones in higher primates. *Arch. Sex. Behav.* 5:269–74.

Solecki, R. S. 1975. Shanidar IV, a Neanderthal flower burial in northern Iraq. *Science* 190:880–81.

Solien, N. L. 1954. A cultural explanation of geophagy. *Fla. Anthropol.* 7:1–9.

Solms, J., and R. Wyler. 1979. Taste components of potatoes. In *ACS Symposium Series*, Vol. 115, Food taste chemistry, ed. J. Boudreau, 175–84. Washington: American Chemical Society.

Solomons, N. W. 1984. Nutrition and parasitism. In *Genetic factors in nutrition*, ed. A. Velázquez and H. Bourges, 225–42. Orlando: Academic Press.

Sorokin, P. A. 1975. *Hunger as a factor in human affairs.* Gainesville: University Presses of Florida.

Sparre, B. 1973. Tropaeolaceae. In *Flora of Ecuador*, ed. B. Sparre and G. Harling. *Opera Botanica*, series B, 2:2–30.

Speck, F. G. 1917. Medicine practices of the northeastern Algonquians. In *Proceedings of the International Congress of Americanists*, Washington, D.C. 19:303–21.

Speth, J. D. 1987. Early hominid subsistence strategies in seasonal habitats. *J. Archaeol. Sci.* 14:13–29.

Sposito, G. 1984. *The surface chemistry of soils.* New York: Oxford University Press.

Spring, B. 1986. Effects of foods and nutrients on the behavior of normal individuals. In *Nutrition and the brain*, Vol. 7, Food constituents affecting normal and abnormal behaviours. ed. R. J. Wurtman and J. J. Wurtman, 1–47. New York: Raven Press.

Srikantia, S. G. 1982. An outbreak of aflatoxins in man. In *Adverse effects of foods*, ed. E. F. P. Jelliffe and D. B. Jelliffe, 45–50. New York: Plenum Press.

Stahl, A. B. 1984. Hominid dietary selection before fire. *Curr. Anthropol.* 25:151–68.

———. 1989. Plant-food processing: Implications for dietary quality. In *Foraging and farming*, ed. D. R. Harris and G. C. Hillman, 171–94. London: Unwin Hyman.

Stein, M. 1986. A reconsideration of specificity in psychosomatic medicine: From olfaction to lymphocyte. *Psychosom. Med.* 48:3–22.

Stone, D., ed. 1984. *Pre-Columbian plant migration.* Papers of the Peabody Museum of Archaeology and Ethnology, Harvard University, no. 76. Cambridge: Harvard University Press.

Story, M., and J. E. Brown. 1987. Do young children instinctively know what to eat: The studies of Clara Davis revisited. *New Eng. J. Med.* 316:103–6.

Subramanian, V.; L. G. Butler; R. Jambunathan; and K. E. P. Rao. 1983. Some agronomic and biochemical characters of brown sorghums and their possible role in bird resistance. *J. Agric. Food Chem.* 31:1303–7.

Swain, T. 1979. Tannins and lignins. In *Herbivores: Their interactions with secondary plant metabolites,* ed. G. A. Rosenthal and D. H. Janzen, 657–82. New York: Academic Press.

———. 1986. The evolution of flavonoids. In *Plant flavonoids in biology and medicine,* ed. V. Cody, E. Middleton, and J. B. Harborne, 1–14. New York: Alan R. Liss.

Symon, D. E. 1979. Sex forms in *Solanum* (Solanaceae) and the role of pollen collecting insects. In *The biology and taxonomy of the Solanaceae,* ed. J. G. Hawkes, R. N. Lester, and A. D. Skelding, 385–97. Linnean Society Symposium Series, no. 7. London: Academic Press.

Takeda, K. 1972. The steroidal sapogenins of the Dioscoreaceae. *Prog. Phytochem.* 2:287–333.

Te Rangi Hiroa, 1970. *The coming of the Maori.* Wellington, N.Z.: Maori Purposes Fund Board.

Tetenyi, P. 1971. *Infraspecific chemical taxa of medicinal plants.* New York: Chemical Publishing.

Theng, B. K. C. 1974. *The chemistry of clay-organic reactions.* New York: John Wiley and Sons.

Thompson, L. U.; C. L. Button; and J. A. Jenkins. 1987. Phytic acid and calcium affect the in vitro rate of navy bean starch digestion and blood glucose response in humans. *Amer. J. Clin. Nutr.* 46:467–73.

Tingey, W. M. 1984. Glycoalkaloids as pest resistance factors. *Amer. Potato J.* 61:157–67.

Tingey, W. M., and S. L. Sinden. 1982. Glandular pubescence, glycoalkaloid composition, and resistance to the green peach aphid, potato leafhopper, and the potato fleabeetle in *Solanum berthaultii. Amer. Potato J.* 59:95–106.

Tjamos, E. C., and J. A. Kuc. 1982. Inhibition of steroid glycalkaloid accumulation by arachidonic and eicosapentaenoic acids in potato. *Science* 217:542–44.

Tobias, P. V. 1981. The emergence of man in Africa and beyond. *Phil. Trans. R. Soc. Lond. B* 292:43–56.

Torii, K. 1986. Effects of dietary protein on the taste preference for amino acids in rats. In *Interaction of the chemical senses with nutrition,* ed. M. R. Kare and J. G. Brand, 45–69. Orlando: Academic Press.

Toth, N. 1985. The Oldowan reassessed: A close look at early stone artifacts. *J. Archaeol. Sci.* 12:101–20.

———. 1987. The first technology. *Sci. Amer.* 256:112–21.

Towers, G. H. N., et al. 1985. Antibiotic properties of thiarubrine-A, a naturally occurring dithiacyclohexadiene polyine. *Planta Medica* 3:225–29.

Treit, D.; M. L. Spetch; and J. A. Deutsch. 1981. Variety in the flavor of food enhances eating in the rat: A controlled demonstration. *Physiol. Behav.* 30:207–11.

Trotter, R. J. 1973. Aggression: A way of life for the Qolla. *Sci. News* 103:76–77.

Tschopik, H. 1946. The Aymara. In *Handbook of South American Indians,* 7 vols., ed. J. H. Steward, 2:501–73. Bureau of American Ethnology Bulletin 143. Washington, D.C.: Smithsonian Institution Press.

Turner, N. J.; R. Bouchard; and D. I. D. Kennedy. 1980. *Ethnobotany of the Okanagan-Colville Indians of British Columbia and Washington.* Occasional Papers British Columbia Provincial Museum, no. 21.

Turner, N. J.; H. V. Kuhnlein; and K. N. Egger. 1987. The cottonwood mushroom (*Tricholoma populinum*): A food resource of the interior Salish Indian peoples of British Columbia. *Can. J. Bot.* 65:921–27.

Turner, P. R. 1972. *The highland Chontal.* New York: Holt, Rinehart, and Winston.

Tyler, V. E. 1987. *The new honest herbal.* Philadelphia: George F. Stickley.

Uehara, S. 1982. Seasonal changes in the techniques employed by wild chimpanzees in the Mahale Mountains, Tanzania, to feed on termites (*Pseudacanthotermes spiniger*). *Folia Primatol.* 37:44–76.

Ugent, D. 1967. Morphological variation in *Solanum × edinense*, a hybrid of the common potato. *Evolution* 21:696–712.

———. 1968. The potato in Mexico: Geography and primitive culture. *Econ. Bot.* 22:108–23.

———. 1970a. *Solanum raphanifolium,* a Peruvian wild potato species of hybrid origin. *Bot. Gaz.* 131:225–33.

———. 1970b. The potato. *Science* 170:1161–66.

———. 1981. Biogeography and origin of *Solanum acaule* Bitter. *Phytologia* 48:85–95.

Ugent, D.; T. Dillehay; and C. Ramirez. 1987. Potato remains from a late Pleistocene settlement in southcentral Chile. *Econ. Bot.* 41:17–27.

Ulrich, R. S. 1984. View through a window may influence recovery from surgery. *Science* 224:420–21.

Unschild, P. U. 1977. The development of medical-pharmaceutical thought in China. *Comp. Med. East West* 5:109–15.

Usai, A. 1969. *Il pane di ghiande e la geophagia in Sardegna.* Cagliari: Editrice Sarda Fratelli Fossataro.

Van Olphen, H. 1977. *An introduction to clay colloid chemistry.* 2d ed. New York: Wiley.

Van Soest, P. J. 1982. *Nutritional ecology of the ruminant.* Corvallis, Oregon: O & B Books.

Vaughan, G. N. 1986. Australian medicinal plants. In *Folk medicine: The art and the science,* ed. R. P. Steiner, 103–9. Washington, D.C.: American Chemical Science.

Vermeer, D. E. 1966. Geophagy among the Tiv of Nigeria. *Annals Assn. Amer. Geogrs.* 56:197–204.

Vermeer, D. E., and R. G. Ferrell. 1985. Nigerian geophagical clay: A traditional antidiarrheal pharmaceutical. *Science* 227:634–36.

Vinson, S. B., and G. F. Iwantsch. 1980. Host suitability for host parasitoids. *Ann. Rev. Entomol.* 25:397–419.

Waddy, J. 1982. Biological classification from a Groote Eylandt Aborigine's point of view. *J. Ethnobiol.* 2:63–77.

Wadsworth, G. 1984. *The diet and health of isolated populations*. Boca Raton, Fla.: CRC Press.

Wagner, W. H. 1980. Origin and philosophy of the groundplan-divergence method of cladistics. *Syst. Bot.* 5:173–93.

———. 1983. Reticulistics: The recognition of hybrids and their role in cladistics and classification. 2 vols. In *Advances in cladistics*, ed. V. A. Funk and N. I. Platnick. 2:63–79. New York: Columbia University Press.

Wai, K.-N., and G. S. Banker. 1966. Some physiochemical properties of the montmorillonites. *J. Pharm. Sci.* 55:1215–20.

Walker, A. 1981. Dietary hypotheses and human evolution. *Phil. Trans. R. Soc. Lond. B* 292:57–64.

Wang, Y.-M., and Y.-J. Hu. 1985. Toxicity and side effects of some Chinese medicinal plants. In *Advances in Chinese medicinal materials research*, ed. H. M. Chang et al., 109–24. Singapore: World Scientific.

Washburn, S. L., and C. S. Lancaster. 1968. The evolution of hunting. In *Man the hunter*, ed. R. B. Lee and I. DeVore. Chicago: Aldine.

Wat, C-K.; T. Johns; and G. H. N. Towers. 1980. Phototoxic and antibiotic activities of plants of the Asteraceae used in folk medicine. *J. Ethnopharmacol.* 2:279–90.

Watts, D. P. 1985. Observations on the ontogeny of feeding behavior in mountain gorillas (*Gorilla gorilla beringei*). *Amer. J. Primatol.* 8:1–10.

Wedekind, K. J.; H. R. Manesfield; and L. Montgomery. 1988. Enumeration and isolation of cellulolytic and hemicellulolytic bacteria from human feces. *Appl. Envir. Microbiol.* 54:1530–35.

Weder, J. K. P. 1986. Inhibition of human proteinases by grain legumes. In *Nutritional and toxicological significance of enzyme inhibitors in foods*, ed. M. Friedman, 239–79. New York: Plenum Press.

Weigel, M. M., and R. M. Weigel. 1988. The association of reproductive history, demographic factors, and alcohol and tobacco consumption with the risk of developing nausea and vomiting in early pregnancy. *Amer. J. Epid.* 127:562–70.

Weil, J. 1986. Beyond the mistique of cocaine: Coca in Andean cultural perspective. In *Plants used in indigenous medicine and diet: Biobehavioral approaches*, ed. N. L. Etkin, 306–28. South Salem, N.Y.: Redgrave.

Weiss, E. A. 1979. Some indigenous plants used domestically by East African coastal fishermen. *Econ. Bot.* 33:35–51.

Weiss, P. 1953. Los comedores peruanos de tierras. *Peru Indigena* 5:12–21.

Werge, R. W. 1979. Potato processing in the central highlands of Peru. *Ecol. Food Nutr.* 7:229–34.

Westley, J. 1980. Rhodanese and the sulfane pool. In *Enzymatic basis of detoxication*. 2 vols., ed. W. B. Jakoby, 2:245–62. New York: Academic Press.

Westoby, M. 1974. An analysis of diet selection by large generalist herbivores. *Amer. Nat.* 108:290–304.

Whitaker, T. W., and W. P. Bemis. 1975. Origin and evolution of the cultivated *Cucurbita*. *Bull. Torrey Bot. Club* 102:362–68.

White, J. L., and S. L. Hem, 1983. Pharmaceutical aspects of clay-organic interactions. *Ind. Eng. Chem. Prod. Res. Dev.* 22:665–71.

Whiting, A. F. 1939. *Ethnobotany of the Hopi.* Museum of Northern Arizona Bulletin 15, 1–120.

Whiting, M. G. 1963. Toxicity of cycads. *Econ. Bot.* 17:271–302.

Whittaker, R. H., and P. O. Feeny. 1971. Allelochemicals: Chemical interactions between species. *Science* 171:757–70.

Wildmann, J., et al. 1988. Occurrence of pharmacologically active benzodiazepines in trace amounts in wheat and potato. *Biochem. Pharmacol.* 37: 3549–59.

Wilkes, H. G. 1977. Hybridization of maize and teosinte in Mexico and Guatemala and the improvement of maize. *Econ. Bot.* 31:254–93.

Willaman, J. J., and H.-L. Li. 1970. Alkaloid bearing plants and their contained alkaloids. *Lloydia* 33, supplement 3A.

Williams, W. 1984. Lupins in crop production. *Outlook on agriculture* 13: 69–76.

Windholz, M., ed. 1983. *The Merck index.* Rahway, N.J.: Merck.

Winterhalder, B.; R. Larson; and R. B. Thomas. 1974. Dung as an essential resource in a highland Peruvian community. *Human Ecol.* 2:89–104.

Woods, S. C.; J. R. Vasselli; and K. M. Milam. 1977. Iron appetite and latent learning in rats. *Physiol. Behav.* 19:623–26.

Woolfe, Jennifer A. 1987. *The potato in the human diet.* Cambridge: Cambridge University Press.

Woolhouse, N. M. 1986. The debrisquine/sparteine oxidation polymorphism: Evidence of genetic heterogeneity among Ghanaians. In *Ethnic differences in reactions to drugs and xenobiotics,* ed. W. Kalow, H. W. Goedde, and D. P. Agarwal, 39–54. New York: Alan R. Liss.

Wrangham, R. 1977. Feeding behavior of chimpanzees in Gombe National Park. In *Primate ecology,* ed. T. H. Clutton-Brock, 504–38. London: Academic Press.

Wrangham, R. W., and J. Goodall. 1987. Chimpanzee use of medicinal leaves. In *Understanding chimpanzees.* Chicago: Chicago Academy of Sciences.

Wrangham, R. W., and T. Nishida. 1983. *Aspilia* spp. leaves: A puzzle in the feeding behavior of wild chimpanzees. *Primates* 24:276–82.

Wrangham, R. W., and P. G. Waterman. 1983. Condensed tannins in fruits eaten by chimpanzees. *Biotropica* 15:217–22.

Wurtman, R. J., and J. J. Wurtman. 1988. Do carbohydrates affect food intake via neurotransmitter activity? *Appetite* 11:S42–S47.

Yang, C. S., and J.-S. H. Yoo. 1988. Dietary effects on drug metabolism by the mixed-function oxidase system. *Pharmac. Ther.* 38:53–72.

Yarnell, R. A. 1964. *Aboriginal relationship between culture and plant life in the Upper Great Lakes region.* Museum of Anthropology, Anthropological Papers no. 23. Ann Arbor: University of Michigan Press.

Yen, D. E. 1974. *The sweet potato and Oceania.* Honolulu: Bishop Museum Press.

Younos, C., et al. 1987. Repertory of drugs and medicinal plants used in traditional medicine of Afghanistan. *J. Ethnopharmacol.* 20:245–90.

Zahorik, D. M. 1977. Associative and non-associative factors in learned food preferences. In *Learning mechanisms in food selection*, ed. L. M. Barker, M. R. Best, and M. Domjan. Waco, Texas: Baylor University Press.

Zahorik, D. M.; S. F. Maier; and R. W. Pies. 1974. Preferences for tastes paired with recovery from thiamine deficiency in rats: Appetite conditioning or learned safety. *J. Comp. Physiol. Psych.* 87:1083–91.

Zohary, D. 1984. Modes of evolution in plants under domestication. In *Plant biosystematics*, ed. W. F. Grant, 579–86. Toronto: Academic Press.

Index

Bark, plant, in human medicine, 259, 286–87
Barley, *See Hordeum vulgare*
Baunei, Sardinia, 87
Beans. *See Phaseolus* spp.
Bees, as potato pollinators, 145
Behavior, human: effects of diet on, 43, 282; and toxic chemicals, 15, 19, 216, 234–40, 274
Behavioral data, in studies of dietary evolution, 217
Bentonite (clay), 94, 95, 96. *See also* Clays; Geophagy; Montmorillonite
Benzodiazepines, in plants, 275
Beverages, relation to food and medicine, 283
Bioassays, use of, 286, 290
Biochemistry, comparative, 228
Biosystematics, plant: definition of, 114; of domesticates, 103, 114, 199; methods, 118. *See also Ajawiri; Sisu; Solanum acaule; S. × ajanhuiri; S. megistacrolobum; S. stenotomum; Yari*
Biotransformation, in detoxication, 46–49. *See also* Bacterial transformations; Detoxication
Birds, pharmacophagy by, 254
Bitter taste: adaptive significance of, 14; Aymara relation to, 180, 189; classification of. *See also* Quinine
Blood pressure, role in learning about medicine, 42, 67, 257, 275, 276
Body temperature, and toxicity, 42, 67. *See also* Fever
Boiling, detoxification by, 73, 74
Bolivia, 27–29, 92, 94
Bone chemistry, and dietary evolution, 220
Botany, history of, 266
Botulism. *See Clostridium*
Brain development, and evolution of human diet, 219, 224, 241, 285. *See also* Intellectual capacity
Brassica oleracea, 117

Breeding systems, plant, 114–15, 123
Buddhist, legends of medicine, 271
Byzantine scholars, 265–66

C_4 and C_3 plants, 220
Caffeine: chemical selection for, 108; perception of, 38, 230; properties of, 51, 275, 276, 283
California, use of acorns in, 85–86
Campylobacter jejuni, 67
Cañahua. See Chenopodium canihua
Cancer: etiology of, 248, 290; herbal treatments for, 279
Capsaicin, 35, 283. *See also Capsicum*
Capsicum spp., 19, 163; in Aymara diet, 185; chemical selection for, 108, 110, 200. *See also* Capsaicin
Caquiaviri, Bolivia, 28, 126, 144, 170; as source of seed potato, 151, 155, 156. *See also* Ajawiri Marka
Carbohydrates: in human diet, 231, 240, 283, 284–85; perception of, 236, 238; staples, detoxification of, 80, 204
Cardiac glycosides, 57, 59; pharmacology, 252, 266, 276; toxicity, 59, 86
Cardiotonic drugs, 276. *See also* Cardiac glycosides
β-Carotene, as dietary antioxidant, 48. *See also* Vitamin A
Cassava. *See Manihot esculenta*
Catnip (*Nepeta catararia*), 254
Cats, pharmacophagous behavior of, 254
Cellulose, digestion of, 227
Central America, geophagy in, 68
Cercopithecus aethiops (monkey), diet of, 222
Chaconine, 69; as chemotaxonomic marker, 129, 130, 131, 132, 133; taste perception of, 178–79, 192–93; toxicity, 169–70
Chagnon, Napoleon, 235

Timothy Johns was appointed assistant professor of human nutrition at McGill University in 1987. He received his Ph.D. from the University of Michigan, with his doctoral dissertation receiving the Distinguished Dissertation Award of 1985 from the Council of Graduate Studies in the United States/University Microfilms International. The expansion of that thesis into a book began while he was a research fellow at the University of California, Berkeley.